Gel Electrophoresis of Proteins

The Practical Approach Series

SERIES EDITORS

D. RICKWOOD

Department of Biology, University of Essex
Wivenhoe Park, Colchester, Essex CO4 3SQ, UK

B. D. HAMES

Department of Biochemistry, University of Leeds
Leeds LS2 9JT, UK

Affinity Chromatography

Animal Cell Culture

Animal Virus Pathogenesis

Antibodies I and II

Biochemical Toxicology

Biological Membranes

Biosensors

Carbohydrate Analysis

Cell Growth and Division

Centrifugation (2nd edition)

Clinical Immunology

Computers in Microbiology

Directed Mutagenesis

DNA Cloning I, II, and III

Drosophila

Electron Microscopy in Molecular Biology

Fermentation

Flow Cytometry

Gel Electrophoresis of Nucleic Acids (2nd edition)

Gel Electrophoresis of Proteins (2nd edition)

Genome Analysis

HPLC of Small Molecules

HPLC of Macromolecules

Human Cytogenetics

Human Genetic Diseases

Immobilised Cells and Enzymes

Iodinated Density Gradient Media

Light Microscopy in Biology

Liposomes

Lymphocytes

Lymphokines and Interferons

Mammalian Development

Medical Bacteriology

Medical Mycology

Microcomputers in Biology

Microcomputers in Physiology

Mitochondria

Mutagenicity Testing

Neurochemistry

Nucleic Acid and Protein Sequence Analysis

Gel Electrophoresis of Proteins

A Practical Approach

SECOND EDITION

Edited by

B. D. Hames

Department of Biochemistry,
University of Leeds, Leeds, UK

and

D. Rickwood

Department of Biology,
University of Essex, Essex, UK

IRL PRESS
—at—
OXFORD UNIVERSITY PRESS
Oxford New York Tokyo

Oxford University Press
Walton Street, Oxford OX2 6DP

Oxford is a trade mark of Oxford University Press

Published in the United States
by Oxford University Press, New York

British Library Cataloguing in Publication Data
Gel electrophoresis of proteins.
1. Proteins. Chemical analysis. Gel electrophoresis.
Laboratory techniques
I. Hames, B. D. (B. David) II. Rickwood, D. (David) 1945–
547.75046
ISBN 0-19-963074-7 (hbk.)   ISBN 0-19-963075-5 (pbk.)

Library of Congress Cataloging in Publication Data
Gel electrophoresis of proteins: a practical approach/edited by
B. D. Hames and D. Rickwood.—2nd ed.
(The Practical Approach Series)
Includes index.
1. Proteins—Analysis. 2. Gel electrophoresis. 3. Polyacrylamide
gel electrophoresis. I. Hames, B. D. II. Rickwood, D. (David)
III. Series.
QP551.G334 1990   574.19'245—dc20   90-38956
ISBN 0-19-963074-7 (hbk.)   ISBN 0-19-963075-5 (pbk.)

Typeset and printed by Information Press Ltd, Oxford, England

Preface

SINCE the first edition of this book, the electrophoretic analysis of proteins in polyacrylamide gels has continued to grow in importance as an essential research technique in the life sciences. Whilst some techniques have changed only marginally, many new procedures and applications have arisen in the intervening years. This second edition seeks to reflect these changes.

Without doubt, one-dimensional polyacrylamide gel electrophoresis is currently the most widely used form of the technique in all areas of the life sciences and so a greatly extended first chapter is devoted to this topic. This chapter also covers many of the recently developed methods for analysing gels, especially the use of different staining and blotting protocols. The subsequent chapters describe in great practical detail the other major gel electrophoretic techniques that are now in common use, including isoelectric focusing, with both conventional and immobilized pH gradients, two-dimensional gel electrophoresis, peptide mapping, and immuno-electrophoresis.

Our hope is that readers of this methods manual will find it to be as instructive and valuable a laboratory companion as many colleagues were kind enough to say they found the first edition.

Leeds and Colchester B. D. HAMES
1990 D. RICKWOOD

Contents

2 Isoelectric focusing

Pier Giorgi Righetti, Elisabetta Gianazza, Cecilia Gelfi, and Marcella Chiari

xiii

Contributors

A. T. ANDREWS

AFRC, Institute of Food Research, Reading Laboratory, Shinfield, Reading RG2 9AT, UK.

T. C. BØG-HANSEN

Research Center for Medical Biotechnology, The Protein Laboratory, University of Copenhagen, Sigurdsgade 34, DK-2400 Copenhagen, Denmark.

J. A. A. CHAMBERS

Department of Biotechnology Research, Pioneer Hibred International, PO Box 38, Johnston, Iowa 50131, USA.

M. CHIARI

Faculty of Pharmacy and Department of Biomedical Sciences and Technology, University of Milano, Via Celoria 2, I-20133 Milano, Italy.

B. D. HAMES

Department of Biochemistry, University of Leeds, Leeds LS2 9JT, UK.

C. GELFI

Faculty of Pharmacy and Department of Biomedical Sciences and Technology, University of Milano, Via Celoria 2, I-20133, Milano, Italy.

E. GIANAZZA

Faculty of Pharmacy and Department of Biomedical Sciences and Technology, University of Milano, Via Celoria 2, I-20133 Milano, Italy.

R. JURD

Department of Biology, University of Essex, Wivenhoe Park, Colchester CO4 3SQ, UK.

D. RICKWOOD

Department of Biology, University of Essex, Wivenhoe Park, Colchester CO4 3SQ, UK.

P. G. RIGHETTI

Faculty of Pharmacy and Department of Biomedical Sciences and Technology, University of Milano, Via Celoria 2, I-20133, Milano, Italy.

S. P. SPRAGG

Department of Chemistry, Birmingham University, Edgbaston, Birmingham, UK.

Abbreviations

ACM	*N*-acroyloyl-morpholine
AMPS	2-acrylamide-2-methyl propane sulphonic acid
ANS	1-aniline-8-naphthalene sulphonate
AP	alkaline phosphatase
BAC	*N,N'*-bisacrylylcystamine
BCIP	5-bromo-4-chloro-3 indolyl phosphate
Bis-ANS	bis(8-*p*-toluidino-1-naphthalene sulphonate)
CA	carrier ampholyte
CIE	crossed immunoelectrophoresis
CPCL	cetylpyridinium chloride
CTAB	cetyltrimethylammonium bromide
DAB	3,3′ diaminobenzidine
DATD	*N,N'*-diallyltartardiamide
DBM	diazo-benzyloxymethyl
DDA	dodecyl alcohol
DDE	didodecyl ether
DDS	didodecyl sulphate
DHEBA	*N,N'*-(1,2 dihydroxyethylene) bisacrylamide
DMAPN	3-dimethylamino-propionitrile
DMSO	dimethylsulphoxide
DNP	dinitrophenol
DPT	diazophenylthioether
DTT	dithiothreitol
EDIA	ethylene diacrylate
EIA	enzyme immunoassay
ELISA	enzyme-linked immunosorbent assay
FIA	fluorescent immunoassay
FITC	fluorescein isothiocyanate
HbF	fetal haemoglobin
HRP	horseradish peroxidase
IEA	immunoelectrophoretic analysis
IEF	isoelectric focusing
IPG	immobilized pH gradient
LGT	low gelling temperature (agarose)
MDPF	2-methoxy-2,4-diphenyl-3(2H)-furanone
MTT	methyl thiazolyl tetrazolium
NBT	nitroblue tetrazolium
NCS	Nuclear Chicago solubilizer
NEPHGE	non-equilibrium pH gradient electrophoresis
OPA	*o*-phthaldialdehyde
PAS	periodic acid − Schiff
PCMB	*p*-chloromercuribenzoic acid
PITC	phenylisothiocyanate
PMS	phenazine methosulphate

PMSF	phenylmethylsulphonyl fluoride
POPOP	1,3-bis-2-(5-phenyloxazole)
PPO	2,5 diphenyloxazole
PVDF	polyvinyl-difluoride
QAE	quaternary amino ethyl
RIA	radioimmunoassay
SDS-PAGE	sodium dodecyl sulphate polyacrylamide gel electrophoresis
SSA	sulphosalicylic acid
TCA	trichloroacetic acid
TEMED	N,N,N',N'-tetramethylethylenediamine

1

One-dimensional polyacrylamide gel electrophoresis

B. DAVID HAMES

1. Introduction

Many years after its first use, polyacrylamide gel electrophoresis continues to play a major role in the experimental analysis of proteins and protein mixtures. Although two-dimensional gel separations of proteins have the highest resolving power, one-dimensional polyacrylamide gel electrophoresis is still the most widespread form of the technique since it offers sufficient resolution for most situations coupled with ease of use and the ability to process many samples simultaneously for comparative purposes. The basic protocols for preparing and running one-dimensional polyacrylamide gels have changed relatively little in recent years but there have been considerable advances in the analysis of proteins separated by polyacrylamide gel electrophoresis; for example, silver staining and a whole range of blotting methodology.

This chapter describes in detail the practicalities of one-dimensional polyacrylamide gel electrophoresis of proteins, together with theoretical considerations where appropriate. Not surprisingly, many of the methods and approaches described here are also applicable to two-dimensional separations which are described later in this book (Chapter 3).

2. Why polyacrylamide gel?

Any charged ion or group will migrate when placed in an electric field. Since proteins carry a net charge at any pH other than their isoelectric point, they too will migrate and their rate of migration will depend upon the charge density (the ratio of charge to mass) of the proteins concerned; the higher the ratio of charge to mass the faster the molecule will migrate. The application of an electric field to a protein mixture in solution will therefore result in different proteins migrating at different rates towards one of the electrodes. However, since all proteins were originally present throughout the whole solution, the separation achieved is minimal. Zone electrophoresis is a modification of this procedure whereby the mixture of molecules to be separated is placed as a narrow zone or band at a suitable distance from the

1

electrodes such that, during electrophoresis, proteins of different mobilities travel as discrete zones which gradually separate from each other as electrophoresis proceeds. In theory, separation of different proteins as discrete zones is therefore readily achieved provided their relative mobilities are sufficiently different and the distance allowed for migration is sufficiently large. However, in practice there are disadvantages to zone electrophoresis in free solution. First, any heating effects caused by electrophoresis can result in convective disturbance of the liquid column and disruption of the separating protein zones. Second, the effect of diffusion is constantly to broaden the protein zones and this continues after electrophoresis has been terminated. To minimize these effects, zone electrophoresis of proteins is rarely carried out in free solution but instead is performed in a solution stabilized within a supporting medium. As well as reducing the deleterious effects of convection and diffusion during electrophoresis, the supporting medium allows the investigator to fix the separated proteins at their final positions immediately after electrophoresis and thus avoid the loss of resolution which results from post-electrophoretic diffusion. The fixation process employed varies with the supporting medium chosen.

Many supporting media are in current use, the most popular being sheets of paper or cellulose acetate, materials such as silica gel, alumina, or cellulose which are spread as a thin layer on glass or plastic plates, and gels of agarose, starch, or polyacrylamide. These media fall into two main classes. Paper, cellulose acetate, and thin-layer materials are relatively inert and serve mainly for support and to minimize convection. Hence separation of proteins using these materials is based largely upon the charge density of the proteins at the pH selected, as with electrophoresis in free solution. In contrast, the various gels not only prevent convection and minimize diffusion but in some cases they also actively participate in the separation process by interacting with the migrating particles. These gels can be considered as porous media in which the pore size is the same order as the size of the protein molecules such that a molecular sieving effect occurs and the separation is dependent on both charge density and size. Thus two proteins of different sizes but identical charge densities would probably not be well separated by paper electrophoresis, whereas, provided the size difference is large enough, they could be separated by polyacrylamide gel electrophoresis since the molecular sieving effect would slow down the migration rate of the larger protein relative to that of the smaller protein.

The extent of molecular sieving depends on how close the gel pore size approximates the size of the migrating particle. The pore size of agarose gels is sufficiently large that molecular sieving of most protein molecules is minimal and separation is based mainly on charge density. In contrast, starch and polyacrylamide gels have pores of the same order of size as protein molecules and so these do contribute a molecular sieving effect. However, the success of starch gel electrophoresis is highly dependent on the quality of the starch gel itself, which, being prepared from a biological product, is not reproducibly good and may contain contaminants which can adversely affect the quality of the results obtained. On the other hand, polyacrylamide gel, as a synthetic polymer of acrylamide monomer,

2

can always be prepared from highly purified reagents in a reproducible manner provided that the polymerization conditions are standardized. The basic components for the polymerization reaction are commercially available at reasonable cost and high purity although for some purposes extra purification may be required. In addition, polyacrylamide gel has the advantages of being chemically inert, stable over a wide range of pH, temperature, and ionic strength, and is transparent. Finally, polyacrylamide is better suited to a size fractionation of proteins since gels with a wide range of pore sizes can be readily made whereas the range of pore sizes obtainable with starch gels is strictly limited. For these and other reasons, poly-acrylamide gels have become the medium of choice for zone electrophoresis of most proteins although starch gels have been widely used for the analysis of isoenzymes. Starch gel electrophoresis has been reviewed by Gordon (1), Smith (2), and Andrews (3). Agarose gels are used for the fractionation of molecules or complexes larger than can be handled by polyacrylamide gels, especially certain nucleic acids and nucleoproteins. In addition, agarose is widely used in immunoelectrophoresis where zone electrophoresis of proteins is coupled to immunological detection and quantitation (Chapter 4).

This chapter is concerned with analytical zone electrophoresis of proteins in polyacrylamide gels plus modifications which allow small-scale preparations of proteins of interest. Detailed quantitative approaches to analytical zone electrophoresis and special techniques for large-scale preparation of proteins by zone electrophoresis are not described.

3. Properties of polyacrylamide gel

3.1 Chemical structure

Polyacrylamide gel results from the polymerization of acrylamide monomer into long chains and the crosslinking of these by bifunctional compounds such as N,N'-methylene bisacrylamide (usually abbreviated to bisacrylamide) reacting with free functional groups at chain termini. Other crosslinking reagents have also been used to impart particular solubilization characteristics to the gel for special purposes (Section 7.5.2). The structure of the monomers and the final gel structure are shown in *Figure 1*.

3.2 Polymerization catalysts

Polymerization of acrylamide is initiated by the addition of either ammonium persulphate or riboflavin. In addition, N,N,N',N'-tetramethylethylenediamine (TEMED) or, less commonly, 3-dimethylamino-propionitrile (DMAPN) are added as accelerators of the polymerization process.

In the ammonium persulphate−TEMED system, TEMED catalyses the formation of free radicals from persulphate and these in turn initiate polymerization. Since the free base of TEMED is required, polymerization may be delayed or even

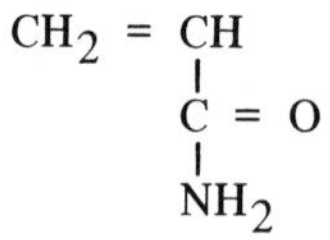

$$CH_2 = CH$$
$$|$$
$$C = O$$
$$|$$
$$NH_2$$

Acrylamide

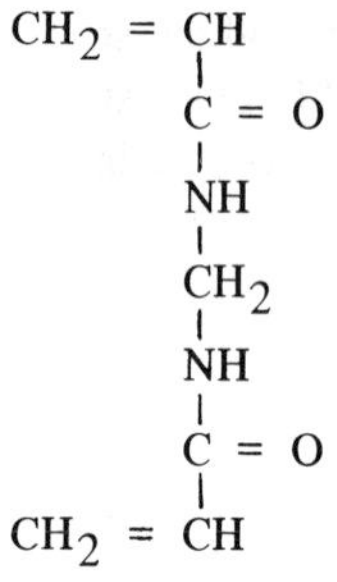

$$CH_2 = CH$$
$$|$$
$$C = O$$
$$|$$
$$NH$$
$$|$$
$$CH_2$$
$$|$$
$$NH$$
$$|$$
$$C = O$$
$$|$$
$$CH_2 = CH$$

N,N'—methylene bisacrylamide

$$-CH_2-CH-[CH_2-CH-]_n CH_2-CH-[CH_2-CH-]_n CH_2-$$

with CO/NH₂, CO/NH/CH₂/NH/CO, CO/NH₂ branches and further repeating units forming the crosslinked network.

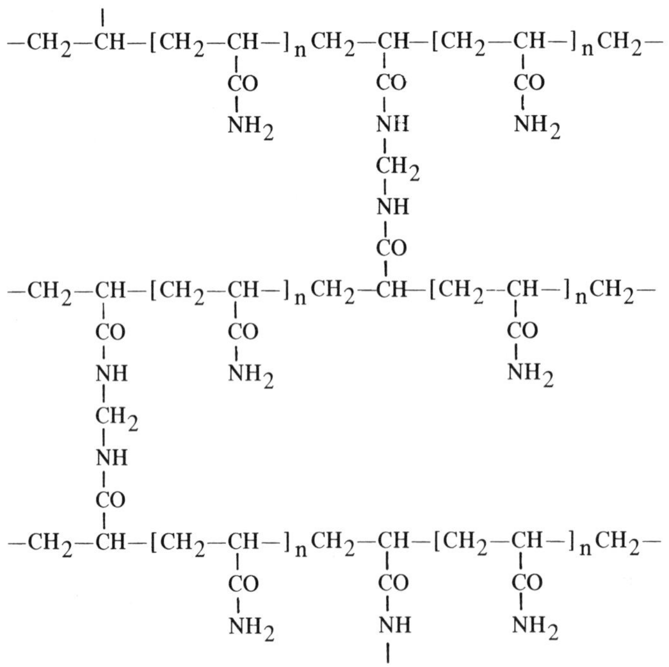

Polyacrylamide gel

Figure 1. The chemical structure of acrylamide, *N,N'*-methylene bisacrylamide, and polyacrylamide gel.

prevented at low pH. Increases in either the TEMED or ammonium persulphate concentration increase the rate of polymerization.

In contrast to chemical polymerization with persulphate, the use of the riboflavin TEMED system requires light to initiate polymerization. This causes photodecomposition of riboflavin and production of the necessary free radicals. Although gelation occurs when solutions containing only acrylamide and riboflavin are irradiated, TEMED is usually also included since under certain conditions polymerization occurs more reliably in its presence.

Oxygen inhibits polymerization and so gel mixtures are usually degassed prior to use.

3.3 Effective pore size

The effective pore size of polyacrylamide gels is greatly influenced by the total acrylamide concentration in the polymerization mixture, effective pore size decreasing as acrylamide concentration increases. Gels with concentrations of acrylamide less than about 2.5%, which are necessary for the molecular sieving of molecules above a molecular mass of 10^6, are almost fluid but this can be remedied by the inclusion of 0.5% agarose (Section 9.7). At the other extreme, polyacrylamide gels will form at over 30% acrylamide at which concentration polypeptides with a molecular mass as low as 2000 experience considerable molecular sieving. As one might expect, the choice of acrylamide concentration is critical for optimal separation of protein components by zone electrophoresis and will be considered in more detail later (Section 4.6).

The composition of any given polyacrylamide gel is now usually described by two parameters, $\%T$ and $\%C$. The $\%T$ value represents the total concentration of monomer used to produce the gel (acrylamide plus bisacrylamide) in grams per 100 ml (i.e. w/v), and $\%C$ is the percentage (by weight) of the total monomer which is the crosslinking agent. For any given total monomer concentration, the effective pore size, stiffness, brittleness, light scattering, and swelling properties of the polyacrylamide gel vary with the proportion of crosslinker used. Polymerization in the absence of crosslinker leads to the formation of random polymer chains resulting only in a viscous solution. When bisacrylamide is included in the polymerization mixture, gelation occurs with random polymer chains crosslinked at intervals to form a covalent meshwork. As the proportion of crosslinker is increased, the pore size decreases. Initial studies (4) suggested that pore size reaches a minimum when the bisacrylamide represents about 5% of the total monomer concentration (i.e. $C_{Bis} = 5\%$) irrespective of the absolute value of $\%T$. However, more recent work has shown that, above about $T = 15\%$, the proportion of crosslinker required for minimum pore size increases with the value of $\%T$ (5).

As the proportion of crosslinker is increased above the value required for minimum pore size, the acrylamide polymer chains become crosslinked to form increasingly large bundles with large spaces between them so that the effective pore size increases again. The variation in pore size is substantial. Thus for a 5% polyacrylamide gel ($T = 5\%$, $C_{Bis} = 5\%$) the pore size is approximately 20 nm but at very high proportions of bisacrylamide crosslinker ($C = 30-50\%$) the pore size can reach 500−600 nm (6).

4. Experimental approach

Before planning the detailed methodology of zone electrophoresis in polyacrylamide gels, it is worthwhile considering the general strategy for the separation to be attempted. In practice, the researcher must decide upon the physical form the gel should best take (gel rods or slabs), whether to use a dissociating or non-dissociating buffer system (and if this should be a continuous or discontinuous system), what pH and buffer ionic strength to use for the separation, and finally what gel concentration would be most appropriate for the sample to be fractionated. The answers to some of these questions will be immediately obvious once one considers the information desired and that obtainable by the methods available. Other questions will require preliminary experimentation in order that they be answered satisfactorily.

4.1 Rod or slab gels

Originally analytical zone electrophoresis in polyacrylamide made use of cylindrical rod gels in glass tubes but now flat slab gels, 0.75–1.5 mm thick, are usually preferred instead. One of the most important advantages of slab gels is that many samples, including molecular mass marker proteins, can be electrophoresed under identical conditions in a single gel such that the band patterns produced are directly comparable (*Figure 2*). In contrast, due to minor differences in polymerization efficiency, gel length, and diameter, etc., rod gels even of the same sample are rarely identical. Additional advantages of slab gels are:

- Any heat produced during electrophoresis is more easily dissipated by the standard slab gel than the thicker rod gels usually used, thus reducing distortion of protein bands due to heating effects.
- Their rectangular cross-section allows densitometry and photography with less risk of optical artefacts and they can be easily dried for storage or autoradiography.
- Less time is required for the preparation of gels for a large number of samples to be electrophoresed under identical conditions. Up to 25 samples can be easily accommodated on a standard-size slab gel.

Given so many advantages one might ask whether there is still a requirement for rod gels for analytical separations. There are several situations where rod gels will continue to be used. First, there are those instances where the investigator wishes to slice the gel after electrophoresis and either determine the radioactivity present in each slice (for radioactive proteins) or elute and assay proteins of interest by their biological activity. Although both rod and slab gels can be sliced, the normal dimensions of the gels used mean that the former have greater sample capacity and may be preferred for this reason. Furthermore, if the investigator wishes to use an automatic fractionation device he may choose rod gels since many of these devices were designed for rod gels. Second, rod gels are often the preferred format when determining the optimum pH or gel concentration for separation of protein

6

B. David Hames

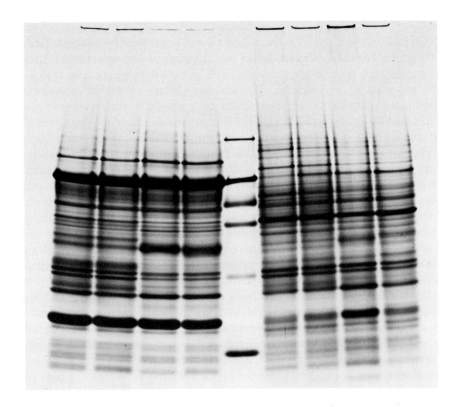

Figure 2. A typical analysis of sample polypeptide composition by SDS-PAGE using the slab gel format. Soluble proteins from the lumen of different regions of rat epididymis were fractionated using the SDS-discontinuous buffer system in a 10 – 15% linear gradient gel run at 20 mA for 5 h at room temperature. The track in the centre contained the following molecular mass marker polypeptides; β-galactosidase, bovine serum albumin, γ-globulin heavy chain, ovalbumin, γ-globulin light chain, and cytochrome c. Polypeptides were visualized after electrophoresis by staining with Coomassie Blue R250. (Photograph courtesy of Dr D.Brooks.)

components since a large number of conditions can be tested with minimum effort, especially using a rod gel apparatus modified for this purpose. Finally, increasing use is being made of two-dimensional electrophoretic techniques whereby the protein mixture is separated in the first dimension in a rod gel and then this is attached along one edge of a slab gel for electrophoretic separation of the component in the second dimension. This high-resolution method is described in detail in Chapter 3.

4.2 Dissociating or non-dissociating buffer system

The vast majority of studies employing zone electrophoresis of proteins in polyacrylamide gel use a buffer system designed to dissociate all proteins into their individual polypeptide subunits. The most common dissociating agent used is the ionic detergent, sodium dodecyl sulphate (SDS), although other cationic detergents particularly cetyltrimethylammonium bromide (CTAB) and cetylpyridinium chloride (CPCl) have also been used (7). The protein mixture is denatured by heating at 100°C in the presence of excess SDS and a thiol reagent (to cleave disulphide bonds). Under these conditions, most polypeptides bind SDS in a constant weight ratio (1.4 g of SDS per gram of polypeptide). The intrinsic charges of the polypeptide are insignificant compared to the negative charges provided by the bound detergent, so that the SDS−polypeptide complexes have essentially identical charge densities and migrate in polyacrylamide gels of the correct porosity strictly according to polypeptide size (8, 9). Thus, in addition to analysing the polypeptide composition of the sample, the investigator can determine the molecular mass of the sample polypeptides by reference to the mobility of polypeptides of known molecular mass under the same electrophoretic conditions (Section 4.7.2). The simplicity and speed of the method, plus the fact that only microgram amounts of sample proteins are required, have made SDS-polyacrylamide gel electrophoresis (SDS-PAGE) the most widely used method for determination of the complexity and molecular masses of constituent polypeptides in a protein sample. Proteins from almost any source are readily solubilized by SDS so that the method is generally applicable.

Urea has also been used as a dissociating agent and works by disrupting hydrogen bonds. High urea concentrations (~ 8 M) are necessary, a thiol reagent is also required for complete denaturation of proteins containing disulphide bonds, and urea must be present during electrophoresis to maintain the denatured state. The advantage of urea for some applications is that it does not affect the intrinsic charge of proteins and so separation of the constituent polypeptides will be on the basis of both size *and* charge, in contrast to the use of SDS. One disadvantage is that this combined size and charge fractionation prevents accurate molecular mass determinations. Furthermore, urea is not as good as SDS in dissociating proteins; up to 50% of a complex protein mixture may fail to enter the gel whereas at least 90% of even crude cell lysates will enter the gel if SDS is the dissociating agent used. However, some proteins require the presence of both urea and SDS if most of the material is to enter the gel (Section 9.2 and ref. 282).

In contrast to the above systems, zone electrophoresis of native proteins under non-dissociating buffer conditions is designed to fractionate a protein mixture in such a way that subunit interaction, native protein conformation, and biological activity are preserved. Separation of the native proteins occurs on the basis of both size *and* charge. Further details of the types of non-dissociating buffer systems available are given in the next section and later in this chapter.

An interesting development which has yet to be fully exploited is electrophoresis under acid conditions in polyacrylamide gels containing urea and non-ionic detergents. The proteins migrate as cations and are differentially retarded depending on their ability to bind the detergent and to form mixed micelles between the detergent and the hydrophobic regions of the polypeptide chains. Under these conditions, solubility depends on the hydrophobicity of the proteins as well as size and charge (10). The continuous buffer systems originally devised following this approach were designed to separate histones (11) as was a subsequent discontinuous system (12). However, this approach to separate two proteins of similar size and charge but differing hydrophobicity should be applicable to a variety of proteins, as shown by the successful fractionation of mammalian haemoglobins (13). Further details are given in Section 9.2.1.

4.3 Continuous or discontinuous (multiphasic) buffer system

Zone electrophoretic systems in which the same buffer ions are present throughout the sample, gel, and electrode vessel reservoirs (albeit possibly at different concentrations in each) at constant pH, are referred to as continuous buffer systems. In these systems the protein sample is loaded directly onto the gel in which the separation will occur, the resolving gel (*Figure 3a,b*) which has pores sufficiently

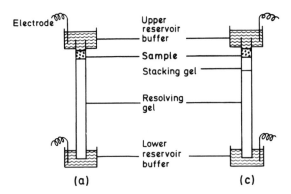

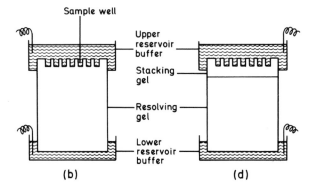

Figure 3. Use of continuous and discontinuous buffer systems with rod and slabs gels. (a) Continuous buffer system used in conjunction with a rod gel; the sample is loaded directly onto the resolving gel. (b) Continuous buffer system used in conjunction with a slab gel; samples are loaded into wells formed directly in the resolving gel. (c) and (d) Discontinuous buffer system used in conjunction with rod and slab gels, respectively; samples are loaded directly onto the stacking gel or into wells formed in the stacking gel, respectively.

small to cause a size fractionation of the sample components during electrophoresis. In contrast, discontinuous (or multiphasic) buffer systems employ different buffer ions in the gel compared to those in the electrode reservoirs. Most discontinuous buffer systems have discontinuities of both buffer composition and pH. In these, the sample is loaded onto a large-pore 'stacking' gel polymerized on top of the small-pore resolving gel (*Figure 3c,d*).

The major advantage of these discontinuous buffer systems over continuous buffer systems is that relatively large volumes of dilute protein samples can be applied to the gels but good resolution of sample components can still be obtained. The reason for this is that the proteins are concentrated into extremely narrow zones (or stacks) during migration through the large-pore stacking gel prior to their separation during electrophoresis in the small-pore resolving gel. The production of thin protein starting zones by discontinuous systems can be understood by consideration of the original discontinuous system of Ornstein (14) and Davis (15) as an example.

Consider a weak acid, such as glycine, at a pH near its pK_a. Only part of the population of molecules will be negatively charged at any one time. If x is the proportion of the total molecules which is dissociated, and hence present as negatively charged ions, then each molecule can be regarded as being charged for x proportion of the time and uncharged the rest of the time. Hence, if the mobility of the charged species is M, the effective mobility $= Mx$ and the velocity of migration $= V. Mx$ (where V is the voltage gradient). Therefore, an ion of lower mobility can migrate as fast as one with higher mobility if the products of voltage and effective mobility are equal. Now, in the Ornstein–Davis system, the sample itself plus the stacking gel contain Tris–HCl buffer whilst the upper electrode reservoir contains Tris–glycine buffer (*Figure 4a*). At the pH of the sample and stacking gel (pH 6.7),

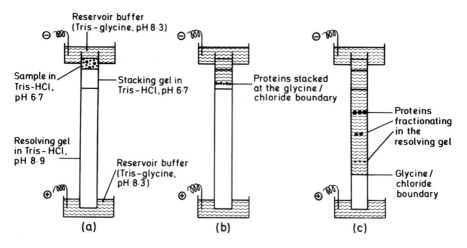

Figure 4. Operation of Ornstein – Davis discontinuous buffer system. **(a)** At the beginning of electrophoresis, **(b)** during stacking, **(c)** during separation in the resolving gel. For a detailed explanation, see text.

glycine is very poorly dissociated so that its effective mobility is low. Chloride ions have a much higher mobility at this pH whilst the mobilities of proteins are intermediate between that of chloride and glycine. The moment the voltage is applied, the chloride ions (the leading ions) migrate away from the glycine (the trailing ions) leaving behind a zone of lower conductivity. Since conductivity is inversely proportional to field strength, this zone attains a higher voltage gradient which now accelerates the glycine so that it keeps up with the chloride ions. A steady state is established where the products of mobility and voltage gradient for glycine and chloride are equal, these charged species now moving at the same velocity with a sharp boundary between them. As this glycine/chloride boundary moves through the sample and then the stacking gel, a low-voltage gradient moves before the moving boundary and a high-voltage gradient after it. Any proteins in front of the moving boundary are rapidly overtaken since they have a lower velocity than the chloride ions. Behind the moving boundary, in the higher voltage gradient, the proteins have a higher velocity than glycine. Thus the moving boundary sweeps up the proteins so that they become concentrated into very thin zones or 'stacks', one stacked upon the other in order of decreasing mobility, with the last protein followed immediately by glycine (*Figure 4b*). The concentration of protein in the moving boundary depends only on the concentration of Tris−HCl in the sample and the stacking gel and not on the initial concentration of proteins in the sample. The thickness of the steady-state protein stack is independent of the initial concentration of protein in the sample and dependent only on the total amount of protein loaded onto the gel. At the protein loads typical of analytical polyacrylamide gel electrophoresis, the protein stacks are only a few microns thick. Since the stacking gel is a large-pore gel, no molecular sieving occurs at this stage.

When the moving boundary reaches the interface of the stacking and resolving gels, the pH value of the gel increases markedly and this leads to a large increase in the degree of dissociation of the glycine. Therefore the effective mobility of the glycine increases so that glycine overtakes the proteins and now migrates directly behind the chloride ions. At the same time, the gel pore size decreases markedly, retarding the migration of the proteins by molecular sieving. These two effects cause the proteins to be unstacked. The proteins now move in a zone of uniform voltage gradient and pH value (now Tris−glycine, pH 9.5, instead of the original resolving gel buffer of Tris−HCl, pH 8.9) and are separated according to their intrinsic charge and size, the latter depending on the molecular sieving effect of the small-pore resolving gel (*Figure 4c*).

Because of the high resolution obtainable with discontinuous buffer systems, the SDS-discontinuous system (a discontinuous buffer system with SDS added to all buffers) is usually the system of choice for high-resolution fractionation of protein mixtures under dissociating conditions. The most commonly used SDS-discontinuous system is that originally described by Laemmli (16) based upon the discontinuous Ornstein−Davis buffer system with SDS present. Indeed slab gels are now used almost exclusively with this Laemmli SDS-discontinuous system. Neville (17) has described an alternative SDS-discontinuous system based on Tris−borate buffer.

Rod gels are often used with either the Laemmli SDS-discontinuous system or a continuous buffer system, usually sodium phosphate buffer containing SDS. The merits of the SDS−phosphate system are its simplicity and the fact that it is less susceptible than the Laemmli system to artefacts caused by contaminants in some commercial sources of SDS. Nevertheless the superior resolving power of the Laemmli SDS-discontinuous buffer system means that this will undoubtedly continue to be the main dissociating buffer system used, especially for complex protein mixtures. The major drawback of this system is that, even in 15% gels, proteins smaller than about 12 000 molecular mass are very poorly separated (see Section 4.7.2). Thus if the polypeptides to be fractionated fall into this size range, alternative dissociating buffer systems must be used (see Section 9.1). Most of these rely on the combined use of urea and SDS to fractionate the oligopeptides−SDS complexes but an alternative protocol, using Tricine as the tracking ion in an SDS-discontinuous buffer system instead of glycine, has recently been reported. This latter system, described in Section 9.1 can fractionate proteins in the size range 1000−100 000 molecular mass.

Unfortunately there is no universal buffer system ideal for electrophoretic separations of native proteins. The choice must be based upon the conditions required to maintain activity of the proteins of interest whilst achieving sufficient resolution of the protein components for the problem under investigation. The major advantage of discontinuous systems is still the higher resolution obtainable compared to the use of continuous buffer systems. However, for the separation of native proteins, continuous buffer systems offer the advantage that the precise buffer composition and buffer pH is known and that the pH remains constant throughout the separation. This may be of overriding importance if the native protein of interest is particularly labile. Based on a knowledge of pK and ionic mobility data of buffer constituents and the theory of discontinuous (multiphasic) zone electrophoresis, several thousand discontinuous buffer systems have been designed for use at any pH in the useful range of pH 2.5−11.0 and are available as a computer output (18). However, in practice, most separations can be achieved using one of the following widely-used buffer systems: a low pH system resolving proteins at pH 3.8 (19), a 'neutral' pH system resolving proteins at pH 8.0 (20), the Ornstein−Davis high pH system resolving proteins at pH 9.5 (15) as modified by Laemmli (16) but lacking SDS. Details are given in Section 6.3.2.

Some native proteins aggregate and may precipitate at the very high protein concentrations reached in the sharply stacked zones of discontinuous buffer systems and then either fail to enter the resolving gel or cause 'streaking'. This phenomenon occurs when aggregated protein accumulates at the gel surface and then slowly dissolves during electrophoresis, causing protein streaks running parallel to the direction of migration. The problem can sometimes be overcome by using a continuous buffer system since this avoids the concentration of sample proteins to such an extent that they precipitate. Fortunately the use of a continuous buffer system can still give good resolution provided that certain conditions are met. First, the sample must be applied in as small a volume as possible to give a thin starting zone.

Depending on the method used to detect the separated protein zones after electrophoresis this usually requires that a concentrated solution of sample proteins is available (about 1 mg/ml). Further zone-sharpening occurs as the proteins enter the single resolving gel in which the mobilities of the proteins in question are considerably less than in free solution. Second, additional zone-sharpening can be obtained by loading the protein sample in a buffer which has a lower ionic strength than that of the gel and electrode buffer (say 10% −20%). The proteins will initially be in a zone of lower ionic strength (lower conductivity) and hence higher voltage and so move still faster in free solution, slowing down as they move into the gel as a result of the sieving effect produced by the gel and the drop in voltage gradient as they enter the more concentrated gel buffer. Virtually any buffer can be used for electrophoresis of native proteins in a continuous buffer format and so for most proteins it is a matter of experimentation to determine which buffer is most appropriate. McLellan (21) has devised several useful buffer systems for electrophoresis at different pH values in which both the anionic and cationic components act as buffering agents. Some common non-dissociating buffers are given later (Section 6.3.2). However, certain classes of proteins such as histones, nuclear non-histone proteins, ribosomal proteins, and membrane proteins are not soluble in the usual non-dissociating buffers so that to analyse these on a charge and size basis requires additional agents such as urea, chloral hydrate, or non-ionic detergents. These are discussed in Section 9.2.

4.4 Choice of pH

Polyacrylamide gel electrophoresis can be carried out at a pH anywhere between 2.5 and 11, but in practice the limits are pH 3 and 10 since some protein hydrolytic reactions (such as deamidation) occur at the extremes of pH.

In SDS-PAGE the SDS−polypeptide complexes are negatively charged over a wide range of pH such that the pH of the SDS-phosphate continuous buffer system is not critical. The pH of the SDS-discontinuous buffer system is important in such separations only in that it permits sample concentration via the stacking phenomenon to occur. In contrast, pH *is* critical in polyacrylamide gel electrophoresis of proteins in non-dissociating buffers where native proteins are separating on the basis of size and charge density. Here changes in the pH alter the net charge of the protein components and hence the separation which can be achieved.

In choosing the pH of a buffer to be used in a continuous buffer system for native proteins, the initial consideration must be the pH range over which the proteins of interest are stable. This pH range may be narrower if one wishes the proteins to retain biological activity (either for detection purposes or because the polyacrylamide gel electrophoretic step is being used preparatively) than if detection is on the basis of standard protein stains where all that is required is to prevent dissociation of the native protein into its substituent subunits. Within this pH range, the selection of pH is a compromise between two opposing considerations. The further the pH of the electrophoresis buffer from the isoelectric points of the proteins to be separated, the higher the charge on the proteins. This leads to shorter times required for electrophoretic separation and hence reduced band spreading due to diffusion. On the other hand, the closer the pH to the isoelectric points of the proteins the greater the charge differences between proteins, thus increasing the chance of separation. Many proteins have isoelectric points in the range pH $4-7$ so a common compromise is to use buffers in the pH $8.0-9.5$ region. Ideally a systematic study should be carried out in which the pH is progressively adjusted nearer the isoelectric points of the proteins until a pH is found which yields optimal resolution and separation for the protein mixture. A similar study can be made with basic proteins which need to be separated at acid pH.

4.5 Choice of polymerization catalyst

For most situations the choice between the ammonium persulphate$-$TEMED and riboflavin$-$TEMED systems is a matter of personal preference, although ammonium persulphate is the usual initiator for resolving gels. Stacking gels of discontinuous buffer systems are frequently polymerized with either ammonium persulphate or riboflavin. If native proteins are being separated which are particularly sensitive to persulphate ions and yet these have been used to polymerize the resolving gel in a continuous buffer system, excess persulphate ions may need to be removed by a period of pre-electrophoresis prior to loading the sample. This is not possible for discontinuous buffer systems and so in these cases it would be wise to polymerize at least the large-pore stacking gel with riboflavin.

The other advantage of using riboflavin as catalyst is that polymerization will not begin until the gel mixture is illuminated. This is useful if there is likely to be some delay in overlayering the mixture with buffer, or adjusting the sample comb in slab gel electrophoresis, or if multiple gels are being made. If riboflavin is used as catalyst in the resolving gel, the time period of illumination should be standardized since this affects the gel porosity and hence protein mobility.

In cases where very acidic buffers are employed, it may be advisable to use a modified catalyst system. Sodium sulphite has been used effectively in these situations (22) as has ascorbic acid and iron sulphate (23).

4.6 Choice of gel concentration

Separation of protein bands via zone electrophoresis in polyacrylamide gels refers to the distance between bands irrespective of band width, whereas resolution refers to separation relative to band width. Resolution between two components is influenced by all the factors that affect band sharpness. Thus the use of small amounts of sample proteins to prevent overloading, the use of a discontinuous buffer system with dilute protein samples, and removal of high concentrations of ions from samples will all maximize resolution. The major factors affecting separation are the pH of the gel (in polyacrylamide gel electrophoresis of native proteins but not in SDS-PAGE; Section 4.4) and the gel concentration. At the extremes, choice of an incorrect gel concentration can lead to total exclusion of the proteins which would be unable to

B. David Hames

penetrate the gel, or conversely, lack of fractionation with the proteins running with
the buffer front. Between these two extremes the proteins will be separated to variable
extents depending on the gel concentration. Although there is a single gel
concentration which is optimal for the resolution of any two proteins (see below)
there can be no gel concentration which will give maximum separation of all the
components from each other in a complex protein mixture. Therefore the gel
concentration chosen for fractionating a heterogeneous mixture is usually one which
gives an adequate display of all the components of interest. A reasonable approach
for initial analysis of a protein mixture using a non-dissociating buffer system would
be to start with a 7.5% acrylamide gel and then attempt a number of gel concentrations between 5% and 15% and choose the most desirable concentration for further
studies. A similar strategy can be attempted for use with SDS-PAGE unless the
molecular mass range of the polypeptide mixture is known, when a suitable gel
concentration can be selected from existing calibration curves of polypeptide
molecular mass versus mobility (see later, *Figure 6*). An alternative approach, which
is strongly recommended for initial analysis of protein mixtures by SDS-PAGE, is
to use a concentration gradient polyacrylamide gel in which the concentration of
acrylamide increases (and hence pore size decreases) in the direction of protein
migration. Gradient gels have two considerable advantages over uniform concentration gels in the analysis of complex protein mixtures. First, they are able to fractionate
proteins over a wider range of molecular masses than any uniform concentration
gel. Second, the gradient in pore size causes significant sharpening of protein bands
during migration. The result is that gradient gels are unsurpassed in the resolution
of protein mixtures covering a wide range of molecular masses where a display of
all the components on one gel is desired. A useful gradient gel for initial SDS-PAGE
analysis is a 5−20% or 6−18% linear gradient slab gel. Details of gradient gel
preparation and use are given in Section 9.3.

If the aim of polyacrylamide gel electrophoresis is to obtain optimal resolution
of any two proteins, rather than to simply display the protein components of interest,
this will only be achieved using the optimal concentration of acrylamide in a uniform
concentration polyacrylamide gel. The gel concentration required will depend on
the size and charge of the proteins under study and can be determined by measuring
the mobility of each protein in a series of gels of different acrylamide concentration
and then constructing a Ferguson plot (24) for each protein of interest, that is a plot
of $\log_{10}$ relative mobility (R_f) versus gel concentration, $\%T$, (percentage total
monomer, i.e. grams acrylamide plus bisacrylamide per 100 ml). The relative
mobility, R_f, refers to the mobility of the protein of interest measured with reference
to a marker protein or to a tracking dye where:

$$R_f = \frac{\text{distance migrated by protein}}{\text{distance migrated by dye}} .$$

Each Ferguson plot can be characterized by its slope K_R and its ordinate intercept
Y_0. Since the Ferguson plot relates to mobility during electrophoresis when only

15

the gel pore size (as determined by %T) is varying, then the slope of the Ferguson plot, K_R, is a measure of the retardation of the protein by the gel; that is, K_R is a retardation coefficient which can be related to molecular size. The ordinate intercept, Y_0 (when %$T = 0$), is a measure of the mobility of the protein in free solution. Rodbard *et al.* (25) have classified four sets of separation problems which are highlighted by such an analysis (*Figure 5*).

Case A is the situation when the proteins have identical charge densities and show identical mobilities in free solution. In polyacrylamide gels these proteins migrate strictly according to size. This is the situation approximated to by most SDS-denatured proteins (Section 4.7.2). In *Case B*, the protein with the greater mobility in free solution (greater charge density) also has the smaller size. Thus the effects of fractionation on the basis of size and charge are synergistic and increased gel concentration leads to increased separation. In *Case C*, the larger protein has the higher free mobility such that size and charge fractionation are antagonistic. This case is commonly found in non-dissociating systems. In *Case D*, the proteins have the same size but different free mobilities (e.g. isozymes such as lactate dehydrogenase, haemoglobins). In this case, increasing acrylamide concentration has no effect upon the relative separation between the two proteins and one should consider using a charge separation method such as electrofocusing or isotachophoresis.

The use of Ferguson plots to determine conditions for the optimal separation and resolution of proteins by polyacrylamide gel electrophoresis is detailed in ref. 3.

4.7 Molecular mass estimation

4.7.1 Native proteins

During the electrophoresis of native proteins in polyacrylamide gel, separation takes place according to both size and charge differences of the molecules. By constructing Ferguson plots (*Figure 5*), the charge aspect is eliminated in that the slope, K_R (retardation coefficient), is a measure only of molecular size (see Section 4.6). Indeed, Hedrick and Smith (26) found that there is a linear relationship between K_R and the molecular mass of native proteins so that, by first using a series of standard native

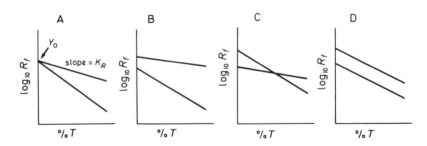

Figure 5. Classification of separation problems as determined by Ferguson plot analysis. See text for details.

proteins of known molecular mass to construct a plot of K_R against molecular mass, one can determine the molecular mass of any sample protein simply by determining its K_R and then referring to the standard curve [although Rodbard and Chrambach (27) have argued that a better relationship exists between $(K_R)^{1/2}$ and the molecular radius for globular proteins]. In the original procedure, the investigator analysed the mobility of the native protein in multiple gels of different $\%T$ to construct the Ferguson plot (e.g. see ref. 3). It is much simpler to use a single slab polyacrylamide gel with a transverse gradient of acrylamide perpendicular to the direction of electrophoretic migration (272, Section 9.4), although this is reportedly a less accurate method (224).

Ferguson plot analysis has been very useful for determining the molecular size of globular native proteins. Nevertheless, the problem of molecular mass determination with native proteins is that it is only valid if the standard proteins used to generate calibration curves have the same shape with the same degree of hydration and partial specific volume. For this reason, molecular mass determination of proteins using polyacrylamide gel electrophoresis is now usually performed in the presence of SDS after reduction of protein disulphide bonds with a thiol reagent. The ionic detergent SDS virtually eliminates conformational and charge density differences amongst proteins and reduces the effect of variability in partial specific volume and hydration. However, it should be noted that the molecular mass obtained using SDS-PAGE is the polypeptide subunit molecular mass and not that of the native protein if it is oligomeric.

4.7.2 Denatured proteins (SDS-PAGE)

i. Basic method

When denatured by heating in the presence of excess SDS and a thiol reagent [usually 2-mercaptoethanol or dithiothreitol (DTT)], most polypeptides bind SDS in a constant weight ratio such that they have essentially identical charge densities and migrate in polyacrylamide gels of the correct porosity according to polypeptide size (Section 4.2). Under these conditions, a plot of $\log_{10}$ polypeptide molecular mass versus relative mobility (R_f) reveals a straight-line relationship (8,9). The approach is therefore to electrophorese a set of marker polypeptides of known molecular mass and use the distance migrated by each to construct a standard curve from which the molecular mass of the same polypeptides can be calculated based on their mobility under the same electrophoretic conditions. When using rod gels the marker polypeptides and sample polypeptides are necessarily run on separate gels. Since rod gels experience considerable variability from gel to gel in terms of distance migrated by any given polypeptide, it is always wise to express all distances migrated as R_f values relative to the dye front. The position of the dye front is marked by insertion of a piece of thin wire into the gel or injection of indian ink before staining to locate the polypeptide bands. However, a better method is to use slab gels for molecular mass estimation since sample and marker polypeptides can be electrophoresed on a single gel and therefore under identical conditions. In this situation it is sufficient simply to measure the distance migrated by all the polypeptides after staining (measured from the top of the resolving gel), and to construct a standard curve by plotting $\log_{10}$ molecular mass against the distance migrated by the marker polypeptides (*Figure 6*). Practical details are given in Section 6.6.1.

It must be emphasized that for any given gel concentration the relationship between $\log_{10}$ molecular mass and relative mobility is linear over only a limited range of

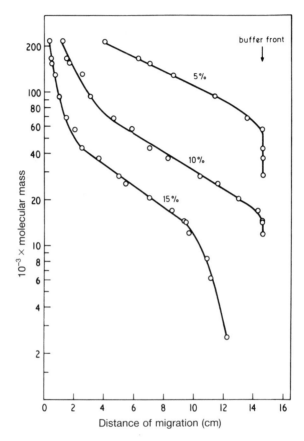

Figure 6. Calibration curves of $\log_{10}$ polypeptide molecular mass versus distance of migration during SDS-PAGE in slab gels using the SDS-discontinuous buffer system. The polyacrylamide gels used were uniform concentration 5%, 10%, or 15%. The polypeptide markers, in order of decreasing molecular mass, were myosin (M_r 212 000), RNA polymerase β'(165 000) and β(155 000) subunits, β-galactosidase (130 000), phosphorylase a (92 500), bovine serum albumin (68 000), catalase (57 500), ovalbumin (43 000), glyceraldehyde-3-phosphate dehydrogenase (36 000), carbonic anhydrase (29 000), chymotrypsinogen A (25 700), soybean trypsin inhibitor (20 100), horse heart myoglobin (16 950), horse heart myoglobin cyanogen bromide cleavage fragments I + II (14 404), lysozyme (14 300), cytochrome *c* (11 700), horse heart myoglobin cyanogen bromide cleavage fragments I (8 159), and II (6 214), and III (2 512). The molecular masses of horse heart myoglobin and its cyanogen bromide cleavage fragments are calculated from the primary sequence given in ref. 38. References to the molecular masses of the other polypeptides are given in *Table 6*.

18

molecular mass. As a general guide for the SDS–phosphate buffer system, the linear relationship holds true over the following ranges: 15% acrylamide, M_r 12 000–45 000; 10% acrylamide, M_r 15 000–70 000; 5% acrylamide, M_r 25 000–200 000. Improved resolution of dilute protein samples can be achieved using the Laemmli SDS-discontinuous buffer system instead, with the linear relationship holding over a similar range for 15% gels. However, with 10% and 5% gels, polypeptides with molecular masses less than about 16 000 and 60 000, respectively, migrate with the buffer front (*Figure 6*). For the fractionation of polypeptides smaller than about 12 000–15 000 molecular mass, see Section 9.1.

ii. Anomalous behaviour of polypeptides.
It is implicit in the linear relationship between $\log_{10}$ molecular mass and mobility for SDS-PAGE that all polypeptides bind a constant mass ratio of SDS. Therefore it is important that an excess of SDS to polypeptide of at least 3:1 is present during protein dissociation; otherwise polypeptides which do obey the linear relationship under the correct conditions may fail to do so. An excess of thiol reagent is also essential to ensure breakage of disulphide bridges which otherwise oppose denaturation and prevent saturation of the polypeptide with SDS. Anomalies can be detected by carrying out the electrophoresis of both sample proteins and molecular mass markers at a number of different gel concentrations and compiling a Ferguson plot for each polypeptide. Ideally, all SDS–polypeptide complexes have identical charge densities and so have an identical Y_0 value (*Figure 5*, case A). In practice this ideal is often not achieved, but most polypeptides which behave 'normally' do approximate to this situation and have an R_f value similar to that of the standard polypeptides at or near $\%T = 0$ (e.g. *Figure 7*) whereas anomalous polypeptides show up as having markedly different Y_0 values.

Many glycoproteins behave anomalously even when SDS and thiol reagent are in excess, probably because they bind SDS only to the protein part of the molecule. The reduced net charge resulting from reduced SDS binding lowers the polypeptide mobility during electrophoresis, yielding artefactually high molecular mass estimates. However, with increasing polyacrylamide gel concentration, molecular sieving predominates over the charge effect and the apparent molecular masses of glycoproteins decrease and approach their real molecular masses. Based on this, Segrest and Jackson (28) have presented a method for molecular mass estimation of a glycoprotein which involves determining the apparent molecular mass, relative to standard proteins, at a number of polyacrylamide gel concentrations to yield an asymptotic minimal molecular mass (*Figure 8*) which approximates to the real molecular mass of the glycoprotein. An alternative approach may be to use concentration gradient gels (Section 9.3).

Malelyated polypeptides, collagenous polypeptides and other polypeptides with high proline contents all give abnormally high molecular masses by SDS-PAGE (see ref. 278), as do very basic proteins (such as histones, Section 9.2.1), and very acidic proteins (279). Alterations of even a single amino acid may change the mobility of proteins in SDS-PAGE (280, 281) up to ±10%.

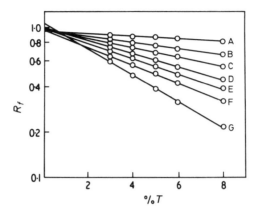

Figure 7. A typical Ferguson plot analysis of standard polypeptides analysed by SDS-PAGE. A, myoglobin; B, chymotrypsinogen; C, lactate dehydrogenase; D, ovalbumin; E, glutamate dehydrogenase; F, bovine serum albumin; G, phosphorylase a. Redrawn with permission from ref. 39 (see also refs 17 and 40).

Finally, polypeptides with a molecular mass below about 12 000 are not well resolved on uniform concentration polyacrylamide gels, even 15% polyacrylamide, when SDS is the sole dissociating agent (*Figure 6*). Gel systems for the separation of these oligopeptides are described in Section 9.1.

iii. SDS-PAGE in non-reducing conditions

Proteins which contain disulphide bonds often have different mobilities and hence different apparent molecular masses when analysed by SDS-PAGE in the presence and absence of reducing agents. This occurs not only when a protein is polymeric and the reduction of inter-polypeptide disulphide bands leads to dissociation into the individual subunits, but also when inter-polypeptide disulphide bands are cleaved resulting in more unfolding of the polypeptide and increased accessibility for SDS binding. Theoretically, therefore, a comparison of protein mobilities under these two sets of conditions can yield useful information on protein structure. However, when complex multi-subunit proteins or macromolecular protein complexes are examined separately by SDS-PAGE under reducing and non-reducing conditions, it can be extremely difficult and even impossible to interpret the banding pattern accurately in terms of protein structure. A novel variation of SDS-PAGE circumvents this difficulty (29). The system used is the standard SDS-PAGE discontinuous buffer system of Laemmli (16), Section 6.3.1, but one half of the sample is denatured in SDS-sample buffer containing 2-mercaptoethanol or DTT while the other half is denatured in SDS-buffer lacking reducing agent. Aliquots of the reduced and non-reduced samples are then loaded into adjacent wells of the slab gel and co-electrophoresed. This format greatly aids interpretation of the effect of reduction, especially since during electrophoresis some lateral diffusion of reducing agent occurs

B. David Hames

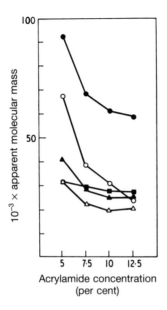

Figure 8. Molecular mass estimation of glycoproteins by electrophoresis in polyacrylamide gels of different acrylamide concentration (28). Human erythrocyte glycoprotein (●); human erythrocyte membrane tryptic glycoprotein (○); fragmented human erythrocyte membrane glycopeptide (■); porcine ribonuclease, higher molecular mass species (▲); porcine ribonuclease, lower molecular mass species (△) (Reproduced from ref. 28 with permission.) The glycoproteins were electrophoresed in uniform concentration polyacrylamide gels at the gel concentrations shown on the abscissa, using the SDS – phosphate buffer system. The apparent molecular mass of each glycoprotein was determined for any given gel concentration by comparison of the electrophoretic mobility of the glycoprotein with that of standard polypeptides of known molecular mass. The anomalous behaviour of glycoproteins in SDS-PAGE is shown by the decreasing glycoprotein apparent molecular mass with increasing acrylamide concentration; non-glycosylated polypeptides give consistent molecular masses at all acrylamide concentrations. Although the asymptotic minimal molecular mass for each glycoprotein shown here approximates to the real molecular mass, sialic acid-containing and desialylated glycoproteins behave quite differently in SDS-PAGE even by this protocol (28) so that the molecular mass obtained for any unknown glycoprotein is still very much only an approximation.

which results in partial reduction of neighbouring 'non-reduced' samples and the creation of arcs of protein which clearly link the fully reduced and non-reduced forms of the protein (*Figure 9*).

iv. SDS-PAGE in concentration gradient gels
Although, in the past, polypeptide molecular mass estimation using SDS-PAGE has usually been carried out in gels of uniform acrylamide concentration, concentration gradient gels are now often used for this purpose (see Section 9.3).

21

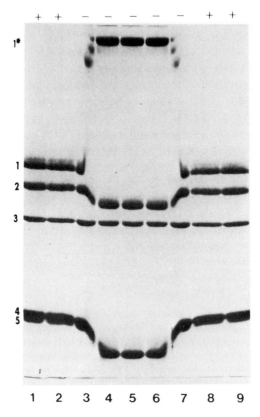

Figure 9. SDS-PAGE profiles of (1) 1gG3 heavy chain (2) 1gG1 heavy chain (3) pig skeletal muscle actin (4) kappa light chain (5) kappa light chain derived from reduced 1gG3k electrophoresed in lanes 1, 2, 8 and 9 with (+) or in lanes 3 – 7 without (–) prior addition of 5% β-mercaptoethanol to the samples. In the absence of reducing conditions, the 1gG3 heavy chain and kappa light chain migrate together as intact 1gG3k (1*). The molecular masses of marker proteins whose migration positions are shown to the left of the gel are 94 000 (phosphorylase a), 68 000 (bovine serum albumin), 42 000 (actin) and 31 000 (DNase I). (From ref. 29 with permission.)

5. Apparatus

5.1 Gel holders and electrophoresis tanks

During electrophoresis, heat is generated by the passage of electric current through the gel. Since the mobility of migrating ions is increased as the temperature rises, it is apparent that the temperature must be controlled if electrophoretic separations are to be reproducible. Furthermore, if the heating effect is significant, there will be a temperature gradient from the centre of the gel to the gel surface which will cause proteins to migrate faster at the gel centre than the surface, resulting in distorted bands. The increased temperature may also inactivate labile proteins.

The actual temperature will depend on both the rate of heat production and its rate of dissipation. The rate of heat production is proportional to the current passed. Hence heat production is minimized by using low-conductivity buffers or, if high-conductivity buffers must be used, by extending the duration of the run using a lower power input. Suitable buffers and electrophoresis conditions are described later (Section 6). Heat dissipation is facilitated by using rod and slab gels as thin as possible to allow the efficient loss of heat whilst being thick enough to allow a reasonable loading capacity. In practice, for one-dimensional zone electrophoresis of proteins on an analytical scale, gels should not be thicker than about 0.6 cm and preferably less. In addition, electrophoretic separations of labile native proteins in non-dissociating buffer systems are best carried out using an apparatus which has some means of cooling the gel. Cooling is often not necessary for simple comparative analyses of protein mixtures by SDS-PAGE using a slab gel apparatus since minor changes in the mobility of sample polypeptides due to small temperature variations are compensated for by changes in the mobility of the molecular mass marker polypeptides. However, with SDS-PAGE as with all other buffer systems, accurate temperature regulation is essential for quantitative studies of protein mobility.

5.1.1 Rod gel apparatus

The cylindrical tubes in which the gels are formed should be made of glass, usually with an internal diameter of 0.4−0.6 cm and a 0.1 cm wall thickness for one-dimensional electrophoretic separations, although gels are often much narrower for some two-dimensional separations (Chapter 3). Perspex tubes should be avoided for analytical zone electrophoresis of proteins since they do not allow heat dissipation as efficiently as glass tubes. The exact length of the tubing employed varies from one laboratory to another and will depend on the experience of the user. In general, rod gel tubes should be about 2−3 cm longer than the final gel. A convenient length is 12 cm, which allows for the use of continuous buffer systems with a single resolving gel 9−10 cm long or discontinuous buffer systems with a resolving gel of similar length but with an additional stacking gel (about 1 cm). Tubing should be of uniform diameter throughout its length and fire polished at the ends.

During gel preparation the rod gel holders must be kept exactly vertical to ensure a horizontal, sharp meniscus. Racks for holding the gel tubes in this way are available commercially or can be made easily in the laboratory from wooden battens and suitably sized clips. For electrophoresis, the rod gels are transferred to an apparatus such as that shown in *Figure 10*. The apparatus is made mainly of Perspex. The gel tubes are suspended by means of silicone-rubber grommets held in holes drilled in the base of the upper buffer reservoir; any holes not required are blocked with rubber stoppers. In the apparatus shown, most of the gel length is immersed in buffer held in the lower buffer reservoir which can be cooled by means of a water jacket. It is important that the platinum electrodes should be the same distance from each gel tube and hence the electrodes are circular in shape and located centrally. Simpler

Figure 10. A typical rod gel apparatus. The model shown (Bio-Rad model 150A) is constructed of Perspex, incorporates a water-jacketed low buffer reservoir, and holds 12 tubes, 0.7 – 0.85 cm o.d. (optional 0.5 cm o.d.) × 15 cm long. A more recent model (Bio-Rad 175) has interchangeable buffer reservoirs to hold a range of tube sizes from 0.4 – 1.7 cm o.d.

apparatuses may lack the water jacket and rely solely on the large heat capacity of the lower buffer reservoir for cooling. More sophisticated apparatuses have interchangeable upper buffer reservoirs capable of holding rod gels of different diameters, plus a number of other refinements. Some commercial vertical slab gel apparatus also has adaptors available to enable it to be used to run rod gels (e.g. Hoefer Scientific, Bio-Rad).

5.1.2 Slab gel apparatus

Apparatus has been designed for protein and nucleic acid fractionation either by vertical or horizontal slab gel electrophoresis. Whilst considerable use is made of

horizontal slab electrophoresis for protein electrofocusing and immunoelectrophoresis, zone electrophoresis of proteins in polyacrylamide slab gels is almost always
now carried out in the vertical format; indeed discontinuous buffer separations can
only be done using vertical gels. One of the most popular designs has been that of
Studier (30). A modification of this apparatus which can be easily constructed in
a laboratory workshop (*Protocol 1*) is shown in *Figure 11*. Essentially the apparatus
consists of two buffer reservoirs, the upper of which is notched, supported on an
integral plastic stand. The gel is formed between two glass plates each about 0.3 cm
thick. One plate is rectangular in shape and the second is the same size but with
a notch cut in one edge. The two plates are placed together to form the gel holder
with a plastic spacer running down each vertical side of the sandwich. Sample wells
are formed in the gel during polymerization using a plastic sample comb; sample
well number and dimensions can then easily be altered by changing the number and
dimensions of the comb teeth. The detailed pouring of such slab gels is described
later. The polymerized gel, held between its glass plates, is attached to the
electrophoresis apparatus by means of strong metal spring clips in such a way that
the notched glass plate is aligned with and adjacent to the notch in the upper buffer
reservoir to allow contact with the top of the gel and the upper reservoir buffer.
The bottom of the gel is immersed in the buffer in the lower reservoir. Since many
samples can be analysed on one slab gel, the platinum electrodes are placed so as
to be equidistant from each sample which means that they are positioned along the
length of both upper and lower buffer reservoirs. Similar apparatus is also
commercially available from a number of sources (e.g. Koch-Light Ltd, Shandon
Southern).

Protocol 1. Construction of a Studier-type slab gel apparatus

Construct the apparatus from Perspex sheeting as follows.

1. Cut a base (20.5 cm × 20.5 cm × 0.9 cm). Also prepare a vertical section
   (21.3 cm × 17.8 cm × 0.9 cm) and cut a notch 3.0 cm deep and 14.0 cm long
   in one of its 17.8-cm-long edges. Mill a groove beneath the notch to take a
   piece of silicone rubber tubing for a seal (see *Figure 11*). Cement and screw
   the un-notched 17.8-cm edge of the vertical section to the base plate at a distance
   of 12.5 cm from one of the base edges.

2. Form the base of the upper buffer reservoir and three sides from 0.3-cm-thick
   Perspex to give a chamber 14.6 cm long, 6.0 cm wide, and 7.0 cm deep
   (outside edge measurements). Cement this to the vertical section of the
   apparatus, which forms the final side of the upper reservoir, flush with the
   top edge.

3. Cut a sheet of 0.3-cm thick Perspex, 29.8 cm long and 6.5 cm wide, mould
   it to form the three sides of the lower reservoir and then cement it to the lower
   part of the vertical section (which forms its final side) and to the base of the
   apparatus (which forms the reservoir base). The final lower reservoir is 17.8 cm
   long, 6.0 cm wide, and 6.5 cm deep (outside edge measurements).

Protocol 1. *continued*

4. Cement two Perspex blocks, 2.5 cm × 0.9 cm × 0.9 cm to the base of the lower reservoir 12 cm apart (from the centre of each block) and abutting the vertical section so as to form supports for the slab gel glass-plate sandwich.

5. Run platinum wire electrodes along the front and back edge of the lower and upper buffer reservoirs respectively, and hold these in position with small blocks of cemented Perspex. Lead the electrodes to male terminals screwed into a Perspex block (3.0 cm × 1.5 cm × 1.5 cm) cemented alongside the upper reservoir.

6. For safety reasons, a lid can be made of 0.3-cm-thick Perspex and incorporating female sockets which interconnect with the male terminals on the apparatus and lead to the power supply. The electrical circuit is then complete only with the lid in position. However, some researchers choose to cool the slab gel with a fan during electrophoresis. In this situation the apparatus terminals are connected to the power pack without a lid but it must be realized that this then constitutes a potential safety hazard.

7. Construct each comb for sample-well formation from a 0.15 cm × 2.5 cm × 13.0 cm Perspex strip with teeth cut 1.25 cm into one 13.0 cm edge. Then cement one face of the uncut half of the strip to a second Perspex strip (0.15 cm × 1.25 cm × 13.0 cm) which serves to hold the comb horizontal during use. A useful comb is for 13 wells (0.7 cm wide with 0.25 cm spaces between teeth) since it is a good compromise between adequate track width, sample-well volume, and the number of samples which can be analysed simultaneously. However, the number and dimensions of the sample wells are entirely dependent on the preferences of the investigator.

8. The glass plates are cut from glass about 0.3 cm thick. This is best carried out by someone experienced in the art of glass cutting! One plate is rectangular in shape (17.0 cm × 19.5 cm) and the other is the same size but with a notch 2.0 cm deep and 14.0 cm long cut in one of the 17.0 cm edges.

9. Cut Perspex side spacers to hold the glass plates apart during gel polymerization. These are 1.0 cm wide and 0.15 cm thick. If the three-spacer method of sealing the glass-plate sandwich is to be used (Section 6.5.1), also cut a third spacer of appropriate size for the bottom of the gel mould. An alternative procedure is to use only the two side spacers which are cut to be long enough to run the whole length of the glass plates. To seal the glass-plate sandwich by this method, grease the two side spacers lightly with petroleum jelly (Vaseline) and then lay these down the sides of the un-notched plate. Then place the notched glass plate face down onto this using gloved hands and taking care not to allow the grease to touch anywhere but the plate edges. Clamp the plate assembly together with strong metal clips which are positioned to press on the sandwich just over the spacer positions. Next stand the gel mould vertically in a Perspex trough

Protocol 1. *continued*

and fill the trough with agarose or acrylamide mixture. A suitable trough for the workshop-built apparatus described has the dimensions 19.0 cm long, 1.3 cm wide and 0.6 cm deep and is milled in a Perspex block 19.5 cm long, 5.0 cm wide and 1.25 cm thick. It has a capacity of 10 ml with the glass-plate sandwich in position. For gradient gels (Section 9.3) the trough can be filled with 10 ml of the most concentrated acrylamide mixture. The polyacrylamide gel is poured after the sealing gel has been polymerized. The advantage of using a gel plug to seal the plate bottom is that after the polyacrylamide gel has polymerized the assembly is ready for electrophoresis without the further manipulation associated with other methods.

It is important to note that much of the gel is uncooled using this apparatus, the investigator relying on the minimal thickness (0.75 – 1.5 mm) of the slab gel to allow efficient heat dissipation. The apparatus works well for SDS-PAGE run at room temperature. However, care must be taken not to exceed the heat dissipative ability of the slab gel by using a too-high power input; otherwise the centre section of the gel increases in temperature relative to the rest of the gel so that polypeptides migrate faster if loaded at the gel centre than if loaded at the gel edge, resulting in an upwardly curved display of polypeptide components known as 'smiling'. This effect can be prevented, whilst allowing separation in SDS-PAGE at higher voltage in less time, by using one of the many cooled vertical slab gel electrophoresis systems which are available commercially, (see, for example, *Figure 12*). For these reasons, most laboratories engaged in gel electrophoresis on a routine basis now employ cooled slab gel apparatus. However, the cheap, easily constructed Studier-type apparatus works sufficiently well for most comparative analyses of sample polypeptide composition and molecular mass estimation by SDS-PAGE and this is still in use by many researchers. Therefore, where appropriate, this chapter will refer to both types of equipment. Electrophoretic separations of labile native proteins in non-dissociating buffer systems are best carried out using a cooled slab gel electrophoresis apparatus at 4 – 8 °C.

5.2 Additional items of equipment required for electrophoresis

Additional items of equipment which may be required apart from standard laboratory glassware and magnetic stirrers are given below.

- A 15 W daylight fluorescent lamp; for photopolymerization of riboflavin-catalysed polyacrylamide gels.
- Microlitre syringes or micropipettes; for loading samples onto the gel. A number of alternatives are available including various pipettors with plastic disposable pipette tips, microcaps (calibrated glass capillary tubes of known volume), and

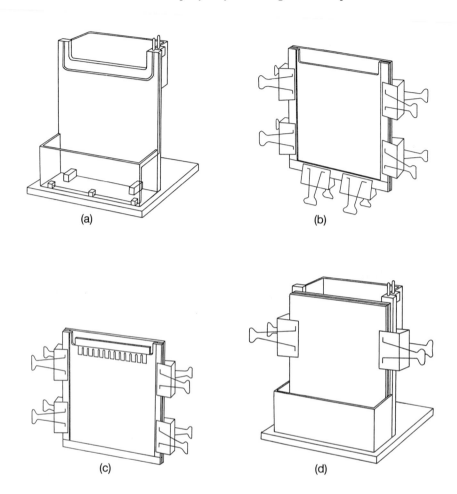

Figure 11. Schematic drawing of a Studier-type slab gel apparatus. (a) the Perspex electrophoresis apparatus consisting of upper and lower buffer reservoirs attached to and on opposite sides of a vertical sheet of Perspex in which is cut a deep notch in its upper edge. A groove cut in the vertical Perspex sheet below the notch and following its contour houses flexible silicone rubber tubing which protrudes above the surface of the groove along its length. This will form a seal when the slab gel plates are clamped in position against the vertical notched face of the apparatus (see *Figure 11d*). The platinum wire electrodes run along the front and back edge of the lower and upper buffer reservoirs, respectively, held in place with small blocks of cemented Perspex. Both electrodes connect to male terminals alongside the upper reservoir. (b) Sealing the glass plate sandwich using three side spacers with the assembly held together with metal spring clips (c) positioning of the sample comb between the glass plates into the gel mixture now filling the mould (d) attachment of the polymerized slab gel to the electrophoresis apparatus using metal spring clips. The notched glass plate of the gel mould has been aligned with the notch in the vertical section of the apparatus (*Figure 11a*) before clamping with the clips. The silicone rubber which sits partially in the groove running below the apparatus notch is compressed by the glass plate assembly and so forms a leak-proof seal.

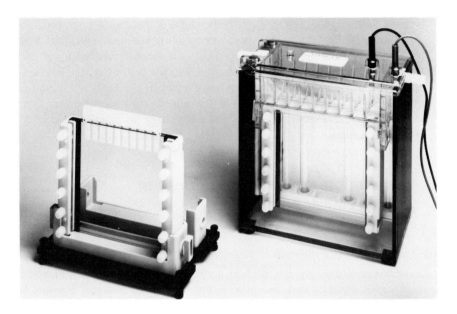

Figure 12. A cooled vertical slab gel apparatus. The model shown is the Protean™ I unit which was made by Hoefer Scientific Instruments and marketed by Bio-Rad Laboratories Ltd. Up to two slab gels can be electrophoresed simultaneously. Each slab gel is formed between two glass plates $16.0 \times 18.0 \times 0.3$ cm, neither of which are notched, which are held apart by PVC spacers, 0.075, 0.15, 0.3, or 0.6 cm thick, placed down each of the two vertical sides of the sandwich. The spacers are held in position by one-piece plastic clamps. The bottom of the sandwich is sealed using polycarbonate cams which press the base against a silicone rubber gasket in the casting stand and locked in position against another silicone rubber gasket in the upper buffer reservoir using the same cam system. The upper and lower reservoirs are filled with reservoir buffer and that in the lower reservoir is cooled by a coolant passing through a glass tube heat exchanger, placed between the two slab gels, and stirred by a magnetic stir bar. Bio-Rad have since modified the design to incorporate several changes including a notched glass plate system and improved side clamps. This new model is the Protean™ II and can electrophorese up to four slab gels simultaneously. Hoefer still market essentially the original design (the SE 600 vertical slab unit) although with longer-wearing cams. Several other types of cooled vertical slab gel apparatus are also commercially available. In many of these, the slab gels are held in place by rubber grommets in the upper buffer reservoir instead of the cam-operated sealing system.

calibrated plunger-type microsyringes. The choice depends upon personal preference when loading rod gels but the control needed for careful sample loading plus the narrow sample wells in slab gels is best coped with by using a 50 μl or 100 μl microsyringe.

- A peristaltic pump; required for the preparation of acrylamide concentration gradient gels (Section 9.3.3) and also useful for the careful overlayering of all slab gels to obtain a flat gel surface.

• A power pack; capable of supplying about 500 V and 100 mA. In many cases it is not too important whether constant current or constant voltage conditions should be used, but for versatility it is worthwhile obtaining a power pack which has these as alternative modes of operation. A number of suitable power packs are available commercially although simple inexpensive power supplies can be built which usually prove to be entirely satisfactory. If electrophoretic destaining of gels (Section 7.2.1) is envisaged, it is worth having a second power supply with a high current ouptut (1−2 A) and low voltage; in practice a car battery charger is adequate.

6. Preparation and electrophoresis of polyacrylamide gels

6.1 Reagents

Acrylamide and bisacrylamide

It cannot be emphasized enough that both acrylamide and bisacrylamide monomers are highly toxic either by skin absorption or by inhalation of monomer powder. The effects range from skin irritation to central nervous system damage. Therefore disposable plastic gloves *must* be worn at all times when handling monomers in either solid or liquid form and mouth pipetting must be prohibited. Because of this toxicity, it is inadvisable to attempt purification of commercial acrylamide to bisacrylamide before use. Although several purification protocols have been published, the purity of most commercial sources of monomers is certainly adequate for most electrophoretic separations. Once polymerization has occurred, the resulting polyacrylamide gel is relatively non-toxic. However, a small proportion of monomer may well remain unpolymerized and therefore it is still wise to wear gloves when handling gels.

If absolutely necessary, acrylamide and bisacrylamide can be recrystallized from a number of solvents. One method is to dissolve 70 g of acrylamide in 1 litre of chloroform at 50°C, filter the solution hot without suction, and store at −20°C to allow recrystallization. The crystals are collected by filtration, washed briefly with cold chloroform and then dried exhaustively under vacuum. Bisacrylamide may be purified by dissolving 10 g in 1 litre of acetone at 50°C, filtering hot, and storing at −20°C. The crystals are collected by filtration, washed briefly with cold acetone and dried.

SDS

Several grades of SDS are commercially available but only highly purified grades should be used. The use of impure SDS leads to marked alterations in the resolution, banding pattern and apparent molecular masses of polypeptide mixtures (e.g. refs. 31, 32). Fortunately SDS in a form specially purified for electrophoresis is now available from a number of sources (e.g. BDH Chemicals Ltd, Bio-Rad Laboratories Ltd) and appears to be satisfactory. However, one would still be advised to select and use SDS from one source only. If necessary, SDS can be recrystallized from ethanol. Approximately 200 g of SDS is dissolved in 3 litres of boiling ethanol, filtered hot without suction, and stored at 4°C to allow the SDS to recrystallize. The crystals are recovered by filtration and dried.

Urea

When urea is used, the main problem is the accumulation of cyanate ions in stock solutions as a result of chemical isomerization. The cyanate reacts with amino groups to form stable carbamylated derivatives thus altering the charge of the proteins. If this reaction does not go to completion, several artefactual species of proteins with differing charges will result. The simplest remedy is to use fresh urea solutions and, where appropriate, to buffer the solutions with Tris, the free amino groups of which neutralize the cyanate ions. Since cyanate ion formation is accelerated with increasing temperature, heating of urea-containing solutions should be avoided if possible. For exacting situations, urea solutions should be deionized, just before use, by passage through a mixed-bed ion-exchange resin.

TEMED

The TEMED used should be a colourless liquid. Most commercial sources are at least 99% pure.

All other reagents should be of the highest purity available.

6.2 Stock solutions

Details of methodology vary with the type of gel and buffer system selected but certain chemicals and solutions are common to many methods and are described here. Once made, most of these are stable for several months under the conditions described, so that to attempt different gel systems the investigator need only prepare fresh buffer solutions.

- *Acrylamide – bisacrylamide* (30:0.8), is prepared by dissolving 30 g of acrylamide and 0.8 g of bisacrylamide in a total volume of 100 ml of water. The solution is filtered through Whatman No. 1 filter paper, and stored at 4°C in a dark bottle. Hydrolysis of acrylamide monomer to yield acrylic acid and ammonia will occur upon prolonged storage. To ensure reproducibility of data, only enough acrylamide – bisacrylamide stock solution to last for about 1 – 2 months should be prepared.

   Some researchers store stock acrylamide solutions in the presence of a little basic ion-exchange resin (e.g. Dowex 1 or Amberlite IRA-400) to bind any acrylic acid which forms.

- *TEMED*, used as supplied. It is stable in undiluted solution at 4°C in a dark bottle.

- *Ammonium persulphate* (1.5%, w/v), 0.15 g of ammonium persulphate is dissolved in 10 ml of water. This solution is unstable and should be made fresh just before use.

- *Riboflavin* (0.004%, w/v); 4 mg of riboflavin are dissolved in 100 ml of water. The solution is stable when stored at 4°C in a dark bottle.

- *SDS* (10%, w/v); prepared by dissolving 10 g of SDS in water to 100 ml. The solution should be both clear and colourless. If the water temperature is too low, not all the SDS will dissolve and heating is necessary. The solution is stable at room temperature for several weeks but precipitates in the cold.

- *Electrophoresis buffers*; the exact buffer required will depend on the buffer system chosen (see below).

6.3 Gel mixture preparation

The recipes tabulated below yield a sufficient volume of gel mixture for 12−15 rod gels or one standard-size slab gel. The volume of stacking gel mixture for discontinuous buffer systems is ample for both rinsing the resolving gel surface prior to stacking gel polymerization and the final stacking gel itself. Higher concentration gels usually polymerize more rapidly than lower concentration gels at any given TEMED concentration. Thus, the volume of TEMED given in the following tables is only a guide and should be adjusted to obtain gel polymerization within 10−30 min.

6.3.1 Dissociating buffer systems (SDS-PAGE)

Details of buffer composition and gel mixture preparation for the SDS−phosphate (continuous) system, essentially as described by Weber and Osborn (9), and the SDS-discontinuous buffer system based on the method of Laemmli (16), are given in *Tables 1* and *2* respectively. Note that the acrylamide concentration of the stacking gel for the SDS-discontinuous system is constant irrespective of the acrylamide concentration chosen for the resolving gel. Buffer systems for the fractionation of small polypeptides by SDS-PAGE are described in Section 9.1.

6.3.2 Non-dissociating buffer systems

Proteins differ widely in their sensitivity to ionic strength, ionic species, and cofactor requirements. Therefore the buffer chosen for the zone electrophoresis of native proteins will depend entirely on the proteins under study.

With regard to continuous buffer systems, almost any buffer between pH 3 and 10 may be used for electrophoresis. In general, only solutions of relatively low ionic strength (and thus only weakly conductive) are suitable as electrophoresis buffers since these keep heat production to a minimum. On the other hand, if the ionic strength is too low, protein aggregation may occur. Obviously the buffer concentration which satisfies these requirements will depend on the particular buffer ions chosen and the proteins under study, but in general the concentration limits for electrophoresis are from about 0.01 M to 0.1 M. Typical buffer systems which have been used are Tris−glycine (pH range 8.3−9.5); Tris−borate (pH range 8.3−9.3); Tris−acetate (pH range 7.2−8.5); Tris−citrate (pH range 7.0−8.5). Usually Tris concentrations are 0.02−0.05 M. For basic proteins one can use β-alanine-acetate (pH range 4.0−5.0) with 0.01−0.05 M β-alanine. The choice of a suitable pH was discussed earlier (Section 4.4). If reducing agents are found to be essential at all times to retain protein activity, dithiothreitol (1 mM) can be added to the gel mixture (33) but may inhibit polymerization. At the concentration needed to be an effective reducing agent

Table 1. Recipe for gel preparation using the SDS – phosphate (continuous) buffer system

Stock solution	Final acrylamide concentration (%)[a]							Reservoir buffer[b]
	20.0	**17.5**	**15.0**	**12.5**	**10.0**	**7.5**	**5.0**	
Acrylamide – bisacrylamide (30:0.8)	20.0	17.5	15.0	12.5	10.0	7.5	5.0	—
0.5 M sodium phosphate, pH 7.2	6.0	6.0	6.0	6.0	6.0	6.0	6.0	200
10% SDS	0.3	0.3	0.3	0.3	0.3	0.3	0.3	10
1.5% ammonium persulphate	1.5	1.5	1.5	1.5	1.5	1.5	1.5	—
Water	2.2	4.7	7.2	9.7	12.2	14.7	17.2	790
TEMED	0.015	0.015	0.015	0.015	0.015	0.015	0.015	—

[a] The columns represent volumes (ml) of the various reagents required to make 30 ml of gel mixture.
[b] Volumes (ml) of reagents required to make 1 litre of reservoir buffer.

33

Table 2. Recipe for gel preparation using the SDS-discontinuous buffer system

Stock solution	Stacking gel (ammonium persulphate as catalyst)	Stacking gel (riboflavin as catalyst)	Final acrylamide concentration in resolving gel (%)[a]							Reservoir buffer[b]
			20.0	17.5	15.0	12.5	10.0	7.5	5.0	
Acrylamide–bisacrylamide (30:0.8)	2.5	2.5	20.0	17.5	15.0	12.5	10.0	7.5	5.0	–
Stacking gel buffer stock[c]	5.0	5.0	–	–	–	–	–	–	–	–
Resolving gel buffer stock[d]	–	–	3.75	3.75	3.75	3.75	3.75	3.75	3.75	–
Reservoir buffer stock[e]	–	–	–	–	–	–	–	–	–	100
10% SDS	0.2	0.2	0.3	0.3	0.3	0.3	0.3	0.3	0.3	–
1.5% ammonium persulphate	1.0	–	1.5	1.5	1.5	1.5	1.5	1.5	1.5	–
0.004% riboflavin	–	2.5	–	–	–	–	–	–	–	–
Water	11.3	9.8	4.45	6.95	9.45	11.95	14.45	16.95	19.45	900
TEMED	0.015	0.015	0.015	0.015	0.015	0.015	0.015	0.015	0.015	–

Final concentration of buffers: stacking gel; 0.125 M Tris–HCl, pH 6.8
resolving gel; 0.375 M Tris–HCl, pH 8.8
reservoir buffer; 0.025 M Tris, 0.192 M glycine, pH 8.3

[a] The columns represent volumes (ml) of the various reagents required to make 30 ml of gel mixture.
[b] Volumes (ml) of reagents required to make 1 litre of reservoir buffer.
[c] Stacking gel buffer stock: 0.5 M Tris–HCl (pH 6.8); 6.0 g of Tris is dissolved in 40 ml of water, titrated to pH 6.8 with 1 M HCl (~48 ml), and brought to 100 ml final volume with water. The solution is filtered through Whatman No. 1 filter paper and stored at 4°C.
[d] Resolving gel buffer stock: 3.0 M Tris–HCl (pH 8.8): 36.3 g of Tris and 48.0 ml of 1 M HCl are mixed and brought to 100 ml final volume with water. This buffer is then filtered through Whatman No. 1 filter paper and stored at 4°C.
[e] Reservoir buffer stock: 0.25 M Tris, 1.92 M glycine, 1% SDS (pH 8.3); 30.3 g of Tris, 144.0 g of glycine, and 10.0 g of SDS are dissolved in and made to 1 litre with water. The solution is stored at 4°C.

34

Table 3. Recipe for gel preparation using non-dissociating continuous buffer systems

Stock solution	Final acrylamide concentration (%)[a]							Reservoir buffer[b]
	20.0	17.5	15.0	12.5	10.0	7.5	5.0	
Acrylamide – bisacrylamide (30:0.8)	20.0	17.5	15.0	12.5	10.0	7.5	5.0	–
Continuous buffer (5 × conc.)	6.0	6.0	6.0	6.0	6.0	6.0	6.0	200
1.5% ammonium persulphate[c]	1.5	1.5	1.5	1.5	1.5	1.5	1.5	–
Water	2.5	5.0	7.5	10.0	12.5	15.0	17.5	800
TEMED[d]	0.015	0.015	0.015	0.015	0.015	0.015	0.015	–

[a] The columns represent volumes (ml) of the various reagents required to make 30 ml of gel mixture.

[b] Volumes (ml) of reagents required to make 1 litre of reservoir buffer.

[c] Riboflavin (0.004% w/v), 2.5 ml, may be used in place of ammonium persulphate/TEMED at low pH whilst the latter is more effective at high pH.

[d] The concentration of TEMED may need to be increased for low pH buffers.

Table 4. Buffers for non-dissociating discontinuous systems

High pH discontinuous (15)
Stacks at pH 8.3, separates at pH 9.5

Stacking gel buffer:	Tris – HCl (pH 6.8); dissolve 6.0 g of Tris in 40 ml of water and titrate it to pH 6.8 with 1 M HCl (~48 ml). Adjust to 100 ml final volume.
Resolving gel buffer:	Tris – HCl (pH 8.8); mix 36.3 g of Tris and 48.0 ml of 1 M HCl and bring to 100 ml final volume with water. Titrate the solution to pH 8.8, with HCl, if necessary.
Reservoir buffer:	Tris – glycine (pH 8.3) at the correct concentration for use; dissolve 3.0 g of Tris and 14.4 g of glycine in water and bring to 1 litre final volume.

Neutral pH discontinuous (20)
Stacks at pH 7.0, separates at pH 8.0

Stacking gel buffer:	Tris – phosphate (pH 5.5); dissolve 4.95 g of Tris in 40 ml of water and titrate to pH 5.5 using 1 M orthophosphoric acid. Add water to 100 ml final volume.
Resolving gel buffer:	Tris – HCl (pH 7.5); dissolve 6.85 g of Tris in 40 ml water and titrate to pH 7.5 with 1 M HCl. Add water to 100 ml final volume.
Reservoir buffer:	Tris – diethylbarbiturate (pH 7.0); dissolve 5.52 g of diethylbarbituric acid and 10.0 g of Tris in water and make to 1 litre final volume.

Low pH discontinuous (19)
Stacks at pH 5.0, separates at pH 3.8

Stacking gel buffer:	acetic acid – KOH (pH 6.8); mix 48.0 ml of 1 M KOH and 2.9 ml of glacial acetic acid and then add water to 100 ml final volume.
Resolving gel buffer:	acetic acid – KOH (pH 4.3); mix 48.0 ml of 1 M KOH and 17.2 ml of glacial acetic acid and add water to 100 ml final volume.
Reservoir buffer:	acetic acid – β-alanine (pH 4.5); dissolve 31.2 g of β-alanine and 8.0 ml of glacial acetic acid in water and make to 1 litre final volume.

(at least 10 mM), 2-mercaptoethanol inhibits polymerization and so cannot be used. An alternative strategy is to add charged thiol reagents such as 1 mM thioglycolate or 10 mM 3-mercaptopropionic acid to the upper (cathodic) buffer reservoir. The reducing agent then migrates rapidly into the gel as soon as electrophoresis starts. Once a suitable gel concentration (Section 4.6) and buffer have been selected, the gel mixture is prepared according to the directions in *Table 3* which assumes that the buffer can be prepared as a five-times-concentrated stock.

Details of three discontinuous buffer systems for the electrophoresis of native proteins are given in *Table 4*. The high pH system has been the most used whereas

Table 5. Recipe for gel preparation using non-dissociating discontinuous buffer systems

Stock solution	Stacking gel (riboflavin as catalyst)	Final acrylamide concentration in resolving gel (%)[a]							Reservoir buffer[b]
		20.0	17.5	15.0	12.5	10.0	7.5	5.0	
Acrylamide–bisacrylamide (30:0.8)	2.5	20.0	17.5	15.0	12.5	10.0	7.5	5.0	–
Stacking gel buffer stock[c]	5.0	–	–	–	–	–	–	–	–
Resolving gel buffer stock[c]	–	3.75	3.75	3.75	3.75	3.75	3.75	3.75	–
Reservoir buffer stock[c]	–	–	–	–	–	–	–	–	1000 (i.e. undiluted)
1.5% ammonium persulphate[d]	–	1.5	1.5	1.5	1.5	1.5	1.5	1.5	–
0.004% riboflavin	2.5	–	–	–	–	–	–	–	–
Water	10.0	4.75	7.25	9.75	12.25	14.75	17.25	19.75	–
TEMED[d]	0.015	0.015	0.015	0.015	0.015	0.015	0.015	0.015	–

[a] The columns represent volumes (ml) of the various reagents required to make 30 ml of gel mixture.
[b] Volumes (ml) of reagents required to make 1 litre of reservoir buffer.
[c] Stock solution prepared as described in Table 4.
[d] When the low pH discontinuous buffer system is used with ammonium persulphate as catalyst, increase the volume of TEMED to 0.15 ml for the resolving gel and adjust the water volume accordingly. Riboflavin is usually more effective than ammonium persulphate/TEMED at low pH whilst the latter is more effective at high pH.

the low pH system is useful for basic proteins. The neutral pH system is poorly buffered and little used but may preserve some enzyme activities which are not stable at extremes of pH. The volumes of reagents required for gel mixture preparation using any of these three buffer systems are given in *Table 5*. A very large number of other discontinuous buffer systems are available as a computer output (18).

6.4 Preparation of rod gels

6.4.1 Continuous buffer system

The precise method for pouring rod gels will vary from researcher to researcher. Nevertheless the overall procedure is fairly standard and is described in *Protocol 2*.

Protocol 2. Pouring rod gels for the continuous buffer system

1. Use glass tubes 2−3 cm longer than the required gel length. The most important point is to ensure that gel tubes are absolutely clean to obtain uniform adhesion of the gel to the glass. This is done by soaking the glass tubes in chromic acid overnight, rinsing them first with distilled water and finally with ethanol before drying.

   Although low percentage gels (below 12%) are usually easily recovered after electrophoresis by rimming (Section 7.1), this becomes increasingly difficult with higher percentage gels. One way to facilitate gel removal by this method is to use tubes siliconized with an anhydrous solution of 0.5% dimethyl-dichlorosilane in carbon tetrachloride (commercially available as 'Repelcote', Hopkin & Williams Ltd). Instructions on the use of siliconizing reagents are supplied by the manufacturer but it is worth emphasizing that they should be used with care and in an efficient fume cupboard since they are highly toxic. The reduced adhesion of siliconized tubes means that low pecentage glass gels in particular may slip out of the tube during electrophoresis. This can be prevented by covering the bottom of the gel tube, before electrophoresis, with a square piece of nylon mesh (about 100 micron size) held in place with a 2−3 mm section of Tygon tubing.

2. Mark the gel tubes at the required gel length with a fine-tipped water-insoluble marker pen (unsiliconized tubes only), or a glass knife, and seal at the other end with paraffin wax film such as Parafilm or Nescofilm.

3. Place the gel tubes in racks which should hold the tubes snugly in an exactly vertical position. Many workers recommend that gel polymerization should be performed at 0−4°C. However, Gelfi and Righetti (34) have shown that polymerization is not ideal at this temperature and leads to non-uniform and irreproducible gels compared to polymerization at higher temperatures. These workers concluded that the optimum temperature for gel polymerization is 25−30°C. For practical purposes, therefore, polymerization at room temperature is probably most convenient whilst for exact reproducibility one should stand the tubes in a water bath at the optimal temperature.

Protocol 2. *continued*

4. Having selected the resolving gel concentration to be used, prepare the gel mixture omitting TEMED, according to the volumes indicated in *Table 1* (SDS−phosphate) or *Table 3* (non-dissociating buffers) in a small thick-walled flask and degas for 1 min using a water pump. As well as aiding reproducible polymerization rates, degassing prevents bubble formation in the gel.

5. Add the correct volume of TEMED and mix rapidly but gently by swirling. Without delay, fill each tube with gel solution to the mark, being careful to avoid trapping any air bubbles. This can be achieved using either a Pasteur pipette or a 10 ml hypodermic syringe fitted with a long blunt needle, lowering the tip to the tube bottom before expelling the contents.

6. Once the gel tubes have been filled, overlayer the gel solution, preferably with buffer of the same composition as in the gel (although water or isobutanol is often used instead) to a height of about 0.5 cm, both to exclude oxygen (which inhibits polymerization) and to ensure a flat gel meniscus. Great care should be taken to ensure that the overlay is added as gently as possible to prevent mixing with the gel solution; if mixing occurs it will result in a diffuse boundary which will lower the acrylamide concentration at the gel top and reduce zone sharpness. There are a number of ways of adding the overlay but perhaps the most convenient is to use a small-volume syringe with needle attached. Fill the syringe with overlay buffer, wipe the needle to remove any droplets, and then gently touch it to the gel-mixture surface, holding the tip firmly against the tube wall. Raise the tip about 2 mm to leave a wet track to the gel-mixture surface. Keeping the tip firmly against the wall, slowly expel the syringe contents whilst raising the tip to keep it 2 mm above the liquid surface. At this stage, a sharp boundary should be visible between the gel mixture and the overlay. Other workers prefer to use a syringe fitted with a needle bent at 90°, or a Pasteur pipette, for applying the overlay. Whatever the method, it is imperative that the overlay is added with as little mixing as possible.

7. Leave the gels undisturbed to polymerize (10−30 min). Any vibration at this stage will lead to an uneven gel surface. During this time the original interface disappears and is replaced with another slightly below; this is the polymerized gel surface. In general, gels which polymerize in less than 10 min or which have not polymerized by 50−60 min should be discarded since polymerization does not proceed uniformly under these conditions and so the resulting gels will give poor and irreproducible separations. Gel mixtures at low pH will take longer to polymerize than at other pH values but should not exceed 80−90 min. The amount of TEMED and/or ammonium persulphate can be varied somewhat to slow down or speed up polymerization to achieve the desired polymerization time, but beware that excess catalyst can actually inhibit polymerization.

39

Protocol 2. *continued*

8. Leave the gels another 10−15 min after the new interface forms and then remove the water or buffer overlay. If the gel is to be used immediately, rinse the gel surface with reservoir buffer and fill the space above the gel with this buffer. The gels may also be stored, provided that dehydration is prevented. In this case, remove the overlay and rinse the gel surface with fresh buffer of the same composition as in the gel. Next, layer more of this buffer onto the gel surface and seal the tube with Parafilm or a rubber bung.

6.4.2 Discontinous buffer system

The two gel layers required for use with the discontinuous buffer system are prepared by first polymerizing the resolving gel in the tube and then polymerizing the stacking gel on top of this. As a general rule, the stacking gel should not be less than twice the height and volume of the sample to be applied. A stacking gel of about 1.0 cm will suffice for most sample volumes used; in the event that an exceptionally large volume of dilute protein need be loaded, the stacking gel dimensions should be adjusted accordingly. Details of rod gel preparation using discontinuous buffer systems are given in *Protocol 3*.

Protocol 3. Pouring rod gels for the discontinuous buffer system

Steps **1−7** inclusive are identical to steps 1−7 of *Protocol 2* for the continuous buffer system except that the gel tubes are marked at two places, once at the desired height of the resolving gel, and then again at a position 1 cm above this to mark the height of the stacking gel. Fill the gel tubes to the first mark with resolving gel mixture prepared according to the details given in *Table 2* (for the SDS-discontinuous buffer system) or *Table 5* (for non-dissociating discontinuous buffer systems).

8. Once the resolving gel has polymerized, pour off the overlay and prepare the stacking gel mixture (see *Table 2* for SDS-discontinuous buffers or *Table 5* for non-dissociating buffers). If ammonium persulphate is to be used as polymerization catalyst, mix all the components except TEMED and then add the correct volume of this reagent just prior to pouring the stacking gel. If riboflavin is used instead of ammonium persulphate, TEMED can be added to the stacking gel mixture any time prior to pouring the gel since polymerization will not occur until the gel solution is illuminated.

9. Rinse the resolving gel surface with a small volume of stacking gel mixture which is applied using a Pasteur pipette, then remove this and discard it. Next, fill each gel tube to the second mark with stacking gel mixture and overlayer with water or stacking gel buffer according to the procedure detailed in step 6 of *Protocol 2*. If riboflavin has been used as polymerization catalyst for the

Protocol 3. *continued*

stacking gel, place a daylight fluorescent lamp 2−5 cm away from the gel to initiate polymerization.

10. After polymerization of the stacking gel, remove the overlay and rinse the gel surface with reservoir buffer. After discarding this, fill the space above the gel with reservoir buffer.

These gels should be used as soon as possible after preparation since the different buffers of the stacking and resolving gels, and the reservoir buffer, essential for the stacking phenomenon to occur, mix by diffusion upon storage. A convenient alternative is to polymerize the resolving gel in place, rinse this with buffer of the same composition as in the resolving gel, and then store with a fresh overlay of this buffer. The stacking gel is polymerized in place just before the gel is required.

6.5 Preparation of slab gels

The many types of slab gel apparatuses available vary in the final gel dimensions (including gel thickness), whether notched or unnotched glass plates are used, and the methods of sealing the plates and attaching them to the apparatus. However, the method for pouring uniform concentration polyacrylamide slab gels and for producing sample wells is essentially the same irrespective of the equipment used. The main difference encountered would be in the method of sealing the plates in which case, for commercial apparatus, the manufacturer's instructions should be consulted.

Disposable plastic gloves should be worn during slab gel preparation to prevent contamination of clean glass plates with skin proteins.

6.5.1 Continuous buffer system

The preparation of slab gels using a continuous buffer system is relatively simple since only a single gel needs to be poured. The overall procedure is described in *Protocol 4*.

Protocol 4. Pouring slab gels for the continuous buffer system

1. As with rod gels, it is most important to ensure that the slab gel plates are perfectly clean to obtain good gel adhesion to the glass. Clean the glass plates by soaking them in chromic acid overnight, rinse them with water, and then with ethanol. Then put the plates down onto clean tissue paper, with the side which is to be in contact with the gel uppermost, and swab it with an acetone-soaked tissue held in a gloved hand. After a final rinse with ethanol, allow the plates to air-dry.

Protocol 4. *continued*

2. Assemble the glass-plate sandwich which will form the gel mould. The glass plates are usually held the correct distance apart by thin plastic spacers which *must* be of uniform thickness both with respect to each other and along their length to ensure good contact with the plates and a gel of uniform thickness. Usually for SDS-PAGE, these are about 1.5 mm thick, although thinner gels can also be used.

    Several methods of sealing the glass-plate sandwich are possible. For the popular type of cooled commercial apparatus, the plates are held together by one-piece plastic clamps which run down each side of the gel mould. The base of the mould is sealed using a cam system which presses the mould against a silicone rubber gasket in a casting stand (*Figure 12*). For the notched plate system of Studier (*Figure 11a*), the most common method of assembly is to use three spacers, one for each vertical side and one for the base of the glass-plate assembly (*Figure 11b*). These are sealed by coating with grease before assembly, or by dripping molten 2% agarose around the edges after assembly, or using a special adhesive tape (e.g. Tape UF T1/AT; Universal Scientific Ltd) which is resistant to immersion in SDS-containing buffers. The plate assembly is clamped together with strong metal clips which are positioned to press on the sandwich just over the spacer positions (*Figure 11b*). The spacer along the bottom edge is removed after the polyacrylamide gel has polymerized and prior to electrophoresis. An alternative method of sealing is described in *Protocol 1*, step 9.

3. Hold the clamped plate assembly vertically during pouring of the gel. Having selected an appropriate resolving gel concentration, prepare the gel mixture by adding the correct volumes of all components (*Table 1* for SDS−phosphate or *Table 3* for non-dissociating buffer systems), except TEMED, to a small thick-walled flask. Again, as with the preparation of rod gels (see above), this is most conveniently achieved at room temperature. After degassing for 1 min using a water pump, add the TEMED, gently mix it in and pour the gel solution without delay between the glass plates to within 0.5 cm of the top. Immediately insert a sample comb between the glass plates and into the gel mixture. For the Studier notched plate system, this is shown in *Figure 11c*. The teeth of the comb should fit snugly against the glass plates. Take special care to ensure that air bubbles are not trapped beneath the comb otherwise irregularly shaped sample wells will be formed.

4. Leave the assembly undisturbed for the gel to polymerize (10−30 min) as evidenced by the appearance of a sharp boundary below areas of gel/air interfaces. After a further 10 min, remove the comb carefully and slowly.

5. If the gel is to be used immediately, rinse out the sample wells formed by removal of the comb with reservoir buffer using a Pasteur pipette, or a syringe

Protocol 4. *continued*

fitted with a needle, and then fill them with reservoir buffer. Straighten any of the gel sections bordering the sample wells which have become distorted during removal of the comb using a syringe needle or microspatula. If the gel is to be stored before use, rinse out the sample wells with buffer of the same composition as in the gel and then fill them with this buffer.

In some situations when a continuous buffer system is required, the resolving gel is too brittle to allow the ready formation of sample wells. This can be overcome by polymerizing a low concentration gel (3.75−5.0%) on top of the resolving gel, and the sample wells formed in this in an analogous manner to that described for a discontinuous buffer system (see below) but with the low concentration gel containing the same buffer as the resolving gel (e.g. *Protocol 16*).

6.5.2 Discontinuous buffer system

The use of a stacking gel polymerized on top of the resolving gel, as required by the discontinuous buffer system, means that sample wells are formed in the stacking gel. Therefore the resolving gel is poured first, and allowed to polymerize to give a flat meniscus, followed by pouring of the stacking gel in which the comb is inserted to form the sample wells. The procedure is described in detail in *Protocol 5*.

Protocol 5. Pouring slab gels for the discontinuous buffer system

1 − 2. Clean and assemble the gel plates as described for steps 1−2 of slab gel preparation using the continuous buffer system (*Protocol 4*). Therefore, the resolving gel is poured first, and allowed to polymerize to give a flat meniscus, followed by pouring of the stacking gel in which the comb is inserted to form the sample wells. The procedure is described in detail in *Protocol 5*.

3. Having selected an appropriate resolving gel concentration, prepare the gel mixture in a small, thick-walled flask by mixing the components (except TEMED) in the volumes listed in *Table 2* for SDS-discontinuous or *Table 5* for non-dissociating buffer systems. Degas the mixture for 1 min using a water pump, add the correct volume of TEMED, and gently mix it in.

4. Pour the resolving gel mixture into the space between the glass plates leaving sufficient space at the top for a stacking gel to be polymerized later and sample wells formed. The stacking gel needs to be at least twice the height of the sample; 2 cm will be sufficient in most cases. Taking into account the depth of the sample wells, this means a space of about 3.5 cm needs to be left above the resolving gel. The volume of resolving gel required will obviously depend on the gel dimensions. Now layer gel buffer, of the same composition as in the resolving gel, onto the gel surface using a peristaltic pump fitted with fine plastic tubing, or with a hypodermic syringe fitted with fine tubing. Position the tip of the

Protocol 5. *continued*

> tubing so as to deliver the buffer just above the gel surface and at the centre of the glass plates.

5. After polymerization (10−30 min), as evidenced by the presence of a sharp interface between the polymerized gel and the overlay, tilt the assembly to pour off the overlay. Although the stacking gel can be polymerized in place and the gel used immediately, routinely the resolving gel is overlayered with buffer of the same composition as in the resolving gel and left overnight before use.

6. Prepare the stacking gel according to the protocol listed in *Table 2* for the SDS-discontinuous, or *Table 5* for non-dissociating buffers.

7. Use a small volume of the stacking gel mixture to rinse the surface of the resolving gel. Pour this off and then fill the remaining space between the gel plates with stacking gel mixture.

8. Immediately insert the comb into the stacking gel mixture, being careful to avoid trapping any air bubbles beneath it. For the notched glass plate system of Studier, this is shown in *Figure 11c*. For riboflavin-catalysed stacking gel polymerization only, place a daylight fluorescent lamp within 2−5 cm of the gel to initiate polymerization. Leave the assembly undisturbed whilst the stacking gel polymerizes.

9. After polymerization, carefully remove the comb to expose the sample wells, rinse them out with reservoir buffer and then fill with this buffer. Any divisions between wells that have become displaced during comb removal can be straightened with a syringe needle or microspatula provided that care is taken to avoid damage. Once the stacking gel has been polymerized in place, use the slab gel immediately.

6.6 Sample preparation

6.6.1 Dissociating buffer system (SDS-PAGE)

It is important not to overload the gel or bands will be distorted. In the case of slab gels an overloaded sample in one track can also distort the electrophoretic pattern of bands in adjacent tracks. At the other extreme, underloading will result in one not detecting minor components, and even major component bands may be too faint after staining for a good photographic record. Therefore it is wise to use standard assays to determine sample protein concentration before electrophoretic analysis is attempted. About 1−10 μg of each polypeptide should be loaded on to the gel to give optimal results, such that for a complex mixture about 50−100 μg is usually sufficient. The volume in which this is loaded is also important. The thickness of the starting zone and hence the volume of the sample applied has a large effect on protein band sharpness using the continuous buffer system (SDS−phosphate) and so the sample volume should be as small as possible. The stacking effect which occurs with the SDS-discontinuous system means that the method is essentially volume independent, but in practice the sample volume is limited in slab gel electrophoresis by the size of the sample wells and the fact that with large volumes large stacking gels are needed and some sideways spreading of polypeptide bands occurs. Best results using 1.5-mm-thick slab gels with 7-mm-wide sample wells are obtained using sample volumes of $10-30$ μl but 60 μl still gives reasonable results. In rod gels, sample volumes should also be kept small for optimal resolution but the amount of sample loaded (in micrograms) and sample volumes can be higher because of the larger cross-sectional area of these gels.

A number of methods can be used for concentrating protein samples too dilute for immediate electrophoretic analysis. These include lyophilization, ammonium sulphate precipitation, and dialysis against a high concentration of polyethylene glycol (M_r 20 000). Alternatively, solid polyethylene glycol may be placed in contact with the outside of a dialysis bag containing the dilute protein sample. Sephadex G-100 or G-200 can be used in a similar manner. When the sample has been concentrated by any of these methods it should be dialysed against 0.01 M sodium phosphate buffer (pH 7.2) for the SDS−phosphate buffer system, or 0.0625 M Tris−HCl (pH 6.8) for the SDS-discontinuous system, to remove salts or low molecular mass polyethylene glycol impurities which may interfere with electrophoresis. Potassium ions in particular must be removed since they precipitate SDS. An appropriate volume of sample protein in 0.01 M phosphate buffer (pH 7.2) or 0.0625 M Tris−HCl (pH 6.8) is then brought to 2% SDS, 5% 2-mercaptoethanol, 10% sucrose (or glycerol), and 0.002% Bromophenol Blue using concentrated stock solutions. Since molecular mass estimations by SDS-PAGE depend on all polypeptides having the same charge density, and hence the same amount of SDS bound per unit weight of polypeptide (see Section 4.7.2), it is important that the SDS is present in excess. Knowing the protein content of the sample, one should calculate whether the SDS is indeed in excess; a ratio of at least 3:1 is required. At least 5% SDS can be added to the sample buffer of the SDS-discontinuous system without deleterious effects. Dithiothreitol can be used instead of 2-mercaptoethanol to disrupt polypeptide disulphide bonds and has the advantages that it is odourless and does not tend to auto-oxidize. The sucrose (or glycerol) is present to increase the density of the sample so that, when applied to the gel, the sample remains as a well-defined overlay and does not undergo convective mixing with the reservoir buffer during the early stages of electrophoresis.

A common method of sample protein concentration, especially for multiple samples, is to precipitate the protein with 10% trichloroacetic acid (TCA) (incubate in ice for 30 min), followed by centrifugation (12 000 g for 5 min), followed by repeated washing with ethanol-ether (1:1 v/v) to remove the TCA. Apart from speed in concentrating large numbers of samples, the method simultaneously removes interfering salts as well as some non-dialysable contaminants. Nevertheless, TCA precipitation should be used with caution since some glycoproteins and histones are soluble in low concentrations of TCA whilst in other cases the protein will precipitate but prove extremely difficult to redissolve completely in the sample buffer. Therefore, before using TCA precipitation routinely, it is wise to check on the efficiency of the procedure for the particular sample under study. After washing with ethanol-ether, TCA precipitates are dissolved directly in 0.01 M sodium phosphate buffer (pH 7.2), 2% SDS, 5% 2-mercaptoethanol, 10% sucrose or glycerol, 0.002% Bromophenol Blue for the SDS−phosphate buffer system or 0.0625 M Tris−HCl (pH 6.8), 2% SDS, 5% 2-mercaptoethanol, 10% sucrose or glycerol, 0.002% Bromophenol Blue for the SDS−discontinuous buffer system. The sample should be blue in colour. If it is yellow then there is still sufficient TCA present to interfere with electrophoresis and the pH must be adjusted by addition of microlitre volumes of concentrated Na_2HPO_4 (for the SDS−phosphate system) or Tris (for the SDS-discontinuous system). However, excessive addition of concentrated buffers can lead to problems during stacking and separation of proteins during electrophoresis and so a better approach is to remove the TCA efficiently initially. If the protein pellet does not completely dissolve with mixing in the neutralized solution it is possible that insufficient SDS is present.

An even simpler method of sample protein concentration which avoids the problems of acid neutralization is acetone precipitation. Five volumes of cold acetone are added to the sample, mixed, and incubated at −20°C for 10 min. The precipitated protein is collected by centrifugation (12 000 g for 5 min) and may be washed by repeated precipitation. After drying, the protein pellet is dissolved directly in the appropriate sample buffer and is then ready for analysis by polyacrylamide gel electrophoresis. As with TCA precipitation, it is important to check that the proteins under study are quantitatively precipitated before using this concentration method routinely.

Prior to electrophoresis and immediately after the addition of SDS, samples for SDS-PAGE are heated in a boiling water bath for 3 min. This ensures denaturation of the protein. Use of lower temperatures may not fully denature some proteins, especially certain proteases which may then proceed to degrade other sample proteins. After heating, the sample is allowed to cool to room temperature. Finally, and of crucial importance, any insoluble material should now be removed by centrifugation (12 000 g for 5 min) or this will cause protein 'streaking' during gel electrophoresis. The sample may be used immediately or stored in the freezer at −20°C. When cooled, SDS crystallizes out of solution and so stored samples must be warmed before use.

Some proteins, particularly nuclear non-histone proteins, require the presence of 8 M urea in the SDS sample buffer if most of the protein is to enter the gel (Section 9.2). Similarly, immunoprecipitates and membrane proteins are often dissolved in SDS sample buffer with urea added to aid solubilization. When urea is present there is no need to add sucrose or glycerol to the sample buffer to increase its density. If the sample must be heated prior to loading, the sample should contain Tris as the buffer to minimize cyanate modification of proteins (Section 6.1). Other problems related to sample preparation are described in ref. 278.

Molecular mass standards

Whenever analytical SDS-PAGE is used, it is wise to include a mixture of polypeptides of known molecular mass. Whilst this is essential for determining the molecular mass of sample polypeptides (see Section 4.7), it is also worthwhile if only undertaking a qualitative assessment of the protein composition of samples since it provides a measure of reproducibility between different gel runs. A number of protein standards that are commercially available are listed in *Table 6*. Each of the standard proteins should be dissolved in the appropriate sample buffer at a concentration of 1 mg/ml, heated at 100°C for 3 min, and stored frozen in small aliquots. When a sample is to be analysed, an aliquot of molecular mass markers can be thawed out, warmed to dissolve any precipitated SDS, and run on a parallel rod gel or in parallel tracks of a slab gel. After electrophoresis, the gel is stained and destained and the distances migrated by the sample and marker polypeptides measured. The molecular mass of radioactive sample proteins can be determined by including radioactive marker proteins, either prepared in the laboratory (41) or purchased from commercial sources, followed by autoradiography or fluorography. Alternatively one can use unlabelled marker proteins and then indicate the position of each in the

Table 6. Molecular masses of polypeptide standards[a,b,c]

Polypeptide	Molecular mass
Myosin (rabbit muscle) heavy chain	212 000
RNA polymerase (*E.coli*) β'-subunit	165 000
β-subunit	155 000
β-Galactosidase (*E.coli*)	130 000
Phosphorylase a (rabbit muscle)	92 500
Bovine serum albumin	68 000
Catalase (bovine liver)	57 500
Pyruvate kinase (rabbit muscle)	57 200
Glutamate dehydrogenase (bovine liver)	53 000
Fumarase (pig liver)	48 500
Ovalbumin	43 000
Enolase (rabbit muscle)	42 000
Alcohol dehydrogenase (horse liver)	41 000
Aldolase (rabbit muscle)	40 000
RNA polymerase (*E.coli*) α-subunit	39 000
Glyceraldehyde-3-phosphate dehydrogenase (rabbit muscle)	36 000
Lactate dehydrogenase (pig heart)	36 000
Carbonic anhydrase	29 000
Chymotrypsinogen A	25 700
Trypsin inhibitor (soybean)	20 100
Myoglobin (horse heart)	16 950[d]
α-Lactalbumin (bovine milk)	14 400
Lysozyme (egg white)	14 300
Cytochrome c	11 700

[a] Several proteases, for example trypsin, chymotrypsin, and papain, have been used as molecular mass standards by various workers but these may sometimes cause proteolysis of other polypeptide standards and so are omitted here.

[b] The polypeptide molecular masses given here are mainly from refs 9, 35, and the references given in Appendix 4, and are the molecular masses in the presence of excess thiol reagent. A more comprehensive list of suitable polypeptides is available from the original sources.

[c] The molecular mass range ~12 000 – 68 000 is reasonably well covered but there are few suitable proteins with subunit molecular masses above this range. This can be overcome by using polypeptides which are crosslinked to form an oligomeric series (36, 37). Kits of these are commercially available.

[d] Calculated from the sequence data given in ref. 38.

stained gel with a small spot of radioactive ink (Section 7.5.1) prior to autoradiography. A calibration curve of $\log_{10}$ molecular mass versus distance migrated is constructed using the data from the standard polypeptides (Section 4.7.2) and then the molecular mass of the sample polypeptides determined from this, knowing their migration distances. To rule out potential anomalies (Section 4.7.2) the analysis should be repeated with at least one different gel concentration and shown to give no significant change in the estimated molecular mass. Ideally, if sufficient material is available, several gel concentrations are used and a Ferguson plot constructed.

6.6.2 Non-dissociating buffer systems

The sample proteins should be present at $1-10$ μg per protein in the volume which is to be loaded onto a slab gel, or about 100 μg for a complex protein mixture. To increase zone-sharpening using a continuous buffer system, the sample proteins should be dissolved in buffer diluted $5-10$-fold over the concentration used in the gel (Section 4.3). Whether diluted sample buffer can be used or not will depend upon the concentration of buffer used in the gel, since if the ionic strength of the sample buffer falls too low, protein aggregation may occur. Since with most vertical slab and rod gel apparatus the sample is usually loaded onto the gel through the upper reservoir buffer (see Section 6.7), the sample should also contain 10% sucrose (or glycerol) to increase its density. A tracking dye (about 0.002%) is also added both to aid in loading the sample onto the gel and to act as marker during electrophoresis to indicate when electrophoresis should be terminated. Bromophenol Blue is widely used as a tracking dye for separations at alkaline pH. Methylene green, methylene blue, or Pyronin Y can be used for separations at acid pH. Samples for analysis using one of the discontinuous buffer systems should be present in $1/4-1/8$ diluted stacking gel buffer stock, 10% sucrose (or glycerol), plus a suitable tracking dye (final concentration 0.002%).

Depending on the protein under study, concentration of dilute samples can be achieved by protein precipitation (with acetone, alcohol, or ammonium sulphate), ion-exchange chromatography, use of Diaflo membrane filters, vacuum ultra-filtration, lyophilization, or use of hydrophilic polymers such as polyethylene glycol or Sephadex (Section 6.6.1). Obviously the trichloroacetic acid precipitation method used for concentrating samples for SDS-PAGE cannot be used for the concentration of native protein samples. If necessary, samples should be dialysed against the sample buffer to be used in the electrophoretic analysis (but lacking sucrose, glycerol, or tracking dye) to remove salts or other low molecular mass contaminants which could interfere with electrophoretic analysis. All samples are then centrifuged (12 000 g for 15 min at 4°C) to remove any insoluble material which would interfere with electrophoresis. Sucrose or glycerol may then be added to 10% final concentration and finally a tracking dye is added.

Usually all steps involved in the preparation of native protein samples are performed at low temperature $(0-4°C)$ both to reduce loss of protein activity through denaturation and to minimize attack by any proteases in the sample. The presence of protease inhibitors, such as phenylmethylsulphonyl fluoride (PMSF) for serine active site proteases, is also helpful in this respect. The problem with all these reagents is that they can only be used if they do not also inactivate the protein of interest. Once ready for electrophoresis the samples may be used immediately, or stored in sample buffer containing glycerol at 4°C, or frozen if the proteins are stable under these conditions.

6.7 Sample loading and electrophoresis

For all discontinuous buffer systems, the sample is loaded before any electrophoresis is attempted. However, when native proteins are being separated using a continuous buffer system, it may be desirable to pre-electrophorese the gel for a period of time (30−40 min) prior to loading the sample. This removes traces of chemicals such as residual ammonium persulphate or acrylic acid which may otherwise reduce the biological activity of the protein of interest.

The method of loading liquid samples is essentially the same for both rod and slab gels using any buffer system and is described in *Protocol 6*. The method given here for slab gels is for the notched glass-plate system of Studier (30) where the slab gel is attached to the electrophoresis apparatus before sample loading. Slab gels formed between the unnotched glass plates of some cooled commercial apparatuses are attached to the apparatus only after sample loading. The method of attachment varies with the commercial apparatus being used (see *Figure 12* legend).

Protocol 6. Sample loading and electrophoresis

1. For rod gels, remove any rubber stoppers, Parafilm, etc., from the base of the gel tubes and mount the gels in rubber grommets located in the upper buffer reservoir of the electrophoresis apparatus (*Figure 10*). For slab gels prepared using the Studier apparatus (30), remove the bottom spacer leaving the two side spacers in place. Lightly grease the silicone rubber tubing which will form the seal below the upper buffer reservoir (*Figure 11a*). Clamp the slab gel prepared in the notched glass-plate sandwich in position using metal clips with the notch of the glass plate adjacent to and aligned with the notch in the upper reservoir (*Figure 11d*).

2. Add reservoir buffer to the lower reservoir of the electrophoresis apparatus. Next, it is essential to remove any air bubbles from the bottom of the gel or these will prevent uniform electrical contact between the gel and reservoir buffer. This can be achieved using a Pasteur pipette with a bent tip. Alternatively for rod gels, the bubbles are easily removed by flicking the top of the gel tube with a gloved finger.

3. Plug any unused holders in the rod or slab gel apparatus and add electrode buffer to the upper reservoir.

Protocol 6. *continued*

4. After checking for leaks, wash the gel surface by directing a gentle stream of electrode buffer into each sample well of a slab gel or onto the gel surface of a rod gel using a Pasteur pipette. This also serves to fill the space above the gel with reservoir buffer, hence removing any air bubbles present.

5. Now carefully load the sample onto the gel surface using a microsyringe or micropipette. The tip of the sample applicator should be held only $1-2$ mm above the gel surface to minimize sample mixing with the reservoir buffer during loading. The dense sample solution will flow onto the gel surface and form a sharply-defined layer. Best results with slab gels are achieved if unused wells are filled with an equivalent volume of blank sample buffer.

6. Connect the electrophoresis apparatus to the power pack (switched off) with the anode $(+)$ connected to the bottom reservoir for SDS-PAGE and electrophoresis of negatively charged proteins in other buffer systems, and the cathode $(-)$ connected to the upper reservoir. For the electrophoresis of positively charged proteins, reverse the polarity of the buffer reservoirs.

7. Connect the power pack to the mains, switch on, and adjust it to deliver the necessary current and voltage. Electrophoresis conditions are quoted as either constant current or constant voltage. Either mode of operation is permissible although only constant voltage conditions give constant protein mobility during electrophoresis. The exact electrophoresis conditions will depend on the buffer and gel conditions employed and usually need to be determined empirically. In general, too-high current has the danger of excessive heating whilst too-low voltage increases electrophoresis time and may decrease resolution as a result of band diffusion. For the SDS−phosphate buffer system, 4 mA constant current is applied per rod gel until the sample has entered the gel and then 6 mA constant current per 9 cm rod gel for about 3 h (5% gels), 4 h (10% gels) or 8 h (15% gels). For the lower conductivity SDS-discontinuous system, $2-3$ mA constant current per rod gel or 100 V constant voltage allows electrophoresis to be completed in similar time periods. Longer rod gels take proportionately longer to run at constant voltage. Similarly, the electrophoresis conditions for slab gels with the SDS-discontinuous system depend on the gel dimensions and gel concentration. For the slab gel format described earlier (Section 5.1.2) one can electrophorese 10% acrylamide gels at room temperature at about 50 V overnight. Higher voltages are necessary for overnight runs with higher percentage slab gels. Alternatively, slab gels can be electrophoresed during the day at $25-30$ mA constant current or by stacking at 120 V followed by electrophoresis at 200 V constant voltage.

   Electrophoresis conditions with non-dissociating buffers vary enormously depending on the conductivity of the buffer system used. In general one should not exceed about 2 mA per rod and $100-200$ V to avoid excessive heating.

7. Analysis of gels following electrophoresis

It is very important that disposable plastic gloves are worn when handling gels to prevent proteins from the skin being transferred to the gels. In addition the gels must not be allowed to come into contact with paper surfaces since polyacrylamide readily sticks to such surfaces.

7.1 Recovery of gels

Slab gels are easily recovered by removing the side spacers and gently levering the glass plates apart at the end away from the notch to avoid damage to the fragile notched end. Leave the gel resting on the unnotched plate. Higher percentage gels (above about 10% acrylamide) may be carefully lifted between thumb and forefinger, using both hands, at the gel base and transferred to a suitable rectangular solvent-resistant plastic tray (ideally at least 5 cm longer and wider than the gel) for soaking in buffers or stains involved in analysis. Some workers find a broad plastic spatula useful in handling lower percentage slab gels.

Rod gels are more difficult to recover. A very effective method is to remove the gel from the intact tube by rimming. The aim of the rimming procedure is to gently pry the gel away from the tube wall, without damaging it, using water pressure. The rod gel is held in one hand whilst a syringe fitted with a long, fine, deliberately blunted needle (e.g. 2.5 inch 23 gauge) is used to squirt water in between the gel surface and the tube wall whilst slowly rotating the gel (*Figure 13*). With siliconized glass tubes the gel usually slides out fairly easily, so easily that the operation should be carried out over a tray (not the sink!). If the gel does not detach itself easily then begin the whole procedure again at the other end of the tube. If required, pressure can be exerted in the final stages of removing the loosened gel by applying a Pasteur pipette bulb to one end of the gel tube. This rimming procedure works well with low concentration polyacrylamide gels (below about 12%) but at gel concentrations above this, especially if the gel is in an unsiliconized glass tube, it may be necessary to use the second method which is to break the tube itself. Perhaps the simplest way to do this is to wrap the gel tube in tissue and place it between the jaws of a workshop vice. The jaws are tightened until a crack is heard and then released. The gel tube fractures into a number of pieces along its length and the gel can be washed free of these without damage. Alternatively one can use a heavy hammer to carefully break the tube, starting from one end, without damaging the gel too much. Ghadge *et al.* (42) have suggested a method which involves inserting a piece of rolled Mylar sheet into the gel tube prior to polymerization. The Mylar sheet is arranged to be slightly longer than the gel tube so that, after electrophoresis, the sheet can be pulled from the tube bringing the gel with it. Finally, Dhamankar *et al.* (43) recommended simply coating the tube walls by dipping into paraffin wax at 70°C prior to gel preparation: after electrophoresis, the gel is recovered simply by dipping the tube into hot water for 15 sec.

Once the rod or slab gel has been recovered it is essential to mark which end is which. In addition it is sometimes necessary to mark the position of the tracking

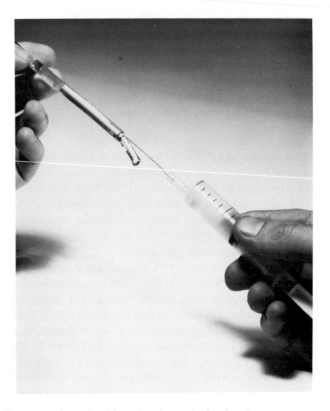

Figure 13. Recovery of a rod gel from its glass tube by rimming.

dye for later R_f determinations. Both of these requirements are met by inserting a small piece of fine-gauge wire (about 0.1 mm in diameter) into the gel at the dye front. Alternatively one can inject a small amount of indian ink at this point.

7.2 Protein staining with organic dyes and quantitation

Protein bands of sufficiently high concentration may be localized by direct photometric scanning of unstained gels at 280 nm. The gel is placed in a suitable quartz trough (or actually run in quartz tubes) and scanned using any suitable modern spectro-photometer fitted with a gel scanning attachment. Various UV absorbing impurities are present in many monomer stocks and these give high background absorbance, as will any unreacted acrylamide monomer. Although much of this problem can be eliminated by using pure reagents and correct polymerization protocols, the method is still unsuitable for most applications because the extinction coefficient for proteins is rather low at 280 nm and light scattering can be a serious problem. These limitations mean that only major components can be quantitated with precision. Because of this, the usual method is to increase the sensitivity of protein detection by reacting the protein bands with an easily visualized reagent.

Various negative stains have been reported (e.g. 44, 45) but for quantitation purposes protocols which give positively stained protein bands against a pale background are preferred. Most of the early work used Amido Black 10B as the general protein stain but this has been largely superseded by Coomassie Blue R-250 (R = reddish hue) which is more sensitive. Another protein stain with almost the same sensitivity as Coomassie Blue R-250 is Xylene Cyanine Brilliant G (Coomassie Blue G-250; G = greenish hue). Wilson (46) has compared the use of these and other organic stains. Readers should note that ICI Ltd, which holds the trademark 'Coomassie', no longer produce either Coomassie Blue R-250 or G-250. Hence manufacturers who still do produce these dyes have necessarily introduced their own trademarks. Many of these are reviewed in ref. 46. For the purposes of this chapter, however, I will continue to refer to these dyes by their original names.

7.2.1 Coomassie Blue R-250

i. Staining and destaining
Usually 0.2−0.5 μg of any protein in a sharp band can be detected using this dye and staining is quantitative to 15−20 μg for at least some proteins. In early procedures it was common to fix proteins in the gel first and then stain them but this has been superseded by methodology in which protein fixation and staining occur simultaneously. Since Coomassie Blue is predominantly non-polar, it is usually used in methanolic solution and excess dye removed from the gel later by destaining. Stacking gels are usually discarded before staining the resolving gels unless the investigator wishes to test for proteins unable to enter the resolving gel. Various recipes have been used for the staining solution containing dye, methanol, and acetic acid, or TCA with the exact proportions of each varying. These have been reviewed by Righetti (47). A common protocol is as follows.

Coomassie Blue R-250 (0.1%) is dissolved in water:methanol:glacial acetic acid (5:5:2 by vol) and filtered through a Whatman No. 1 filter paper to remove any insoluble material before use. Place each rod gel in a test-tube filled with stain and left for at least 4 h at room temperature. The minimum time required for staining depends on the gel thickness and the gel concentration, increasing as the gel concentration increases. The rate of staining is increased at higher temperatures, for example 40−50°C, but in practice it is often more convenient simply to allow the gels to stand in stain overnight at room temperature and destain the following day. Gel slabs are placed in a plastic tray containing stain solution and are stained fully if left overnight at room temperature if only 1.5 mm thick. Some workers find it useful to suspend the gel slab on a nylon mesh screen held on a Perspex frame during staining and destaining. It has been reported that high concentrations of SDS interfere with Coomassie Blue staining but this is avoided if at least 10 volumes of stain solution are used with each gel; for example, 500 ml stain solution per slab gel.

After staining is complete, excess stain must be removed from the gel to allow protein bands to be seen clearly. Two different methods of destaining may be employed; diffusion destaining or electrophoretic destaining. The former is the more widely-used method. In diffusion destaining the gel is transferred to 12.5% isopropanol, 10% acetic acid, and the destain solution is simply renewed as stain leaches out of the gel over a period of about 48 h at room temperature (faster at higher temperatures). Although most background staining has been removed by this time, longer destaining may be necessary to obtain completely clear backgrounds. Faster destaining is possible using 30% methanol, 10% acetic acid, but destained gels should not be stored in this since protein bands are eventually destained also. An alternative to the replacement of destaining solution is to include a few grams of an anion-exchange resin to absorb the stain as it leaches from the gel. Commercial diffusion destainers are available for both rods and slabs (e.g. Bio-Rad Laboratories Ltd) and these usually contain a cartridge of a suitable resin or charcoal for absorbing the dye. Gels stained with Coomassie Blue R-250 are best stored in 7% acetic acid.

Electrophoretic destaining is possible because Coomassie Blue is an anionic dye. Although electrophoretic destaining can be carried out transversely or longitudinally, the former is preferred because of the shorter gel electrophoretic distance and hence shorter time period involved. Transverse electrophoretic destaining apparatus for both rod and slab gels is available commercially but these can be easily manufactured in the average departmental workshop. The gels are held between porous plastic sheets with horizontal electrodes placed either side. The destaining apparatus is filled with 7% acetic acid and connected to a power pack capable of delivering a high current $(0.25-1.0 \text{ A})$; either a commercially available model or even a car battery charger (delivering ~ 12 V, 3 A). Because of the high power input the electrode buffer reservoirs should be fitted with efficient cooling coils. Destaining occurs within $15-30$ min. It is important to note that if destaining is continued too long some destaining of the protein bands may occur and weakly stained bands may be destained completely.

Staining methods have also been described for Coomassie Blue R-250 where destaining is not necessary in order to observe the protein banding pattern. In these methods the dye is dissolved in TCA (48,49) in which it is relatively insoluble and so forms dye−protein complexes preferentially. Therefore, staining is rapid; dense protein bands appear within seconds and almost full intensity is reached after about 45 min although the staining method is reported by some workers not to be quite as sensitive as those involving methanol-based stains followed by destaining. The advantage of these methods is their speed. Furthermore, TCA is a much better fixative than acetic acid and so may be preferred if one suspects loss of protein from the gel during staining. However, most workers have continued to use the methanol-based Coomassie Blue R-250 stain for routine detection and quantitation of protein bands.

ii. Quantitation of stained proteins

A number of instruments specifically designed for scanning are available as well as gel scanning attachments for standard spectrophotometers. Using the latter instrumentation, each destained rod gel is placed in a shallow glass trough containing 7% acetic acid. Individual tracks may also be cut from slab gels and placed against one trough wall for scanning. Most of these apparatuses move the gel at a constant speed perpendicular to a narrow, fixed, parallel light beam, the transmission of which is detected by a photomultiplier. The peak absorption for Coomassie Blue R-250−protein complexes varies between 560−575 nm depending on both the protein and the solvent. The output of the spectrophotometer is recorded as a series of tracings on chart paper. Unfortunately the optics of many of these apparatuses is such that sharp bands clearly separated by the naked eye often do not give clearly separated peaks by densitometry (*Figure 14*).

A recent development in scanning densitometers is the use of a laser light source. These machines are now commercially available and are claimed to give far higher resolution when scanning gels than conventional densitometers. In practice, however, the limitation to resolution is often the electrophoretic separation itself and not the scanning procedure. Therefore, the purchase of an expensive laser densitometer does not necessarily promise better data.

Some densitometers incorporate automatic integration of the densitometric record which allows automatic integration of the areas under each peak. By this means, the amounts of a single component in different gels may be estimated. Indeed, densitometer output is increasingly being computerized by integrated or stand-alone microcomputer hardware that allows the data to be recorded on disc, displayed on the computer screen or printed, and analysed at will including the direct comparison of two or more scans from different gels. For machines lacking automatic integration one can cut out the chart paper and weigh the peaks. Alternatively, for well-resolved stained bands, these may be cut out, macerated, and the Coomassie Blue eluted overnight by shaking at room temperature with 25% (v/v) dioxane or 25% (v/v) pyridine in water (50). The pyridine shifts the dye absorption maximum from about 560 nm to 605 nm. Quantitation of dye eluted by absorbance at 605 nm allows quantitation of proteins in bands in the range 1−100 μg. This method may be particularly useful for quantitating proteins in spots after two-dimensional gel electrophoresis (Chapter 3). Wong *et al.* (51) have reported the use of 80% methanol for elution instead of the more noxious dioxane and pyridine.

Different proteins bind Coomassie Blue to different extents and so quantitative determination of the amount of a particular protein by staining requires a standard curve for that particular protein. This applies whether quantitation is by scanning or elution. However, if another protein is used as a standard, this allows the *relative* amounts of the specific protein to be determined in multiple samples, although the linearity of dye absorbance with mass still needs to be determined for the proteins under study. The overall extent of staining for any protein varies with the time period used for staining and destaining and upon the gel thickness and gel concentration both of which affect the diffusion rate of dye molecules. Clearly for optimal reproducibility of data, these variables should be kept as constant as possible.

iii. Photography of stained protein bands

For storage purposes, rod gels can be placed in sealed tubes and slab gels in sealed

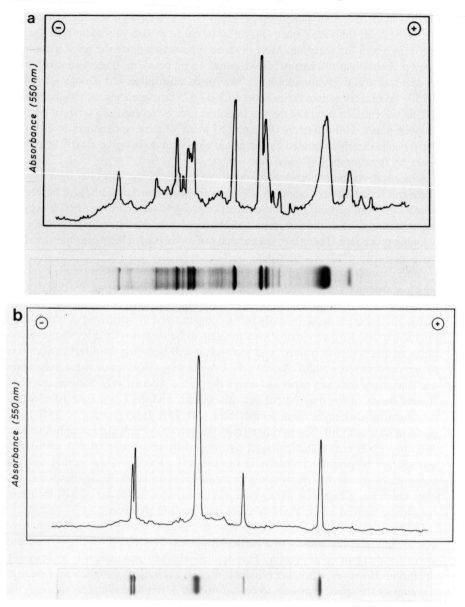

Figure 14. Densitometric analysis of Coomassie Blue stained protein bands. (a) A human placental membrane sub-fraction obtained by wheat germ agglutinin precipitation (kindly donated by Dr A.G. Booth) was analysed on a 6 – 18% polyacrylamide linear gradient slab gel, and then stained with Coomassie Blue R250. A photograph of the stained gel track is shown and above this is the densitometric scan of this track at 550 nm. (b) A track of several molecular mass marker polypeptides (RNA polymerase α, β, β' subunits, bovine serum albumin, and soybean trypsin inhibitor) analysed on a 6 – 18% polyacrylamide linear gradient slab gel, stained and scanned as in (a).

plastic bags each containing 7% acetic acid. Stained bands remain visible almost indefinitely under these conditions. However, gels stored in this wet state take up large amounts of space and are easily mislaid. The best procedure with slab gels is to photograph them and then to dry them down. Do not leave the photography until after drying since gels lose resolution during the drying step. Both photographs and dried gels are easily labelled and stored in notebooks. Rod gels are also best photographed for record purposes. They can be dried down after longitudinal slicing into thin strips (52) but this is not commonly practised since the major applications of dried gels (autoradiography and fluorography; see Section 7.5.1) are best served using the slab rather than the rod gel format.

Wet rod or slab gels are photographed by being laid directly onto an illuminator with an opal white screen (avoiding trapped air bubbles) and kept wet during photography by addition of 7% acetic acid. A fine-grain, panchromatic film should be used, for example Ilford Pan F, and a medium-red filter will increase band contrast. It is worth including a gel title in the photograph to identify the samples later. Also, ensure that the entire gel is photographed to enable R_f values of proteins to be calculated from photographs if necessary in future data analysis.

iv. Gel drying

Successful drying of slab gels is possible only with gels not thicker than about 1.5 mm; thicker gels usually crack upon drying and must therefore be sliced horizontally into thin sheets. Apparatus to do this is available commercially but the best method is to use thin slab gels for the initial electrophoresis so that slicing is avoided.

If the polyacrylamide gels are dried down without any support they shrivel. However, by drying down onto a filter-paper backing, the gel dimensions are preserved as the gel attaches itself to the paper support. A number of gel driers have been described which can be made in the laboratory, the simplest being that of Maizel (53). However, for reliability it is well worth purchasing a commercial gel drier. Many types are available which differ in the size of gel they will accept and the presence or absence of a built-in heater to increase the rate of drying and the ability to choose the drying temperature. A good drier should be large enough to take at least two standard-size slab gels with a built-in heater coupled to a timer (e.g. Bio-Rad, Hoefer, LKB; see *Figure 15a*). For rapid drying, the temperature control of most driers is pre-set at 80°C. Two slab gels 1.5 mm thick will usually dry in less than 1 h using this type of apparatus. Best results are obtained if the gel is soaked overnight in a solution containing 3% glycerol prior to drying. The solution should contain isopropanol or methanol in sufficiently high concentration to prevent swelling of the gel to greater than its original size, or cracking may result. Some concentration gradient gels (e.g. 6−18% acrylamide) can be dried successfully if they are first soaked in 30% methanol, 3% glycerol, but difficulties may be experienced at higher gel concentrations. The best commercial driers allow the drying temperature to be reduced so that drying occurs more gently over a longer time period. This may help reduce cracking of some highly-crosslinked gels and concentrated gradient gels. In addition, most gels being prepared for fluorography are best dried at 60°C to protect the heat-labile fluor impregnated in the gel.

(a)

A

B

C

(b)

Figure 15. (a) A commercial gel dryer. The model shown is the model 583 slab dryer from Bio-Rad Laboratories. It has pre-programmed drying cycles to provide optimum drying for different gel types. The dryer heats gels from the top while vacuum pulls moisture from the bottom. Other commercial models vary in the arrangement of heating elements. (b) A typical arrangement of apparatus for gel drying. A, gel dryer (Bio-Rad model 483); B, vacuum pump; C, dry ice cold trap all mounted on a movable trolley for convenience. (From ref. 67 with permission.)

The precise method for drying gels obviously varies according to the gel drier used. A typical protocol would be to place sheets of pre-wetted filter paper (e.g. 3MM) into the gel drier support (beneath which is the heating block) and then align the slab gel or rod gel slices on this, being careful not to trap air bubbles under the gel since this can result in cracking during drying. This is overlaid with a sheet of pre-wetted cellophane, then a porous plastic sheet, and finally the silicone sheet (attached to the apparatus) forming a leak-proof seal. The gel drier is connected to a vacuum pump fitted with a dry ice cold trap (*Figure 15b*) or a cold finger cooled with liquid nitrogen. After vacuum is applied, the heating block is turned on. The exact time required for drying depends on the size and concentration of the gels being used. The silicone rubber top sheet of many commercial gel driers is transparent so that the progress of drying can be assessed without breaking the vacuum seal. Indeed, if air is allowed to enter the assembly before drying is complete the gel will crack irreparably. The resulting dried gel is sandwiched between its filter paper backing and a protective cellophane sheet. When radioactive gels are dried, the cellophane can be replaced with Saran Wrap (cling film) to prevent contamination of the porous plastic sheet but, since this is non-porous, the gel is more susceptible to cracking.

An alternative method to heat and vacuum for drying gels is to use a simple plastic frame to dry the gel between two sheets of cellophane membrane. Such frames are available commercially (e.g. Biotech Instruments Ltd). The slab gel is equilibrated in 3% glycerol to prevent cracking then placed between two cellophane sheets which have been previously wetted in this solution. Air bubbles must be scrupulously removed both from the gel/cellophane interface and along the gel edges or gel cracking will result. Drying takes place overnight at room temperature or in about 1 h using a fan heater. The dried gel is preserved in a transparent form that can be photographed or scanned densitometrically (although, as mentioned above, some resolution is lost during the drying process so that one would be well advised to photograph or scan the gels before drying).

7.2.2 Xylene Cyanine Brilliant G (Coomassie Blue G-250)

This stain, which has similar sensitivity to Coomassie Blue R-250, is preferred by some workers. A typical procedure is to prepare a 0.05% solution of the dye in methanol and then add an equal volume of 25% TCA. Fresh batches of dye should be prepared every two weeks or so since the dye promotes decomposition of the TCA to release chloroform. If a very rapid localization of a protein band is needed, the gel is incubated in the dye solution for only 5−10 min. Strong bands become visible within this time or a few minutes after transferring the gel to 5% TCA. For more thorough staining, the gel should be left in the staining solution for several hours, followed by destaining in 5% TCA.

A useful modification of the procedure (54) is to arrange the pH of the solution such that the dye is in its almost colourless leuco form. On binding to protein, the dye pK is shifted and a blue colour develops. Therefore when gels are incubated in the staining solution, blue protein bands appear against an almost colourless background. No destaining is necessary. One procedure (46) is described in *Protocol 7.*

Protocol 7. Staining with Coomassie Blue G-250 in perchloric acid

1. Prepare 0.04% Coomassie Blue G-250 in 3.5% perchloric acid.
2. Stir at room temperature for 1 h and filter through Whatman No. 1 paper. Then filter through a Millipore filter (0.45 μm pore size). This solution is stable indefinitely at room temperature.
3. Immerse the gels in the stain such that the ratio of water in the gel to stain volume is approximately 3:5. Very strong protein bands become visible within 10 sec, most bands are visible within 10 min and staining is essentially complete by 90 min. At 37°C the rate of staining is increased approximately 2-fold.
4. The sensitivity of detection can be increased about 3-fold if, after staining, the gels are placed in 5% acetic acid whereupon the background changes to pale blue. The gels can be stored in 0.005% Coomassie Blue G-250 in 6% perchloric acid.

7.3 Silver staining

7.3.1 Advantages and drawbacks

Silver staining was introduced in 1979 by Switzer *et al.* (55) as a novel staining procedure up to 100 times more sensitive than Coomassie Blue R-250, able to detect as little as 0.38 ng/mm^2 of bovine serum albumin. An example of the marked increase in sensitivity possible using silver staining is shown in *Figure 16.* Since the original description of silver staining, over 100 variations of this methodology have been published, many of which have been reviewed elsewhere (56). For example, Ohsawa *et al.* (57) reported an improved silver stain which can detect as little as 10 femtograms of protein after gel electrophoresis. In addition to the published methodology, a number of silver stain kits are commercially available (e.g. Bio-Rad).

Although silver staining is the most sensitive non-radioactive protein detection method currently available, it has a number of drawbacks.

- High backgrounds can result, especially if the water used is insufficiently pure.
- The method is significantly more expensive to operate on a routine basis than, say, Coomassie Blue staining.
- It is a fairly laborious staining procedure with several steps each of which must be performed carefully.
- Some proteins stain either poorly or not at all, sometimes appearing as 'negatively' stained against a darker background (see below). Thus it is always difficult to be certain that all proteins have been detected.
- Some silver stains detect not only proteins but also DNA, lipopolysaccharides, and polysaccharides, so that a stained band does not necessarily identify the reacting component as a protein.

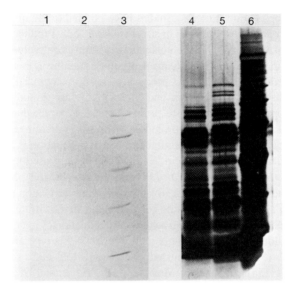

Figure 16. The increased sensitivity of silver staining. Lanes 1 – 3 are identical to lanes 4 – 6 except that lanes 1 – 3 were stained with Coomassie Blue and lanes 4 – 6 with a colour-based silver stain (see *Protocol 9*). (From ref. 67 with permission.)

Since the sensitivity of detection using simple organic stains, such as Coomassie Blue R-250, is sufficient for many applications, it is probably advisable to use silver staining only when its far superior sensitivity is required and so outweighs its practical drawbacks.

7.3.2 Types of silver staining procedures

Prior to silver staining, the proteins need to be fixed. This fixation step has two roles. First, it immobilizes the proteins in the gel, or at least greatly retards their diffusion from the gel. Second, it removes substances which may interfere with the staining procedures such as detergent, reducing agent or reactive buffer components like glycine. A variety of fixatives have been used, including glutaraldehyde (58), ethanol or methanol and acetic acid (59, 60), and TCA (61). The gel is then exposed to silver nitrate solution and finally the silver stain is allowed to develop. A key factor for success in silver staining after SDS-PAGE is the removal of SDS prior to staining. Several washing procedures with different solvents have been suggested but silver stains usually work irrespective of the precise washing procedure used, provided that it is sufficiently thorough to remove SDS.

The basic mechanism of silver staining is the reduction of silver nitrate to metallic silver at a protein band, leading to the deposition of silver grains. Plots of the optical density of silver-stained proteins against their concentration give different slopes for each protein, indicating that the degree of staining is to some extent influenced by specific groups within proteins. Studies have pointed to the importance of basic and sulphur-containing amino acids in silver staining (62). Proteins without cysteine residues or with very low cysteine contents may stain negatively with some silver stains. However, the precise mechanism of silver staining is not yet known.

Two broad categories of staining procedure can be recognized; those involving chemical development of the stain and those involving photodevelopment. The chemical development methods can be further divided. First, the diamine silver stains rely upon the use of ammonium hydroxide to form silver diamine complexes. After fixation the gel is soaked in this solution and image production is then achieved by acidification, usually with citric acid. Typical and widely-used procedures of this kind are the methods of Oakley *et al.* (58) and Wray *et al.* (63). Second, there are non-diamine chemical development silver stains which rely upon the immersion of the fixed gel in silver nitrate at acidic pH. After the silver nitrate has reacted with protein sites, image formation is then achieved by reduction of ionic silver to metallic silver by formaldehyde at alkaline pH. Procedures of this type are the methods of Sammons *et al.* (59), which is reportedly at least 10-fold more sensitive than that of Oakley *et al.* (58), and that of Morissey (60). Additional treatment of the gel with reagents such as dithiothreitol, potassium permanganate, potassium ferricyanide or potassium dichromate is argued to increase the sensitivity of the stain although this additional treatment does not appear to be obligatory (e.g. ref. 59). Treatment with potassium permanganate and $CuCl_2$ during silver staining may allow the detection of protein bands that otherwise remain undetected (64). Staining is stopped by acidification, usually with citric or acetic acid. The stain may then be further enhanced by treating the gel with sodium carbonate (59, 60).

The photodevelopment stains use light energy to reduce the silver ions to metallic silver. Since light can reduce silver at acid pH, photodevelopment stains can use a single staining solution once the gel has been fixed, unlike chemical staining procedures which normally require at least two solutions.

Each type of silver-stain method has its followers. The method of Oakley *et al.* (58) is widely used despite the tendency for deposits to form on the gel surface and a yellowish gel background which can sometimes form when the glutaraldehyde fixative reacts with glycine present in some electrophoresis buffers. The non-diamine chemical development stains (e.g. 59, 60) generally give less deposits on gel surfaces and are cheaper in terms of silver nitrate usage. The photodevelopment methods are rapid, being able to produce an image within 15 min of terminating electrophoresis (65) but are generally less sensitive than the chemical development methods and often produce negatively-stained proteins which are therefore difficult to quantitate. As noted above, the extent of silver staining varies from protein to protein. However, for five specific proteins tested, Merril *et al.* (65) found that photodevelopment silver staining was 5- to more than 20-fold more sensitive than Coomassie Blue R-250 staining whereas a chemical development silver stain was 10- to more than 200-fold more sensitive. Overall, at the present time, most researchers using silver stain prefer one of the non-diamine chemical development methods such as that of Sammons *et al.* (59). A modified protocol for this procedure is described in the following section.

7.3.3 Protocols for monochromatic and colour-based silver stains

Most proteins have a brown or black colour when silver stained. However, some silver stain protocols may yield other colours that may help in the identification of certain proteins. For example, Goldman *et al.* (66) found that some lipoproteins stain blue while some glycoproteins stain yellow, red, or brown. Coloured images may occur if the silver grain size varies, going from red/yellow through blue to black as grain size increases. Charged amino acid side-chains play a major role in colour development. However, other factors are also undoubtedly involved, including the distribution of silver grains in the gel so that the amount of protein in a band may influence colour.

A monochromatic silver stain may be preferred for some analyses such as staining complex protein mixtures in one-dimensional polyacrylamide gels. A reproducible method of this kind, the diamine silver stain of Wray *et al.* (63), is given in *Protocol 8*. However, in many cases the colours observed using a polychromatic silver stain provide useful additional visual information to aid the identification of particular

---

Protocol 8. Procedure for monochromatic silver staining[a]

1. Prepare the following stock solutions using deionized water: Use glassware throughout since plastic can leach organic material which interferes with silver staining.
   Solution A: 0.8 g of silver nitrate (Sigma) in 4 ml of distilled water.
   Solution B: mix 21 ml of 0.36% NaOH (Sigma) with 1.4 ml of freshly-prepared 14.8 M ammonium hydroxide.

2. All the following procedures should be carried out in a glass (not plastic) dish. First, soak the gel in 50% methanol for at least 1 h.

3. Prepare the silver stain by adding solution A dropwise to solution B with constant stirring or vortexing and then making to 100 ml with deionized water. This silver stain should be prepared immediately before use. Stain the gel in this solution for 15 min with gentle agitation, preferably on a platform shaker.

4. Wash the gel in deionized water for 5 min with gentle agitation.

5. Soak the gel in developing solution until bands appear; developing solution is 2.5 ml of 1% citric acid mixed with 0.25 ml of 37% formaldehyde (Fisher) and made to 500 ml with deionized water. It must be prepared just before use. Bands usually appear in less than 10 min.

6. Wash the gel with deionized water and place it in 50% methanol to stop further development.

[a] As described by Wray *et al.* (63).

---

proteins. Furthermore, because of the colour contrast, it is sometimes possible to detect minor proteins which would otherwise go unnoticed in an overloaded gel. A reliable polychromatic stain, from Sammons *et al.* (59), is described in *Protocol 9*.

Protocol 9. Procedure for polychromatic silver staining[a,b]

1. Prepare the following stock solutions:

    Silver stain: 1.9 g of silver nitrate (Sigma)
                            1 litre of deionized water

    Reducing solution: 30 g of NaOH in 1 litre of deionized water. Immediately before use, add 7.5 ml of 37% formaldehyde[c] (Fisher)

    Colour enhancer: 70.5 g of Na_2CO_3 in 10 litres of deionized water.

    Since plastic bottles leach traces of organic components which can interfere with silver staining, these stock solutions are best stored in glass bottles. Carry out all the following steps with the gel in a Pyrex glass dish (not plastic). At all stages, the efficiency of agitation is important and therefore each incubation with reagent is best carried out on a platform shaker set at a speed high enough to ensure good circulation of the solution without damaging the gel.

2. Prepare gel fixative (1 litre of 50% ethanol, 5% acetic acid) just before use. Soak the gels in the fixative using a gel volume:fixative volume of 1:5.5[d]. For optimal results, allow fixation to occur overnight at room temperature.

3. The following day, wash the gel in three changes of deionized water (1 h each wash).

4. Incubate the gel with agitation in the silver stain for 1 h with a ratio of gel volume:stain volume of 1:3.

5. Rinse the gel thoroughly (but briefly; 10–15 sec) by immersing it in deionized water to remove silver on the gel surface. If this washing is not carried out thoroughly, black silver grains will eventually form on the gel surface.

6. Add formaldehyde to the stock of reducing solution as stated in step 1 and immediately add this mixture to the gel on the platform shaker in the ratio of gel volume:solution volume, 1:5.5. Shake for 8–10 min.

7. Pour off the reducing solution and add the colour enhancer in the ratio of gel volume:solution volume, 1:5.5. Incubate the gel in this solution with shaking for 1 h[e].

8. Replace the colour enhancer solution with fresh colour enhancer and shake for a further 1 h[e].

9. Replace the colour enhancer with fresh colour enhancer and shake overnight.

10. The protein bands should appear against a uniform yellow background. Black

Protocol 9. *continued*

and white photography using Pan-X film will not be interfered with by the background[f].

[a] Modified from ref. 59 by Dunbar (67).
[b] This stain is also available commercially (Gelcode from Health Products or Pierce Chemical).
[c] The age and source of formaldehyde is important; take careful note of which batches work best and keep to that source if possible.
[d] The ratios of gel volumes to reagent volumes are critical throughout the procedure. Should the volume of the gel be too small for efficient staining with the reagent volumes needed, stain multiple gels.
[e] The colours of the protein bands appear in the first hour after addition of the colour enhancer (step 7) but the further incubations (steps 8 and 9) give full colour development.
[f] If the background is judged to be too dark, use warm reducing solution before addition of the formaldehyde and incubate the gel in this solution for $2-3$ min longer (step 6).

Using this procedure, proteins stain five basic colours; black, blue, brown, red, or yellow. Nucleic acids stain a distinctive dark colour. The other advantage of this procedure is that it is substantially more sensitive than monochromatic silver stains.

7.3.4 Factors important for success

Protocols 8 and *9* describe two of the most reliable and reproducible silver stain methods currently available (67). Nevertheless, care must be taken to follow a few important guidelines.

- Plasticware should be avoided both for storing stock solutions to be used in silver staining and for the gel staining steps since these can leach organic components which either interfere with silver staining or stain artefactually; glass bottles and Pyrex glass dishes, respectively, should be used.

- Water quality is also critically important—use the purest water available.

- Certain reagents need to be prepared just before use in order to obtain optimal staining results (these are indicated in *Protocols 8* and *9*).

- Good circulation of staining solutions around the gel is essential for the procedure to work well—always carry out the staining steps with the gel tray on a platform shaker to ensure that agitation of the staining solutions is adequate for good equilibration.

7.3.5 Maximizing the sensitivity of silver staining

Obtaining the highest sensitivity with silver stains requires the use of pure reagents, especially the water used to prepare the solutions. This should have a conductivity less than 1 μmho/cm. If still greater sensitivity is required, it is possible to recycle the stained gels through the chemical development procedure again from the silver nitrate stage to add more silver to the original deposits. This may allow the detection of trace amounts of previously undetected proteins. Many photographic image intensity methods can also be used to intensify silver-stained gels. In these methods, intensification may either add more silver to the original deposits or may add other metals such as copper, mercury, uranium, or chromium. Procedures which intensify by adding more silver usually intensify proportionally, that is, the images increase in density proportional to the original image. However, this is often not the case when other intensifying metals are used. Berson (68) has described a blue toning procedure which results in a 3- to 7-fold increase in sensitivity over the use of the chemical development procedure alone.

7.3.6 Variability of protein staining

Silver staining can be used to stain proteins separated by gel electrophoresis in either dissociating or non-dissociating buffer conditions. As mentioned earlier, different proteins stain to different extents. Histones are particularly difficult to silver stain and some silver stains have been reported not to stain, for example, calmodulin or troponin. This is probably related to the absence or low content of cysteine residues in these proteins. Merril *et al.* (69) also observed an acidic protein which failed to stain until the gel was recycled through the silver nitrate stage of the chemical development method. This variability in staining different proteins is more problematic with native proteins than with denatured proteins. One useful modification is to stain gels initially with Coomassie Blue then to silver stain. This reduces the protein to protein variability in silver staining considerably and furthermore has higher sensitivity than silver staining alone (e.g. ref. 70), possibly because the dye molecules bound to protein bands provide extra sites for silver deposition. Andrews (3) recommends staining with Xylene Cyanine Brilliant G (although Coomassie Blue R-250 can also be used), destaining and then silver staining.

As stated above, the fact that not all proteins are equally sensitive to silver staining is a disadvantage when considering its use as a general protein stain. However, several workers have deliberately amplified this differential sensitivity by varying the staining conditions so that the protocol selectively stains specific classes of proteins (e.g. 71, 72).

7.3.7 Linearity of protein staining

As a general guideline, chemical development silver stains are usually linear over an approximately 40-fold range in concentration from about 0.02 ng/mm^2 for most proteins. Above protein concentrations of about 2 mg/mm^2, the silver image becomes saturated and often results in bands where the centres are less stained than the periphery. Nevertheless, the linear response range for silver staining varies somewhat depending on the particular method used and so this must be determined for each protocol prior to analyses which require quantitation. Photodevelopment silver staining is the most difficult to quantitate because of its tendency to give both positively- and negatively-stained proteins. Furthermore, as with most organic dyes including Coomassie Blue (Section 7.2.1), different proteins stain to different extents and so determination of the *absolute* amount of a particular protein requires a standard curve for that protein. If another protein is used as a standard, the staining intensity of the protein of interest in different samples yields only the *relative* amounts of that protein present.

B. David Hames

7.3.8 Drying silver-stained gels

Before drying gels which have been silver-stained by the method of Sammons *et al.* (*Protocol 9*), the gel should be soaked overnight in at least 5 vol. of 1% glycerol − 10% acetic acid. The gel should be photographed before drying, since it will be very dark after drying.

7.4 Fluorescent protein labels

Methods have been devised for pre-staining proteins with a fluorescent dye before electrophoresis and then detection of the fluorescent protein bands after electrophoresis by scanning. Athough dansylation using dansyl chloride was the first method to be described for labelling proteins in this way, later analytical studies have preferred 'fluorescamine' [4-phenylspiro-[furan-2(3H),1-phthalan]-3,3′-dione], trade name 'Fluram', since, unlike dansyl chloride, neither free fluorescamine nor its hydrolysis products are fluorescent, leaving the labelled protein as the only fluorescent component. The major advantages of this reagent is the increased sensitivity of detection which it affords (~5 ng of protein can be detected once it is labelled with fluorescamine) and the ability to monitor the progress of electrophoretic separation simply by exposing the gel to UV light in a darkened room. The disadvantage of fluorescamine is that it converts an amino group to a carboxyl group when it reacts with proteins and so will alter the mobility of proteins in non-SDS-containing buffers. In fact, fluorescamine has also been reported to alter the mobility of some proteins in SDS-PAGE (73), although other workers using the same proteins found that these still obeyed the usual relationship between R_f and $\log_{10}$ molecular mass (74). Barger *et al.* (75) introduced another fluorescent label called MDPF [2-methoxy-2, 4-diphenyl-3(2H)-furanone] which has the advantage that MDPF-labelled proteins in polyacrylamide gels retain their fluorescence for several months whereas that due to fluorescamine is reported to decay over a period of 24 h. Using MDPF the fluorescence signal is linear with protein from 1 − 500 ng and labelled proteins obey the linear relationship between $\log_{10}$ molecular mass and R_f when analysed by SDS-PAGE. Details on the use of fluorescamine and MDPF and the equipment required for scanning are given elsewhere (76). The reagent *o*-phthaldialdehyde (OPA) (77) has also been used for pre-labelling proteins and is reported to be about 10-fold more sensitive than fluorescamine. Simple protocols for using these reagents are given in Chapter 5, Section 3.2. However, possibly because of the requirement for special scanning equipment when quantitating fluorescent proteins and the applicability of Coomassie Blue and silver staining to most situations, the technique of pre-labelling proteins with fluorescent molecules prior to electrophoresis has not found widespread use in analytical polyacrylamide gel electrophoresis.

Protein bands can also be detected *after* electrophoresis by labelling with fluorescent molecules such as 1-aniline-8-naphthalene sulphonate (ANS) (78) and OPA (3). These may be particularly applicable for use after preparative gel electrophoresis for rapid localization of protein bands since the staining procedure does not necessarily require

67

denaturing conditions. The simplest protocol is for ANS. The whole gel is immersed in 0.003% (w/v) ANS in 0.1 M phosphate buffer (pH 6.8). Within 5–10 min, fluorescent bands begin to appear when viewed with a UV lamp. However, the procedure is rather insensitive in that only bands containing about 20 μg of protein or more are visualized. Bis-ANS [bis(8-*p*-toluidino-1-naphthalene sulphonate)] has been reported to be far more sensitive than ANS itself for the detection of proteins after SDS-PAGE, especially if the fluorescence is enhanced with a cation such as potassium (as KCl) when 0.5 μg of protein can be detected (79). For native proteins, OPA is perhaps the reagent of choice since a band containing only 0.25 μg protein can be detected by UV illumination at 340–360 nm after only 25 min under conditions where most enzymic activity is apparently retained (80). Proteins can also sometimes be visualized directly in the gel after electrophoresis by UV light-induced fluorescence without the addition of any reagent. This fluorescence is due to the excitation of tryptophan residues and so proteins which lack this amino acid are not detected.

7.5 Detection of radioactive proteins

Proteins can be radiolabelled either during synthesis, using labelled amino acids, or post-synthetically by iodination, reductive methylation, etc (Appendix 3). A major advantage of radiolabelling is that detection methods for radiolabelled proteins following gel electrophoresis are far more sensitive than staining methods for unlabelled proteins. Two basic approaches are possible. The first is to place the gel next to X-ray film whereupon the radioactive emissions cause the production of silver atoms within silver halide crystals and these are visualized after developing the film (*Figure 17a*). Important variations of this method exist which serve to increase the sensitivity of detection whilst, in some cases, maintaining the linear relationship between film image absorbance and sample radioactivity (see Section 7.5.1). The film image may be recorded photographically or densitometrically by using a spectrophotometer fitted with a scanning attachment. The latter method also enables quantitation by measurement of the peak area, although, as with scanning stained bands, the resolution of closely migrating bands is poor. The second approach involves slicing the gel after electrophoresis and determining the radioactive protein content of each slice by scintillation counting (*Figure 17b*). This is the approach most commonly used for quantitation of radioactive proteins separated in rod gels or of proteins analysed in dual isotope experiments. However, the resolution of the method is dependent upon the size of the fractions produced which for gel slicing techniques is not usually less than 0.5 mm. Far higher resolution is obtained using X-ray film detection methods on intact gels and so these should be used initially, especially with more complex protein mixtures, to identify the number of polypeptides present.

7.5.1 Methods using X-ray film

The optimum exposure time has to be determined empirically for each gel since it will depend on the isotope used, the amount of radioactivity present in each protein

B. David Hames

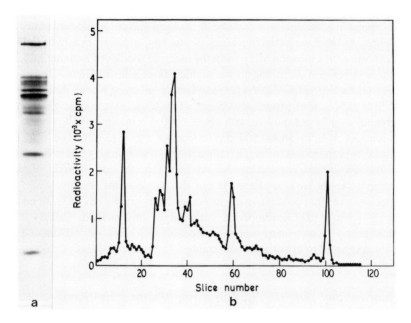

Figure 17. Detection of radioactive protein bands by (a) autoradiography or (b) gel slicing followed by scintillation counting of the slice contents. Detailed protocols are given in the text. The sample shown here is a subfraction of [^{35}S]-methionine-labelled spore proteins of *Dictyostelium discoideum*.

band, and whether the researcher wishes to detect only major components or also more minor ones. Several exposures of different durations will probably be required to detect all or most components in a protein mixture since, whilst a long exposure will detect minor bands, band spreading of the photographic image occurs for major components during long exposures and so bands become indistinct and may overlap.

i. Direct autoradiography

^{32}P- and ^{125}I-labelled proteins can be easily detected in wet gels simply by sealing the slab gel or rod gel slices with Cling film (to prevent dehydration of the gel and wetting of the film) and placing this against one of the 'direct' types of X-ray film such as Kodak Direct Exposure Film (formerly Kodirex or No-Screen). However, resolution is much improved if the gel is first dried down onto a sheet of Whatman 3MM filter paper (Section 7.2.1). Usually the gel is stained with Coomassie Blue and destained before drying so that the photographic image obtained after autoradiography can be compared with the stained protein banding pattern. To facilitate alignment of the stained, dried gel and the developed X-ray film, it is wise to place a few spots of radioactive ink [~ 1 μCi, (40 kBq) ^{14}C per ml ink] on the filter paper at the gel perimeter prior to autoradiography. This ink is also useful for labelling the gel (and hence the X-ray film) to avoid errors later. The stained

69

dried gel with radioactive ink marker spots is placed in direct contact with the X-ray film, clamped in a radiographic cassette or between glass or hardboard plates using metal clips, and then wrapped in a black plastic bag or placed in a light-tight box. After exposure (in a location away from the external sources of ionizing radiation), the film is developed according to the manufacturer's instructions.

Direct autoradiography of dried polyacrylamide gel using Kodak direct exposure X-ray film gave a film image absorbance of 0.02 A_{540} units (just visible above background) in a 24 h exposure with about 6000 d.p.m./cm^2 of ^{14}C or ^{35}S, 1600 d.p.m./cm^2 of ^{125}I, or about 500 d.p.m./cm^2 of ^{32}P (81, 82). The film image absorbance is proportional to sample radioactivity. Unfortunately, ^{3}H is not detected since the low-energy β-particles fail to penetrate the gel matrix. However, fluorography (see below) can detect ^{3}H and also increases the sensitivity of ^{14}C and ^{35}S detection over that possible using direct autoradiography (81, 83). In addition, Laskey and Mills (82) developed a variation of this method, called 'indirect autoradiography' which increases the detection efficiency of ^{125}I and ^{32}P considerably compared to direct autoradiography. The methodologies of fluorography and indirect autoradiography have been reviewed by Laskey (84).

ii. Fluorography (for ^{3}H, ^{14}C, ^{35}S) using PPO as scintillator

In fluorography the gel is impregnated with a scintillator, dried down, and exposed to the X-ray film at $-70°$C. Although low-energy β-particles produced by ^{3}H-labelled proteins are unable to penetrate the gel matrix and expose the X-ray film directly, they are able to interact with the scintillator molecules in the gel to convert the energy of the β-particle to visible light which then forms an image on blue-sensitive X-ray film. Thus absorption of β-particles by the sample and gel matrix is overcome by the increased penetration of light. The low temperature of exposure ($-70°$C) is necessary to stabilize latent image formation during long exposures to the light generated by fluorography (or by indirect autoradiography using intensifying screens; see later). This stabilization can increase the sensitivity of detection by 4-fold or more. There is no improvement in sensitivity for direct exposure of X-ray film to ionizing radiation.

The original fluorographic procedure used 2,5 diphenyloxazole (PPO) as scintillator. The basic protocol is given in *Protocol 10*. It can be used on either unfixed or stained gels. Exposure to dimethylsulphoxide (DMSO) removes some Coomassie Blue from protein bands such that faintly stained bands may not be visible at the end of the procedure. Furthermore the precipitate of PPO which is formed in the gel obscures the staining pattern. For these reasons a photographic record should be made of stained gels before attempting fluorography. Alignment of stained bands in the gel with the final fluorographic image can be made using radioactive ink spots marked on the gel perimeter after drying and prior to clamping the dried gel and X-ray film together (Section 7.5.1*i*).

Unfortunately, using untreated X-ray film, the absorbance of the fluorographic image is not proportional to the amount of radioactivity in the sample, small amounts of radioactivity producing disproportionately faint images. This can be overcome

B. David Hames

Protocol 10. The basic fluorographic technique[a]

Caution: Wear rubber gloves throughout the procedure to avoid skin contact with DMSO.

1. Directly after electrophoresis, or after staining/destaining, soak the gel in about 20 times its volume of DMSO for 30 min followed by a second 30-min immersion in fresh DMSO[b]. These separate batches of DMSO can be stored for use (in the same sequence) with other gels. Prolonged use is to be avoided since all water must be removed or the PPO will not enter the gel.

2. Immerse the gel in 4 vol. of 20% (w/w) PPO in DMSO for 3 h[c].

3. Immerse the gel in 20 vol. of water for 1 h. The PPO is soluble in DMSO but insoluble in water. Therefore, PPO precipitates within the gel matrix turning the gel opaque.

4. Dry the gel under vacuum[d].

5. Place X-ray film in contact with the gel and expose at −70°C. The film originally used was Kodak X-Omat R, XR1, but this is no longer available. Alternative films such as Fuji RX film can be used instead. The relative sensitivities of some films for fluorography are listed in refs 82 and 91. Direct exposure X-ray films used in autoradiography are inefficient at recording the visible light produced by the fluorographic technique and so are not used for fluorography.

6. After exposure, unwrap the film before it warms up (to avoid physical fogging) and develop it according to the manufacturer's instructions.

[a] Adapted from refs 81 and 83.
[b] Methanol is used instead of DMSO for gels which contain less than 2% acrylamide plus 0.5% agarose. After electrophoresis the gel is soaked in 20 vol. of methanol for 30 min followed by a second 30-min immersion in fresh methanol. Then the gel is immersed in 10% (w/w) PPO in methanol for 3 h and finally dried under vacuum without heat.
[c] Excess PPO may be recovered from the DMSO by adding 1 vol. of PPO in DMSO to 3 vol. of 10% (v/v) ethanol. After 10 min, the suspension is filtered and the PPO precipitate is washed with 20 vol. of water, then air-dried.
[d] If gel cracking during drying is a problem, the gel can be successfully dried by soaking in methanolic solutions (up to 30% methanol) containing 3% glycerol (see Section 7.2.1) after step 3 and prior to gel drying.

---

by exposing the film to an instantaneous flash of light (≤ 1 msec) prior to using film for the fluorographic detection of radioactive bands (81). The procedure for controlled pre-exposure of X-ray film is detailed in *Protocol 11*. When the film is pre-exposed to between 0.1 and 0.2 A_{540} units, the absorbance of the fluorographic image becomes proportional to sample radioactivity and the sensitivity of the method increases still further compared to direct autoradiography such that 400 d.p.m./cm^2 of ^{14}C or ^{35}S, or 8000 d.p.m./cm^2 of ^{3}H are detectable (*Table 7*). It is essential to

Protocol 11. Pre-exposure of X-ray film for quantitative fluorography[a]

1. Perform the pre-exposure using a single flash from an electronic flash unit. It is essential that the duration of the light flash is short (≤ 1 msec) since flashes of longer duration only increase the background 'fog' absorbance without hypersensitizing the film. To overcome slight variations in charging the capacitor, use the unit only after the charging lamp has been illuminated for at least 30 sec. Tape three filters to the window of the flash unit to reduce and diffuse the light output:

   (a) an infrared-absorbing filter; placed nearest the flash unit to protect the other filters from the heat generated;

   (b) a coloured filter; to reduce light output. A 'Deep Orange' Kodak Wratten No. 22 is suitable for the above flash unit whereas an 'Orange' Kodak Wratten No. 21 is used for weaker flash units;

   (c) porous paper (e.g. Whatman No. 1 filter paper); to diffuse the image of the bulb so that the film is evenly exposed.

   Make minor adjustments to the illumination intensity by varying either the distance between the film and the light source (usually $60-70$ cm) or the diameter of the aperture in an opaque mask. Back the film with yellow paper during pre-exposure; the surface which has been facing the light source is that which is eventually applied to the gel.

2. Determine the degree of pre-exposure by using any conventional spectrophotometer to measure the increase in background 'fog' absorbance at 540 nm of pre-exposed film compared to unexposed film.

3. Storage of films after pre-exposure is not recommended.

[a] Method derived from ref. 81.

Table 7. Sensitivities of methods for radioisotope detection in polyacrylamide gel[a]

Isotope	Type of method	d.p.m./cm² required for detectable image ($A_{540} = 0.02$) in 24 h	Enhancement over direct autoradiography
^{125}I	indirect autoradiography	100	16
^{32}P	indirect autoradiography	50	10.5
^{14}C	fluorography	400	15
^{35}S	fluorography	400	15
^{3}H	fluorography	8000	>1000

[a]From ref. 84 with permission. The data are for exposure at $-70°C$ using X-ray film pre-exposed to $A_{540} = 0.15$ above the background absorbance of unexposed film.

---

note that only under these conditions of film pre-exposure does the absorbance profile of the fluorographic images represent the true distribution of radioactivity in

B. David Hames

the sample. Therefore, a pre-exposure of 0.15 A_{540} units is used routinely for all quantitative fluorography. The absorbance of any film image produced by a radio-active sample protein is quantitated using a scanning spectrophotometer at 540 nm. The linearity of film image absorbance to radioactivity should be checked with each batch of film, and over the range required in the experiment, by measuring the actual amounts of radioactivity in selected gel slices using liquid scintillation counting (Section 7.5.2). Increasing the pre-exposure of X-ray film above 0.2 A_{540} units increases the sensitivity of ^{3}H, ^{14}C, and ^{35}S detection by fluorography still further, but the absorbance of the film image ceases to be proportional to the radioactivity present (81).

Two practical aspects which affect the interpretation of fluorographic images should be noted. Firstly, as stated earlier, one can stain the gel prior to fluorography with, say, Coomassie Blue. However, staining is known to decrease the intensity of the film image, probably by quenching the emission fluorescence by energy transfer. Coomassie Blue G-250 is fairly resistant to elution with DMSO and so a dramatic (up to 5-fold) reduction in film image intensity can be observed (85). Although most Coomassie Blue R-250 is removed from the gel by DMSO treatment, there is still a significant reduction in image intensity. Since these stains do not stain all proteins to the same extent (see Section 7.2) the reduction in image intensity probably varies from band to band depending on its staining intensity. Therefore the fluorographic banding pattern of stained gels may not accurately reflect the actual pattern of radioactivity in the gel. Quenching is also a factor one must bear in mind with silver-stained gels. Whilst quenching of ^{14}C and ^{35}S is minor with non-diamine silver stains, ^{3}H quenching is severe with all silver staining methods since the silver deposited on the protein absorbs the weak β emissions. If a diamine silver stain was used, ^{3}H-labelled proteins are unlikely to be detected even after destaining.

A second cautionary note is necessary with respect to the use of concentration gradient gels (Section 9.3). The absorbance of the fluorographic image produced by a given level of ^{3}H or ^{14}C radioactivity decreases with increasing acrylamide concentration. Thus, between 5−15% acrylamide, the image intensity is reduced approximately 4-fold for both isotopes (85). This may lead to misleading interpretations of the relative radioactivity of bands in gradient gels.

The PPO−DMSO method described above is no longer the sole procedure available. As an alternative, Skinner and Griswold (86) described the use of PPO in glacial acetic acid as the fluorographic reagent. In this procedure, gels are immersed in 22% PPO in glacial acetic acid directly after destaining.

iii. Fluorography using aqueous fluors
Chamberlain (87) introduced the use of sodium salicylate as the fluor in fluorography, instead of PPO. One advantage of salicylate over PPO is that it is considerably less expensive. The other main advantage is that salicylate is water-soluble, so that fluorography using this fluor avoids lengthy equilibration of the gel in DMSO followed by washing with water. It typically takes 0.5−1 h, instead of about 5 h required with the DMSO−PPO system. Gels at neutral or alkaline pH can be added

73

directly to 1 M sodium salicylate for 30 min but stained gels first need to be soaked for 30 min in water to remove acetic acid. During the gel drying step, acetate sheets must be used instead of Mylar which may otherwise become glued to the dried gel. According to the data published so far, fluorography using salicylate with pre-exposed X-ray film appears to be as sensitive as that using DMSO−PPO, with similar linearity of fluorographic image absorbance to sample radioactivity (87). However, there is reduced sharpness of the film image compared to the use of PPO so that the resolution obtained is significantly lower. It must also be noted that, like DMSO, salicylate is potentially hazardous to health.

iv. Commercial fluorography reagents
Fluorographic reagents are also available from commercial suppliers, for example En³Hance (New England Nuclear) and Amplify (Amersham). These are reported to be safer and more rapid in use than conventional reagents but their sensitivity is disputed. Thus Roberts (88) found that En³Hance and Amplify are 5−8-fold less sensitive than PPO/DMSO for fluorography.

v. Indirect autoradiography (for ^{32}P or ^{125}I)
The method involves placing a pre-flashed X-ray film against the gel (which is either wet or dry, stained or unstained, but otherwise untreated) and then a calcium tungstate X-ray intensifying screen is placed against the other side of the film, the entire sandwich then being placed at $-70°C$ for film exposure (82). Emissions from the sample pass to the film producing the usual direct autoradiographic image, but emissions which pass completely through the film are absorbed more efficiently by the screen where they produce multiple photons of light which return to the film and superimpose a photographic image over the autoradiographic image. The resolution obtained is only slightly less than using direct autoradiography but sensitivity is much increased. Using Kodak X-Omat R film pre-exposed to 0.15 A_{540} units and an intensifying screen, the detection efficiency for ^{32}P was found to be increased 10.5-fold and for ^{125}I 16-fold when compared to direct autoradiography, such that about 100 d.p.m./cm² of ^{125}I or 50 d.p.m./cm² of ^{32}P blackens the film detectably (ref. 84; *Table 7*). The density of the film image is still proportional to the sample radioactivity allowing quantitative work. As with fluorography, pre-exposure of the film above 0.2 A_{540} units increases the detection efficiency still further but the relationship between film image and radioactivity ceases to be linear. The use of a second intensification screen improves the sensitivity of detection still further but this arrangement leads to reduced resolution (89).

vi. Choice of X-ray film
The X-ray film used for fluorography or indirect autoradiography (see Section 7.5.1) must respond efficiently to the blue to UV light emitted by the fluor or intensifying screen, respectively. Therefore 'screen-type' or 'medical' X-ray films must be used, where the screen refers to the fact that the film is sensitive to light emitted by X-ray intensifying screens. Kodak X-Omat R, the most sensitive 'screen-type' film tested by Laskey and Mills (81, 82) has been discontinued and was replaced by Kodak

X-Omat AR. However, many other screen type films are also available, including Fuji RX. This was found to be almost (75%) as sensitive as Kodak X-Omat R (82) and has the advantages of a longer shelf-life, smaller grain size and higher image quality. It gives a linear response relative to the amount of radioactivity when pre-exposed to an A_{540} of 0.15 (85). Films which have been designed for direct autoradiography such as LKB Ultrofilm are not suitable. The source of the calcium tungstate intensifying screen (which can be used repeatedly) is also important. The performance of some intensifying screens is listed in reference 82 but several of these brands have changed names since the original study. Fuji intensifying screens are widely used and seem to perform well.

vii. Image intensification

If particular autoradiographic or fluorographic images are considered too faint, it is possible to intensify them 10-fold or greater using [^{35}S]thiourea which reacts with silver grains in the development film to form silver [^{35}S]sulphide. An autoradiograph can then be taken to amplify the originally faint image. The method is given in *Protocol 12*.

Protocol 12. Intensification of X-ray film images[a]

1. Wash the X-ray film with large volumes of the following solutions (~ 1 litre of each per 18×24 cm sheet).
   - distilled water   2 min
   - distilled water   2 min
   - 20% methanol   2 min
   - 50% methanol   2 min
   - 20% methanol   2 min
   - distilled water   10 min
   - distilled water   10 min

   Dry the film.

2. Prepare a solution of [^{35}S]thiourea in 0.1 M ammonia and adjust to pH 8.0 with dilute HCl (use 1 μCi [^{35}S]thiourea per cm^2 of film [b]. Filter the solution through Whatman No. 1 paper before use[c]. **Note:** thiourea may be a weak thyroid carcinogen so handle with care.

3. Incubate the film with agitation in the thiourea solution for 2 h at room temperature[d].

4. Wash the 'activated' film in large volumes of the following solutions:
   - distilled water   2 min
   - distilled water   2 min
   - 20% methanol   5 min
   - 50% methanol   2 min
   - 20% methanol   5 min

Protocol 12. *continued*

- distilled water   5 min
- distilled water   5 min

Dry the film.

5. Expose the dried 'activated' film to a fresh sheet of X-ray film which has *not* been pre-flashed. Note that when intensifying direct autoradiographs, the original autoradiographic image will be on one side of the film only. Check with a Geiger counter to determine which is the correct side to expose to the new X-ray film.

---

[a] Modified from refs. 90 and 91.
[b] The specific activity and concentration of the [^{35}S]thiourea are not critical.
[c] Filtration of the thiourea solution is essential. This solution can be used at least three times with little loss of performance. It should be re-filtered after three uses if it is to be used further.
[d] If the original film has a high background, it should be treated with a subtractive reducer (92) before activation.

---

viii. Dual isotope detection using X-ray film

McConkey (93) has discussed the approaches to using X-ray film for double-labelling studies and described a method for this purpose. The gel is prepared for fluorography and exposed to X-ray film which is sensitive to the light flashes produced and so detects both ^{14}C and ^{3}H (see above). The gel is then placed in contact with Kodak No-Screen X-ray film or its equivalent which is insensitive to the light produced by fluorography and detects only the ^{14}C by direct autoradiography. Comparison of the images produced on both films allows one to identify proteins labelled with either or both isotopes. Walton *et al.* (94) have published an alternative procedure. The gel is first prepared for fluorography and exposed to sensitive X-ray film to detect ^{14}C and ^{3}H. Then the gel is painted black using marking ink and a second film exposed (direct autoradiography). Provided that the ^{3}H/^{14}C ratio exceeds about 40:1, ^{3}H is detected almost exclusively on the fluorograph whilst ^{14}C only appears on the direct autoradiograph because photons emitted by the ^{3}H are prevented from reaching the film by the black ink. Cooper and Burgess (95) have reported a variation of this method for detecting proteins labelled simultaneously with ^{32}P and ^{35}S. The gel is exposed simultaneously to two X-ray films. The first records ^{35}S directly whereas the other is shielded by aluminium foil and records ^{32}P emissions. Finally, another method being developed is the use of colour negative film (96). This consists of three photographic emulsions each sensitive to one of the three additive primary colours (red, green, and blue). Since the layers are exposed through different thicknesses of gelatin emulsion, emissions of different energies penetrate to different depths and so produce different ratios of exposure in the three layers and so different final colours.

ix. Electronic data capture

Light photons generated by fluorography can now be detected by using electronic image intensification systems (97, 98). This methodology has considerable future potential for the rapid quantitation of both single and multiple radiolabels.

7.5.2 Methods based upon gel fractionation and counting

Despite the sensitivity of detection methods using X-ray film, fractionation of the gel and scintillation counting of the solubilized fraction is still in use for some studies involving quantitation of labelled protein components following polyacrylamide gel electrophoresis. One reason for this is the ability of modern scintillation spectrometers to distinguish between emissions of different energies thus allowing the quantitation (rather than just detection) of samples differentially labelled.

i. Fractionation methods

Slab gels and rod gels which have been dried down onto filter paper (Section 7.2.1) may be sectioned with scissors, but this is inaccurate since it is difficult to obtain slices of even width and so gels are usually sliced whilst still wet. However, one situation where cutting dried gels may be advantageous is if one wishes to quantitate individual bands or measure the isotopic ratio of doubly-labelled bands detected by autoradiography or fluorography. If the gel used for autoradiography or fluorography was unstained, the band may be located by photographing the autoradiograph (to provide a permanent record) and then cutting out the band image of interest from the film. The film is now laid over the unstained gel where it serves as a template to mark the position of the radioactive band which is then cut out from the gel. Gel slices containing ^{32}P or ^{125}I are counted efficiently by simply adding conventional scintillation fluid to the dry gel slice. Gel slices containing ^{14}C, ^{35}S, or ^{3}H, cut from gels prepared for fluorography using PPO, can be counted in the same way without gel hydrolysis. In fact, Laskey (84) has noted that in the latter case there is no need to add scintillant prior to counting since the gel slice already contains PPO. If necessary, dried gel slices can be hydrolysed directly by the methods given below, or radioactive proteins can be leached out of the gel for counting after rehydration of each gel slice with water.

Numerous sectioning devices exist for wet polyacrylamide gels (see ref. 99) only some of which can be applied to both gel rods and slabs. Using a Mickle gel slicer (Joyce-Loebl) the gel rod or gel slab track is held on a carriage which moves past a guillotine knife in calibrated small steps. The smallest practical section size with many polyacrylamide gels is about 0.5 mm. Slices are removed manually during the operation. However, one of the simplest and most widely used fractionators consists of a number of razor blades separated by aluminium, stainless-steel, or plastic spacers, bolted together. A suitable device is easily built in a laboratory workshop but certain commercial models do tend to give more reproducible slices (e.g. Bio-Rad Laboratories Ltd.; *Figure 18*).

Rod gels need to be frozen for easy slicing since low gel concentrations are viscous

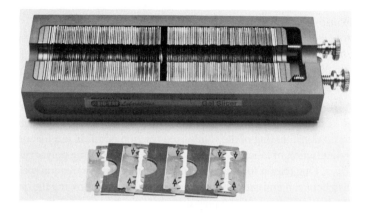

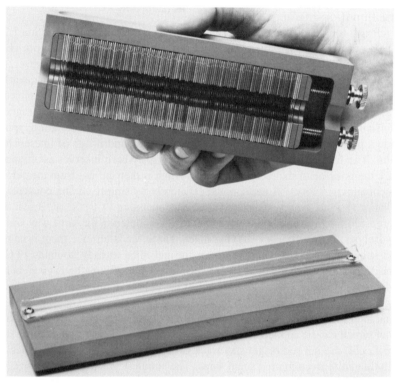

Figure 18. A manual gel slicer (Bio-Rad). (a) The gel slicer consists of a series of razor blades separated by 0.1-cm-thick metal spacers; here, four razor blades and three spacers have been removed and are displayed in front of the gel slicer. (b) The gel to be sectioned can be held in a trough cut from plastic tubing, as shown here.

whilst high gel concentrations are rubbery, both being difficult to cut. The temperature of the frozen gel is important since if the gel is too cold it is extremely difficult

B. David Hames

to slice and tends to crack during the attempt. The easiest method is to place the gels at $-70°C$ for about 1 h. Alternatively the gel is placed in an aluminium-foil trough which is then placed on dry ice until the gel is just frozen. The frozen gel, kept straight during freezing, is placed on a plastic sheet, or in a trough cut from plastic tubing (*Figure 18*), and the slicer lined up along the length of the gel. The gel is then sliced by exerting firm downward pressure whilst rocking the slicer transversely across the gel to help the razor blades penetrate. A syringe needle, a scalpel, or a pair of fine-tipped forceps is used to transfer gel slices from between the blades into counting vials. Next the slicer is washed under the tap, shaken free of excess water, and drained on a pad of filter paper ready for reuse. If gel cracking is a problem during slicing, the gel should be allowed to thaw slightly before attempting to slice. Cracking can also be prevented by incorporating 10% glycerol in the gel mixture prior to polymerization.

Slab gels may be fractionated by first cutting the gel into individual tracks. Attempts to do this using a scalpel usually result in the gel tearing. A better method is to use a long knife with a sharp, flat cutting edge and slice downwards into the gel along the length of the track in a single movement. The slab gel track can then often be sliced transversely using a rod gel slicer (see above) but freezing is not usually necessary, presumably because of the thickness of the gels used (0.75−1.5 mm).

ii. Counting methods

Liquid scintillation counting is the preferred method of isotope quantitation. For maximum counting efficiency the proteins need to be eluted from the gel slices. Two approaches are possible. Either the gel is solubilized or the gel is made to swell and protein leaches out.

Bisacrylamide crosslinked polyacrylamide gels can be solubilized by heating with hydrogen peroxide as described in *Protocol 13*.

Protocol 13. Solubilization of bisacrylamide-crosslinked polyacrylamide gels using hydrogen peroxide

1. Allow the slice (1 mm) to dry in a vial at room temperature overnight or at 50°C for about 2 h.

2. Add 0.25 ml of 30% (w/v) H_2O_2 to each vial and then tightly cap it. Tilt the vial so that the slice is immersed in liquid and incubate at 50°C to allow solubilization.

3. Cool the vials to room temperature before opening. Then add 5 ml (for vial inserts) or 10 ml of a water-miscible scintillation cocktail, consisting of NCS (Amersham) and toluene scintillation fluid (5.0 g of PPO, 0.5 g of POPOP per litre of toluene) mixed in the ratio of 1:5 v/v. A considerably cheaper alternative is to use a Triton X-100:toluene scintillation cocktail (1 vol. Triton X-100:2 vol. toluene scintillation fluid). In this case, dissolve the gel slice in 0.5 ml of 30% (w/v) H_2O_2 and then add 4.5 ml of the Triton X-100:toluene scintillation cocktail. Either method yields a completely clear counting mixture.

Although the loss of ^{14}C as ^{14}CO$_2$ does not exceed about 5% with this protocol, this is avoided in the modification of Goodman and Matzura (100) by using alkaline peroxide to trap any ^{14}CO$_2$. The procedure is as described above except that the reagent used is 1 volume of concentrated NH$_4$OH (specific gravity 0.88) added to 99 volumes of 30% (w/v) H$_2$O$_2$ that has been previously cooled to 4°C. The mixture is kept on ice and used immediately. The temperature of incubation was originally quoted as 37°C but 50°C is often used. Although the NCS:toluene scintillation cocktail still yields a clear counting mixture, the Triton X-100:toluene cocktail gives a slightly cloudy emulsion. Chemiluminescence is sometimes a problem. In this case vials should be counted twice, 24 h apart, or until constant values are obtained. Usually counts due to chemiluminescence decay to background within 48 h.

Another approach to gel solubilization has been to replace bisacrylamide with a crosslinker more susceptible to chemical hydrolysis. A number of such labile crosslinkers are available. Ethylene diacrylate (EDIA) has an ester linkage in place of the amide linkage of bisacrylamide so that ethylene diacrylate crosslinked gels can be solubilized with alkali (101). Unfortunately this alkali lability renders ethylene diacrylate unsuitable for the many types of protein electrophoresis which are carried out at high pH. *N,N'*-bisacrylylcystamine (BAC) is a disulphide-containing analogue of bisacrylamide such that use of this crosslinker renders polyacrylamide gel soluble in 2-mercaptoethanol or dithiothreitol (see ref. 102 for details). Obviously this sensitivity to thiol reagents restricts the use of BAC to those situations where thiols are absent. With modifications to the catalyst system used, this method can be used for the preparation and analysis of soluble gels up to $T = 20\%$ (102,103). For some years, the most generally useful, labile crosslinker appeared to be *N,N'*-diallyl-tartardiamide (DATD) which contains a 1,2 diol structure that can be oxidized by periodic acid. Gels polymerized using DATD are therefore readily solubilized by incubation with 2% (w/v) periodic acid at room temperature for $1-2$ h. When the solubilized gel slices are mixed with a water-miscible scintillation cocktail, high counting efficiencies are obtained for both ^{3}H and ^{14}C (104). Unfortunately, substituting DATD for bisacrylamide on a mole per mole basis produces gels exhibiting reduced physical strength and greater porosity. These disadvantages can be remedied to some extent by substantially increasing the proportion of crosslinker (e.g. ref. 105) and acrylamide concentration, respectively. However, it has become clear that not only is DATD a very poor crosslinker, it actually inhibits the polymerization reaction so that the resulting gel contains large amounts of unreacted monomers (106). Since acrylamide monomer is a potent neurotoxin, under no circumstances should these gels be handled without gloves. Indeed, given this serious safety consideration, Gelfi and Righetti (106) have recommended that DATD should not be used in future for gel electrophoretic studies. Despite this warning, some researchers continue to use this crosslinking agent and work to optimize its use is still proceeding in some laboratories (e.g. ref. 107).

Another solubilizable crosslinking agent is *N,N'*-(1,2-dihydroxyethylene) bisacrylamide (DHEBA) which contains both a periodate-sensitive 1,2 diol structure and two base-cleavable amido methylol bonds (108). Nevertheless the latter bonds are more stable than the ester bonds of ethylene diacrylate so that the common alkaline pH electrophoresis buffers can be used—except for borate buffers which form negatively-charged complexes with *cis* 1,2 diol structures and so could lead to a charged gel. Gels can be stained with Coomassie Blue, destained, sliced, and then each slice placed in a scintillation vial insert with 1 ml of 25 mM periodic acid. After capping tightly, these are incubated at 50°C for 48 h to solubilize the gel. The dye is also bleached during this process so that, after the addition of scintillation fluid, each sample can be assayed for radioactivity directly by scintillation counting. This cleavable crosslinker is likely to find increasing use in the future, since DHEBA exhibits similar polymerization kinetics to bisacrylamide (106) and the resulting gels show good physical and molecular sieving properties.

The precise polymerization kinetics of polyacrylamide gels prepared with each of the above crosslinkers have been examined by Gelfi and Righetti (106). Readers intending to use any of these cleavable crosslinkers are advised to consult this study for detailed reaction characteristics. Their overall order of reactivity is $Bis \simeq DHEBA > EDIA \simeq BAC \gg DATD$.

The alternative approach, of leaching the protein from the gel in order to increase the efficiency of scintillation counting, can be carried out according to several protocols. In the original procedure, each gel slice was swollen by soaking in 0.5 ml NCS for 2 h at 65°C, followed by addition of 10 ml toluene scintillation fluid (109). High percentage polyacrylamide gels may resist swelling unless the NCS is diluted (9 parts NCS:1 part water). A variation on this is to add 10 ml scintillant (14.0 g PPO, 0.21 g POPOP, 143 ml NCS, 3.75 litres toluene) to each gel slice and incubate at 37°C overnight before counting (110). Aloyo (111) suggests adding 10 ml scintillation cocktail (6.0 g PPO, 10 ml NCS, 10 ml hyamine hydroxide, per litre toluene) directly to each wet 1 mm gel slice, vortexing briefly and measuring the radioactivity after 48 h at room temperature. The yields and counting efficiencies by this method compare favourably with the alkaline peroxide solubilization method. Finally, ready-mixed scintillation cocktails for recovery of labelled proteins from gel slices by leaching are commercially available.

7.5.3 Determination of specific radioactivity of proteins

Martin *et al.* (112) have described a method for determining the specific radioactivity of leucine in proteins isolated by polyacrylamide gel electrophoresis. The protein is hydrolysed in HCl followed by radioactivity determination and estimation of leucine content using amino-acyl synthetase. Another method based upon the use of radiolabelled dansyl chloride (113) allows one to determine the specific radioactivity or amino acid content for several amino acids in any protein separated by polyacrylamide gel electrophoresis.

7.6 Detection of glycoproteins

Glycoproteins can be readily located in polyacrylamide gels and distinguished from other non-glycosylated proteins. In the past, the most usual detection method has been the periodic acid – Schiff (PAS) stain using fuchsin (273). However, the most sensitive variation of this technique uses dansyl hydrazine and allows as little as 40 ng of carbohydrate to be detected (274). Another procedure, the thymol-sulphuric acid method (275), has similar sensitivity. More recently, a number of other methods have become available, including radiolabelling of the carbohydrate moiety prior to electrophoresis (either by *in vivo* labelling with sugar precursors or *in vitro* labelling) then glycoprotein detection by fluorography/autoradiography after electrophoresis, and the use of radiolabelled, fluorescent, or enzyme-conjugated lectins. These and other glycoprotein detection methods are referenced in Appendix 2. The most sensitive staining protocol for glycoproteins in polyacrylamide gels is a modification of the silver staining method of Oakley *et al.* (58), claimed to be at least 60-fold more sensitive than the PAS method and able to detect 0.4 ng of bound carbohydrate (114).

7.7 Detection of phosphoproteins

Phosphoproteins can be detected according to any of several protocols (Appendix 2) but the most sensitive method is to label *in vivo* with ^{32}P, then detect phospho-proteins after gel electrophoresis using autoradiography. Lipid and nucleic acids will also be labelled and should be removed from the sample prior to electrophoresis. Alternatively, labelled nucleic acids can be removed by acid hydrolysis after gel electrophoresis (115); if desired, the proteins are first stained using Coomassie Blue (Section 7.2.1) then the gel is equilibrated in 7% TCA overnight followed by hydrolysis in 7% TCA at 90°C for 30 min. The labelled nucleotides are removed by soaking the gel in several changes of 7% TCA at room temperature over a period of 24 h. The gel is shaken gently on a reciprocating shaker during this time. If the gel has been stained prior to acid hydrolysis, it is now destained, dried, and autoradiographed as usual (Section 7.5.1). Labelled lipid migrates with the buffer front and most is extracted by the staining and destaining steps.

Methods also exist for the detection of non-radioactive phosphoproteins (see Appendix 2). The entrapment of liberated phosphate (ELP) method is the most specific of these procedures (118). It depends on the entrapment within the gel of insoluble calcium phosphate produced by the alkaline hydrolysis of protein phosphoester bonds in the presence of calcium chloride, modification to an ammonium molybdate complex and then staining with methyl green. Protein bands containing as little as 1 nmol of phosphate given a bright green band in the gel. A recent variation of the ELP method, based on the formation of insoluble rhodamine β-phosphomolybdate, increases the sensitivity 2 – 3-fold (119). A silver-based staining method (276) is even more sensitive, but always includes the possibility of staining non-phosphoproteins also. Stains-all (277) stains phosphoproteins blue and unconjugated proteins red but the method is relatively non-specific (since glycoproteins, DNA and RNA are also stained blue) and less sensitive than the ELP procedure. Further references to these and other methods are given in Appendix 2.

7.8 Detection of proteins using immunological methods

Particular proteins can often be detected after electrophoresis, even if this was carried out in the presence of SDS, if monospecific antibodies are available. Several methods have been devised for this purpose. Thus antigens may be visualized by incubating the washed gel with specific antibody followed by incubation with anti-IgG coupled to horseradish peroxidase (116). After further incubation with 3,3′-diaminobenzidine and hydrogen peroxide, the antigen-antibody-antibody-peroxidase complexes become visible as stained bands. Alternatively, antigens can be detected by exposing the gel to radioactive antibody followed by autoradiography. A double-antibody method is preferred, with the radiolabel on a second antibody directed against the first, or using ^{125}I-labelled *Staphylococcus aureus* protein A (which binds to the Fc portion of the IgG molecule) instead of a second antibody (117). Careful washing is needed to obtain good ratios of signal to background. Another method involves casting agarose containing antibody directly onto the polyacrylamide slab gel surface following electrophoresis. The protein antigen diffuses out of the polyacrylamide gel and reacts with the antibody in the agarose gel to produce an immune precipitation pattern called an immunoreplica, which can be visualized by staining or by using ^{125}I-labelled protein A (120,121). However, of all the immunodetection procedures available, undoubtedly the most widely used at present rely upon transfer of the separated proteins from the gel to a filter (electroblotting or Western blotting) prior to reaction with the labelled antibody. This approach is discussed more fully in Section 7.10. Other references to the immunological detection of proteins in slab gels are given in Appendix 2.

7.9 Detection of enzymes

If the intention is to detect an enzyme activity after gel electrophoresis, precautions must be taken to minimize loss of activity during electrophoresis. For example, when using a continuous buffer system, pre-electrophoresis is recommended to remove unreacted monomers and persulphate, for all separations the pH chosen should be one at which the enzyme is stable and electrophoresis should be carried out in the cold.

7.9.1 After standard gel electrophoresis

Enzymes can be assayed after gel slicing followed by diffusion elution into a suitable buffer (Section 8.2). However, this is laborious for large numbers of samples and, furthermore, slight differences in the migration of two enzymes may be missed. An alternative method is to stain for specific enzyme activity *in situ*. The major limitations of the method are that detection methods exist for only a relatively small number of enzymes and, secondly, the results are not easily quantifiable. It is therefore mainly useful in analytical comparisons of a particular enzyme in multiple samples run on slab gels or possibly to locate enzyme for preparative purposes, although other proteins are, of course, not visualized and may contaminate the preparation.

It should also be noted that assays for enzyme activity need not necessarily be restricted to gels electrophoresed in non-dissociating buffers. Increasingly, conditions have been reported for the renaturation of denatured proteins followed by assay for biological activity. For example, Scheele *et al.* (122) detected enzymatic activity for 15 human pancreatic proteins after denaturing two-dimensional gel electrophoresis by excising and renaturing the proteins. Nevertheless, an important limitation is that only single subunit enzymes can be detected. As an alternative to elution of the denatured proteins followed by renaturation in free solution, they can be transferred to a paper or membrane support matrix by Western blotting. Once immobilized, SDS can be efficiently removed simply by washing the filter. Western blotting and protein renaturation are discussed further in Sections 7.10 and 10.2.4.

Dehydrogenases can be localized by incubating the gel in a solution of a tetrazolium salt which then acts as the terminal electron acceptor and becomes reduced to yield a coloured formazan. Nitroblue tetrazolium (NBT) and methyl thiazolyl tetrazolium (MTT) have been widely used for this purpose, although other suitable tetrazolium salts are also available. Phenazine methosulphate (PMS) is sometimes included to serve as hydride-ion carrier between the reduced coenzyme or enzyme prosthetic group and the tetrazolium salt. Since tetrazolium salts are light-sensitive, the incubation has to be carried out in the dark. Staining in the absence of dehydrogenase is sometimes observed (123) and so controls should be included to ensure that only the enzyme reaction is being studied. Hydrolases are most readily detected using chromogenic or fluorogenic substrates which yield coloured or fluorescent products after enzyme action. Transferases and isomerases are not amenable to this method and usually required coupling enzymes to be included to the detection mixture which convert the enzyme product into one which can be acted upon by a dehydrogenase also included in the mixture. The dehydrogenase product, and hence the original enzyme, is located as described above. Since the coupling enzymes penetrate polyacrylamide gel only poorly, the coloured product is formed only on the gel surface. Zone spreading and the volume of reagents can be minimized by incorporating the detection mixture, including the coupling enzymes, in an indicator gel, often agarose, which is poured onto the surface of the polyacrylamide gel or preformed and then the gels placed in contact. Ultrathin agar overlays on polyester sheets have also been used (271). If this type of strategy is adopted, a drop of sample should be applied to the indicator gel as a positive control to ensure that the coupling enzymes have survived the agarose treatment.

Overall there is a trend away from using potentially carcinogenic reagents such as benzidine and *o*-dianisidine in enzyme stains, replacing these with reagents such as eugenol or tetrabase. The use of fluorescently-labelled substrates such as umbelliferyl or dansyl derivatives is also becoming more popular. Detection methods for individual enzymes are referenced in Appendix 2.

7.9.2 Activity gels

This technique involves electrophoresis using a polyacrylamide gel in which is embedded the macromolecular substrate of the enzyme of interest. The substrate is added to the gel mixture just prior to polymerization. Electrophoresis of the protein mixture occurs under denaturing conditions, the proteins are then allowed to renature *in situ* and the enzyme visualized by incubation with an appropriate reaction mixture. The technique has been particularly used for the analysis of enzymes involved in DNA or RNA metabolism (124) but has obvious future applications for other proteins. Conditions need to be carefully optimized for each enzyme, especially for the renaturation step.

7.10 Blotting techniques

One of the major recent advances in the analysis of proteins after polyacrylamide gel electrophoresis has been the development of techniques for the transfer of the separated proteins from the gel to a thin support matrix, most commonly a nitrocellulose membrane, to which they bind and are immobilized. This transfer procedure is referred to as a blotting. Although the transfer can be achieved by solvent flow, easily the most widely used procedure employs electrophoretic transfer since it is more rapid and sensitive. This technique has become known as electroblotting or Western blotting. Under ideal conditions, over 90% of the protein can be transferred.

The most important advantage of electroblotting is that it removes the separated proteins from a gel matrix which hinders protein analysis to the surface of a filter sheet where the protein molecules are readily accessible. This is especially important when using gradient gels which, because of the gradation in pore size, vary in the accessibility to detection reagents throughout the gel. However, the advantages also extend to uniform concentration gels. Processing times for staining and destaining, or incubation with other detection reagents, are much shorter than for gels. Thus, using an appropriate filter matrix, staining and destaining of proteins can be performed within 5 min whereas immunodetection of specific proteins takes as little as 6 h. In addition, a single blot can be used for several successive analyses. For example, up to 10 different assays can be performed on certain nylon membranes or DBM or DPT paper (see Section 7.10.4) if the probe is removed each time. Indeed, some analyses can be carried out easily only by using Western blots rather than gels. An additional advantage is that the support matrix for the Western blot is usually far easier to handle than the polyacrylamide gel and certainly more durable. If desired, Western blots can even be dried and stored for long periods prior to analysis. Given the considerable advantages of Western blotting it is not surprising that this field has expanded rapidly. Gershoni and Palade (125), Renart and Sandoval (126), and Towbin and Gordon (127) have reviewed the earlier blotting methodology. The latter review is particularly recommended and includes a good survey of the wide variety of methodological variations and applications of Western blotting. Because of space limitations, only a relatively brief overview can be attempted here.

7.10.1 Blotting matrices

A variety of support matrices have been used for electroblotting. Nitrocellulose membrane is the most widely used and is commercially available from several suppliers (e.g. Schleicher & Schuell, Millipore, Amersham International, Bio-Rad). Pure nitrocellulose membranes have good protein binding capacity ($\sim 80-100$

μg/cm^2) and should be chosen in preference to mixed ester membranes which contain cellulose acetate since these appear to have reduced capacity. Even using pure nitrocellulose, some proteins, especially those of low molecular mass, may bind only weakly and so be lost during transfer (125). One improvement is to reduce the pore size of the filter. Nitrocellulose membranes with 0.45 μm pore size are typically used but 0.22 μm has been recommended for lower molecular mass proteins. Lin and Kasamatsu (128) reported that only very small polypeptides ($< \sim 14$ kd M_r) show significant losses under these conditions but decreasing the pore size still further to 0.1 μm allows even these small polypeptides to be recovered.

The mechanism of protein binding to the nitrocellulose is not fully understood but is certainly non-covalent and probably involves hydrophobic interactions. Some proteins may thus be lost during processing of the membrane, especially if detergents such as Triton X-100 are included to reduce non-specific binding. To minimize these losses, the proteins can be fixed on the nitrocellulose membrane after transfer simply by soaking the filter for 15 min in 0.5% glutaraldehyde in phosphate-buffered saline or for 1 h in 25% isopropanol, 10% acetic acid (129). However, since fixation will destroy biological activity, its use may not always be desirable.

Various nylon membranes have been introduced for blotting which have the advantage of being far stronger, and hence more robust, than conventional pure nitrocellulose sheets. To combat this, several companies have marketed composite nitrocellulose membranes which are also considerably stronger than conventional pure nitrocellulose filters. Nevertheless, nylon membranes have advantages for certain applications. In general they have substantially higher protein binding capacity. For example, one such nylon matrix, Zetaprobe (Bio-Rad), has about 6-fold higher protein binding capacity (~ 480 μg/cm^2) than nitrocellulose. Because of its highly cationic nature, it has been particularly recommended for electroblotting of SDS-PAGE gels to maximize binding of the highly anionic SDS$-$polypeptide complexes. Indeed, charged nylon filters such as Zetaprobe consistently give better results than nitrocellulose in the electroelution of SDS-PAGE gels. The binding also appears to be much stronger than in the case of nitrocellulose. Thus Zetaprobe may be preferred for proteins which bind poorly to nitrocellulose (e.g. low molecular mass proteins) or for multiple probing protocols. Nitrocellulose is not recommended for multiple probing beyond 2 or 3 cycles since proteins are lost during the washes between probes. Because of the larger amounts of protein bound, nylon membranes may also be preferred when immunodetection (Section 7.10.3) is to be employed although, because of the higher affinity for proteins, higher non-specific binding of antibodies can occur. Partly because of this, nitrocellulose is still the most widely-used electroblotting matrix for immunodetection analyses. The cationic nature of charged nylon membranes also makes them less suitable than nitrocellulose for the analysis of basic proteins.

Uncharged nylon membranes are also available (e.g. Hybond N; Amersham International) which should give higher binding of basic proteins. They are also reported by some workers to give lower backgrounds with immunodetection methods than charged nylon membranes. Nevertheless, a major disadvantage of all nylon membranes is the lack of simple general staining procedures, in contrast to the situation with nitrocellulose. Common anionic chemical stains (e.g. Coomassie Blue, Amido Black) give such high non-specific backgrounds that they cannot be used. However, more sophisticated general staining procedures for proteins immobilized on nylon membranes have been reported (see Section 7.10.3).

Other available matrices are diazo-modified papers, particularly diazo-benzyl-oxymethyl (DBM) paper (130), diazophenylthioether (DPT) paper (131), cyanogen bromide-activated paper (132), and classical ion-exchange papers. Since DBM and DPT papers are not particularly stable, they are prepared as amino-derivatives (ABM and APT) and diazotized just before use. Their preparation and use is described in the papers cited above. In practice, since their preparation involves extremely toxic chemicals, they are best purchased from commercial sources as their ABM or APT derivatives, leaving only the activation to be performed in the laboratory. However, the resolution obtained with these rather coarse matrices is generally less than with nitrocellulose or nylon, they have lower protein-binding capacities ($25-50$ $\mu g/cm^2$), and the need for activation makes them less convenient to use. In addition, glycine, which is often a buffer component when electroblotting into nitrocellulose, cannot be used with DBM or DPT papers. If glycine was used in the gel electro-phoresis buffer, it needs to be removed by soaking the gel in, say, phosphate buffer prior to electroblotting. Despite these drawbacks, since the proteins are covalently bound, a diazotized paper may be the matrix of choice if a large number of detec-tion methods are to be applied sequentially to the same blot.

Cyanogen bromide-activated paper also covalently binds transferred proteins but has many of the same limitations as DBM and APT papers. Conventional ion-exchange papers (e.g. DEAE-paper) can be useful for preparative work but the protein binding capacity is low (~ 15 $\mu g/cm^2$) and, since the proteins are bound only by charge interactions, care must be taken in the choice of transfer and processing solutions to avoid unacceptable losses of bound protein. In addition, this paper matrix is mechanically weak which can cause serious problems in processing steps.

Another type of matrix is polyvinyl-difluoride (PVDF). Blotting membranes of PVDF are available commercially (e.g. Immobilon, Millipore). They are hydrophobic in nature and, unlike nylon membranes, are reported to be compatible with commonly-used protein stains as well as standard immunodetection methods. Finally, Janssen-Pharmaceutical has developed a blotting membrane called PCGM-1 which has been recommended for amino and sequence analysis of separated proteins. The proteins can be blotted from SDS-PAGE gels and then acid-hydrolysed directly whilst still immobilized on the filter.

7.10.2 Electroblotting methodology

A schematic diagram of a typical electroblotting apparatus is shown in *Figure 19a*. Several models are available commercially (e.g. Pharmacia-LKB, Bio-Rad, *Figure 19b*). In essence, for nitrocellulose filters, the procedure is described in *Protocol 14*. Wear disposable plastic gloves throughout to avoid contaminating the gel or blotting filter with skin proteins.

One-dimensional polyacrylamide gel electrophoresis

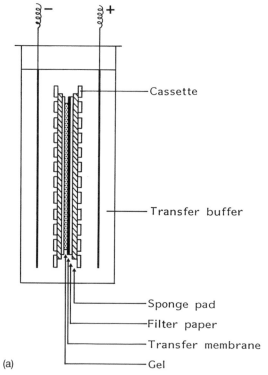

- Cassette
- Transfer buffer
- Sponge pad
- Filter paper
- Transfer membrane
- Gel

(a)

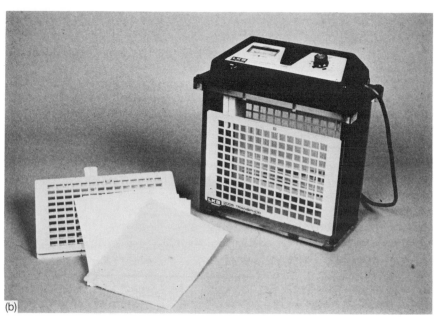

(b)

Protocol 14. A typical procedure for electroblotting

1. Cut the nitrocellulose membrane and pieces of Whatman 3MM paper to the size of the gel and pre-soak these in transfer buffer.
2. Cut off the corner of the gel to mark its orientation and equilibrate it in transfer buffer for about 30 min.
3. Recover the gel onto a glass plate. Place a sheet of the soaked 3MM paper onto the gel, squeezing out any air bubbles by rolling a pipette over the surface or by using gloved fingers.
4. Place a soaked porous pad (such as foam sponge or a ScotchBrite™ scouring pad) on top of the Whatman paper followed by the plastic holder of the transfer apparatus.
5. Invert the assembly and gently remove the glass plate. Place a sheet of the soaked nitrocellulose membrane onto the gel and roll to exclude air bubbles.
6. Place a soaked sheet of Whatman 3MM paper and then a ScotchBrite™ pad on top as before and close the plastic cover of the transfer sandwich. The method of doing this depends on the apparatus used but tight contact between the gel and nitrocellulose membrane is essential for good transfer of proteins.
7. Place the sandwich vertically into the electrophoresis tank filled with transfer buffer (*Figure 19a*) and begin electrophoresis.

The arrangement of platinum electrodes in the apparatus is designed to give a uniform voltage gradient over the whole gel, often by means of a zigzag pattern. The transfer buffer is typically of low ionic strength but its exact composition varies according to the nature of the gel and the proteins being analysed, the blotting matrix being used and the detection method to be employed. Towbin *et al.* (133) used 0.7%

Figure 19. (a) Schematic diagram of electroblotting apparatus. For simplicity only a single cassette assembly is shown immersed in the transfer buffer in the electroblotting apparatus although commercial apparatuses typically allow electroblotting of several gels simultaneously. The gel (stippled) is placed adjacent to the transfer membrane (black) with a sheet of filter paper either side (white). A sponge pad (hatched) is placed either side of this sandwich so that when the cassette is closed it exerts an even pressure over the gel. The cassette (shown here in section at the outmost edges of the sandwich) typically consists of two linked plastic grids which hold the gel sandwich together whilst permitting maximum exposure of the gel sandwich to the buffer. Although shown here as single vertical wires, the electrodes are usually mounted on separate grid-like plastic panels with the platinum electrode wire wound round the grid to produce a uniform electric field (see below). (b) An example of a commercial electroblotting apparatus. The model shown is the Transphor unit from LKB (LKB 2005) which incorporates a power supply in the lid. In other commercial models the power supply is a separate entity. The electrode grid panels are readily visible in place in the main body of the apparatus. On the left-hand side is a gel cassette with sheets of filter paper and sponge pad. Most commercial models come equipped with an optional cooling coil which can be used to cool the gel and buffer during transfer and so allow the use of high field strengths.

acetic acid to transfer proteins to nitrocellulose. If enzymic activity is to be assayed after transfer, the buffer must be chosen to preserve that activity. Typical buffers are 20 mM Tris − 150 mM glycine (pH 8.3), 7.5 mM Tris − 1.2 mM boric acid and 25 mM sodium phosphate (pH 6.5). Buffer components which contain amino groups, such as glycine, cannot be used for electroblotting with DBM and DPT paper since they react with the diazo groups. If glycine-containing buffers were used during gel electrophoresis, the glycine needs to be removed before electroblotting by washing the gel with buffer, for example, phosphate buffer.

Methanol (20% by vol) may be added to the transfer buffer (e.g. ref. 133) since it minimizes swelling of the gel during blotting and increases the binding capacity of nitrocellulose for protein. However, methanol also reduces the efficiency of elution of proteins so that electroelution has to be carried out for a long time (> 12 h) to obtain acceptable transfer of large proteins (> 100 kd) (134). If methanol is omitted, swelling of the gel during electroblotting can be avoided by soaking the gel in transfer buffer for about 1 h before electrophoresis (3).

A common modification is the addition of 0.1% SDS to the transfer buffer. This increases the efficiency of protein transfer but may also reduce protein binding to nitrocellulose.

The pH of the buffer chosen will determine the correct orientation of the electrodes relative to the gel/filter interface. Thus, when electroblotting SDS-PAGE gels, usually in neutral or slightly alkaline buffers, the polypeptides behave as anions so that the transfer filter must be arranged to be on the anodic side of the gel. Conversely, in 0.7% acetic acid the proteins behave as cations and so the filter is placed on the cathodic side of the gel.

Burnette (134) found that a voltage gradient of 6 − 8 V/cm applied overnight was required for the essentially complete transfer of all proteins larger than 100 kd molecular mass from 1.5 mm-thick gels to nitrocellulose. During blotting, buffer components elute from the gel and so increase the conductivity of the transfer buffer. Thus, if blotting is carried out at constant voltage, the current will rise and can exceed 1 A. For this reason, commercial electroblotting apparatus can often accommodate currents up to 5 A. If the current is likely to rise above 1.0 − 1.5 A, a unit fitted with a cooling coil may be needed to keep the temperature within acceptable limits. Alternatively, using standard electrophoresis power packs, one can use constant current (e.g. 200 mA) and allow the voltage to drop during transfer. The main problem encountered is that high molecular mass proteins elute from the gel more slowly than low molecular mass proteins so that the protein profile or the blot may be size-biased. Several approaches have been attempted to minimize this problem including the addition of 0.1% SDS to the sample buffer (135), proteolytically nicking the sample proteins before transfer (136) and using an electroblotting apparatus that generates a gradient in the electric field (137). The simplest method is to increase the elution time to a value where high molecular mass proteins are transferred sufficiently well.

Waterborg and Harrington (138) have reported that when one attempts to use standard electroblotting techniques for transferring histones separated by the

Triton−acid−urea gel system, irreproducible results are obtained. They suggest a procedure which displaces the Triton from such gels with SDS prior to electroblotting.

Finally, another variation called semi-dry blotting has been introduced. Specially designed apparatus (e.g. Pharmacia-LKB) uses buffer-soaked filter paper wicks and allows up to 12 gels to be electroblotted simultaneously by placing the gels and blotting filters in vertical stacks. Similar apparatus is also available from several other suppliers.

7.10.3 Analysis of Western blots

i. General protein stains
Nitrocellulose electroblots can be stained successfully with all of the general protein stains used for polyacrylamide gels including Amido Black, Coomassie Blue R-250, and Fastgreen. Amido Black is often preferred to Coomassie Blue because it destains quickly using only low concentrations of methanol to leave very low backgrounds whereas Coomassie Blue tends to give much higher backgrounds. Nitrocellulose begins to disintegrate after a short time in methanol-acetic acid whereas it is more resistant to isopropanol. Therefore Gershoni and Palade (129) recommend that nitrocellulose blots should be stained for 1 min in 0.1% Amido Black, 25% isopropanol, 10% acetic acid and then destained in this solvent. Soutar and Wade (139) suggest staining in 0.1% Amido Black, 45% methanol, 10% acetic acid for 10 min and destaining in 10% acetic acid until the background is clear (10−15 min). Unfortunately, all these stains have major disadvantages.

- They are relatively insensitive, requiring as much as 0.5 μg of protein for staining.
- They cannot be used to detect proteins immobilized on most nylon membranes or DBM, DPT, or DEAE paper since the matrix itself binds the stain very tightly and gives unacceptably high backgrounds.
- The alcoholic solutions required for staining and destaining can cause nitrocellulose to shrink so that it becomes difficult to compare, say, tracks stained for protein with tracks stained with a specific immunodetection procedure. Similarly, if molecular mass standards are electroblotted and the filter track containing these is cut off and stained, the shrinkage makes it quite inaccurate to deduce the molecular mass of immunodetected bands on the rest of the filter.

Hancock and Tsang (140) reported that indian ink is more sensitive than any of these stains and can detect as little as 80 ng of some proteins. Silver staining is also possible and can detect nanogram quantities of proteins on nitrocellulose blots (e.g. ref. 141).

A variety of new general protein detection methods specifically for use with electroblots have also recently been reported. One is the colloidal gold stain (142) which relies upon gold binding selectively to the immobilized proteins on a blot. The stain is rapid (2−5 h), simple to perform and sensitive (1−5 ng protein can be detected). Furthermore, the gold stain can be enhanced by a 15-min incubation with a silver lactate solution so that as little as 400 pg of protein per band can be detected. The added advantage over the use of classical anionic dyes is that, since alcoholic solutions are not involved, the blot does not shrink and so can be easily compared with other immunoblotted sections of the filter. Unfortunately, however, the colloidal gold stain is relatively expensive to use routinely and is suitable only for nitrocellulose blots.

Colloidal iron sols also have also been used to stain proteins. The positively-charged colloidal iron particles bind to negatively-charged SDS-denatured proteins and so can be used to detect these polypeptides on either nylon or nitrocellulose membranes. The colour of the iron-stained bands can be intensified using acidic potassium ferri-cyanide which gives deep blue stained bands with low background (143). Although substantially less sensitive than colloidal gold staining, the method is applicable to nylon filters which colloidal gold staining is not. Both types of stain are available commercially in kit form.

Several other general protein staining techniques are based upon tagging all the immobilized proteins with a small molecule which can then be recognized by incubation with an easily-located reagent. These methods are applicable to all types of blotting matrix and are extremely sensitive. The first procedure consists of soaking the electroblot in 2,4 dinitrofluorobenzene to add dinitrophenol (DNP) to all the immobilized proteins. The proteins are then detected by incubation with rabbit anti-DNP antibodies followed by horseradish peroxidase (HRP)-labelled goat anti-rabbit IgG (144). This method is reported to be 100-fold more sensitive than Amido Black staining. Kittler *et al.* (145) described a similar procedure, incubating the electroblot with pyridoxal 5′ phosphate to produce 5′ phosphopyridoxal derivatives of the immobilized proteins which were then detected using a mouse monoclonal anti-5′ phosphopyridoxyl antibody. Yet another method uses the high affinity binding of avidin to biotin. The immobilized proteins are biotinylated and then exposed to an avidin−HRP conjugate and an HRP chromogenic substrate (146). Again the procedure is rapid (3−5 h) and the sensitivity of detection is of the order 10−50 ng of protein. An advantage of the 'biotin-blot' technique is that it can be used for nylon membranes. Indeed it is probably the most sensitive total protein staining method currently available for this type of blotting matrix.

ii. Use of molecular weight standards
In the same way that polypeptides of known size are useful in conventional SDS-PAGE to estimate the size of separated sample proteins, early blotting experiments often included one or more lanes of polypeptide size markers which were blotted along with the samples. Molecular mass determination then involved cutting the lane(s) of size markers from the blot, staining it and then realigning this with the rest of the filter which had been probed using other methods such as immunoblotting. Although the shrinkage problems of using anionic dyes can be avoided by using the colloidal gold stain or by methods such as biotinylating the proteins (see above) these still require the marker strip to be cut from the main filter and stained separately.

This extra manipulation can now be avoided by using marker proteins that have been biotinylated prior to electrophoresis. These can be prepared in the laboratory (147) but are also available commercially (e.g. Bio-Rad). Under ideal preparation conditions, each protein should have only a single biotin molecule covalently bound. Since biotin is a small molecule (M_r 244), it does not affect the mobility of the proteins by SDS-PAGE so that these can still serve as accurate size standards. The advantage of using pre-biotinylated marker polypeptides is that the filter can be analysed as a whole, without cutting, using a standard immunoblot protocol. Avidin-conjugated horseradish peroxidase (avidin−HRP) or avidin conjugated alkaline phosphatase (avidin−AP) is mixed with the labelled second antibody. The second antibody locates the position of the specific antigen bands whilst the avidin−HRP or avidin−AP locates the marker polypeptides all in the same incubation.

Pre-stained marker polypeptides are also commercially available for electroblotting (e.g. Bio-Rad). The amount of blue dye bound to each protein molecule can vary substantially. Thus proteins tend to migrate as broad bands and are not recommended for accurate molecular mass determination. However, they are useful for approximate localization of sample proteins of interest and are ideal for monitoring the efficiency of protein transfer during electroblotting since they remain visible both during electrophoresis and on the membrane after blotting.

Detection of radioactive proteins
Electroblotting of proteins radiolabelled with ^{14}C or ^{35}S permits more efficient autoradiography since the gel matrix is no longer present to quench the β emissions of radioactive decay. The minimum levels of ^{14}C or ^{35}S that can be detected in 24 h by autoradiography of electroblots is reported to be about 4×10^2 d.p.m./cm^2. For the same reason, in contrast to polyacrylamide gels where fluorography is essential for detecting ^{3}H (Section 7.5.1), this isotope can be detected as electroblots by direct autoradiography. However, the method is insensitive in that approximately 2×10^4 d.p.m./cm^2 are required for detection in 24 h (88).

The efficiency of detection of all these isotopes is enhanced if fluorography is used. Once dried, nitrocellulose electroblots can be processed for fluorography simply by being dipped once in 20% PPO in toluene or ether and, after removing the excess, allowing the filter to air-dry. Fluorographic detection of ^{35}S or ^{14}C and ^{3}H using Fuji RX film gives a minimum detectable level of radioactivity after 24 h exposure of 1×10^2 d.p.m./cm^2 and 5×10^2 d.p.m./cm^2 respectively (88). It is important to remove all water from the filter before dipping it into the PPO solution since otherwise this prevents entry of PPO and gives non-uniform impregnation. This is easily checked by viewing the PPO-impregnated filter under UV light prior to fluorography. If the filter looks patchy it should be dried thoroughly in air and then re-soaked in fresh PPO solution until a uniform fluorescence is achieved.

iii. Immunodetection methods
Immunological detection of specific proteins in an electroblot ('immunostaining' or 'immunoblotting') is possible using all the transfer matrices described here but, again,

nitrocellose is the most widely-used. This procedure can be applied to denatured polypeptides separated by SDS-PAGE as well as native proteins. The extent of protein renaturation during transfer is probably important for recognition by certain monoclonal antibodies. Some renaturation may occur if the transfer buffer lacks SDS although a low concentration of SDS is included by some workers to improve the efficiency of transfer (e.g. ref. 135). Dunn (148) has shown that incubating the gel in 20% glycerol in Tris−HCl, pH 7.4, prior to electroblotting can improve the recognition of antigens by some antibodies but that a much larger improvement can be gained by transferring the proteins in alkaline pH buffer without SDS, irrespective of whether a prior renaturation step is included. The recommended buffer is carbonate buffer (pH 9.9) instead of the more usual Tris−glycine buffer (pH 8.3).

Since immunostaining is based upon specific antibody binding to particular proteins or antigens, it is important to minimize non-specific binding of the antibody to the filter matrix. When using nitrocellulose filters, non-specific binding sites are readily blocked by incubating with 1% BSA for 1−2 h at 23°C. A non-fat dry milk-based reagent called Blotto has also been used (149). It is claimed to be equally or more effective than BSA yet considerably cheaper for routine use. However, there are complex carbohydrates in milk which can absorb out antibodies that recognize carbohydrate determinants so that, in these cases, BSA may be preferred. If the intention is to stain all proteins after the immunodetection profile has been recorded, one can block nitrocellulose binding sites using 0.05% Tween-20 instead of a protein-based blocking agent (150), although this may not always be advisable since some researchers have noted that sample proteins can be removed from nitrocellulose filters by non-ionic detergents. Towbin *et al.* (151) have described a reversible staining method in which the nitrocellulose-bound proteins are sequentially incubated with heparin and toluidine blue. After recording the banding pattern, this general protein stain is removed by incubation in high ionic strength buffer containing methanol and the filter is then processed normally for immunodetection. Proteins can also be stained while still in the polyacrylamide gel prior to electroblotting then blotted and specific proteins detected by immunostaining (e.g. 152), but using such a method there is always the problem of alignment between the stained gel profile and the immunostained blot profile.

Nylon filters are considerably more difficult to block effectively than nitrocellulose. For example, Zetaprobe requires 12 h at 50°C with 10% BSA. If DBM, DPT, or cyanogen bromide-activated paper is used, residual active groups must be blocked by incubation with 10% ethanolamine (pH 9) for 2 h at 37°C (128) before immunostaining is carried out.

A typical immunodetection protocol for nitrocellulose filters is given in *Protocol 15*.

Protocol 15. Outline method for immunodetection

1. First block non-specific binding sites on the electroblot by incubating it with excess protein (e.g. BSA).

2. Incubate with the specific antibody.

Protocol 15. *continued*

3. Wash away unreacted antibody.
4. Incubate with a labelled second antibody directed against the first antibody.
5. Wash away any excess antibody.
6. Finally, detect the bound second antibody.

In general, each of the reactions with antibody should be carried out in the presence of the blocking component. It is also important to ensure that the reactions are carried out in a reasonably large volume of solution to ensure good equilibration with the filter. This is in contrast to nucleic acid blots where a small volume of radioactive probe solution sealed with the filter in a plastic bag is sufficient. In general, it is also worth avoiding high concentrations of probes or long incubation times which can give high backgrounds.

The second antibody can be tagged in any one of a number of ways. Early workers often used radioiodinated antibody (most readily using ^{125}I: see Appendix 3) in a conventional radioimmunoassay (RIA) protocol with detection by autoradiography. A suitable protocol for this procedure is described by Dunbar (67). Fluorescent-labelled second antibody, for example using FITC (153) or MDPF (154), has also been used for fluorescent immunoassay (FIA) with detection of bound antibody by UV illumination. Although these procedures, especially using radioiodinated antibody, are still in use, much recent progress has been made with reagents consisting of antibody conjugated to a marker enzyme, usually horseradish peroxidase (HRP) or alkaline phosphatase (AP) in an immunoenzyme assay (EIA) protocol. These enzymes will convert a suitable soluble substrate into a coloured precipitate which then marks the site of antibody binding. Many different colour development reagents have been used. The most common are 3,3' diaminobenzidine (DAB), 4-chloro-1-naphthol and 3-amino-9-ethylcarbazole (see ref. 127 for review). DAB is slightly more sensitive than the other two chromogenic reagents but it is a carcinogen and so its use is not recommended. In general, the sensitivity of the HRP method is $100-500$ pg protein. However, when used with the substrates 5-bromo-4-chloro-3 indolyl phosphate (BCIP) and nitroblue tetrazolium (NBT), the AP method is even more sensitive such that $10-50$ pg protein can be detected (155). The other advantage of the AP method is that, unlike the HRP method, the coloured reaction product will not fade when exposed to light. Both the HRP and AP methods are, of course, subject to non-specific colour formation if there are endogenous peroxidase or phosphatase enzymes in the sample. However, they are rapid, very sensitive and generally applicable procedures. A recent variation is to tag the second antibody with biotin (67). The biotinylated second antibody is recognized by avidin which in turn (because of its multiple binding sites) also binds biotinylated peroxidase. The peroxidase is then detected as usual using a colour development reagent such as BCIP. An alternative to using enzyme immunoassay is the colloidal gold conjugate immune detection method ('immunogold staining') (e.g. refs. 156, 157). The second antibody is mixed with colloidal gold spheres which complex tightly to the antibody molecules. Detection of the first antibody therefore simply involves incubating the treated electroblot with gold-conjugated second antibody and then washing away excess antibody. Bound antibody appears as rose-coloured bands. Unlike the HRP and AP methods, this procedure is clearly unaffected by endogenous peroxidase or phosphatase activities. If desired, the sensitivity can be increased by incubating the blot for 15 min with silver lactate which deposits a layer of silver on the gold spheres to give black bands ('immunogold silver staining'). After enhancement, protein bands containing as little as 1 pg protein can be detected. Although the original rose-coloured bands are not light sensitive, the silver lactate reagent will precipitate upon exposure to light so that care is needed in its use. Immunogold and silver enhancement reagents are also available commercially (e.g. Bio-Rad).

Labelled protein A from *Staphylococcus aureus* can be used instead of the second antibody, either radioiodinated or gold-conjugated for example (67, 126, 134, 158). However, protein A does not recognize IgG of all species or all IgG subtypes and is inherently less sensitive than the use of antibodies because binding is not polyvalent. Indeed, in general, the use of a specific antibody is $10-50$ times more sensitive than when using protein A, other conditions being equal.

Most proteins separated on polyacrylamide gels containing urea and/or SDS, or electroblotted in buffer containing SDS, appear to retain sufficient structural identity to be recognized by specific polyclonal antibodies on the blot matrix, but monoclonal antibodies directed against single binding sites may fail to react if this site is particularly sensitive to denaturation. Even gels that have been stained with Coomassie Blue R-250 and destained can be electroblotted to nitrocellulose and used for immunodetection of specific proteins provided that the gels are first incubated with gentle shaking at 4°C for at least 1 h in 1% SDS, 25 mM Tris, 0.192 M glycine buffer (159). Interestingly, the stain is retained by the transferred proteins so that the immunodetected proteins can be correlated with the general profile of blue stained proteins. Likewise, silver-stained proteins retain their antigenic properties and so can also be used for immunodetection (141).

The above procedures aim to locate particular proteins by their reaction with specific antibodies. In addition, immunological techniques exist which utilize the sensitivity of this methodology to allow the detection of all the blotted proteins. These methods are discussed in Section 7.10.3*i*.

iv. Detection of glycoproteins

A general procedure for staining glycoconjugates on blots has been reported (160). The blotted protein is oxidized with sodium periodate, generating aldehyde groups on the sugars which in turn form a stable Schiff base with an enzyme−hydrazide complex such as alkaline phosphatase−hydrazide. The procedure is rapid, simple and sensitive. By performing the periodate oxidation prior to cell breakage it has been used to identify cell surface glycoproteins (161). Particular types of glycoproteins have been detected by lectin overlays using radiolabelled lectins (129) or peroxidase-labelled lectins (162), lectins which are detected using HRP-conjugated anti-lectin antibody (163) or lectins which have the ability to bind HRP directly (164). Glycoproteins can also be readily modified on electroblots, for example using glycosidase treatment, prior to lectin binding in order to examine glycoprotein identity and structure (129).

v. Detection methods for other specific proteins

Lipoproteins have also been detected immunologically on electroblots (165). Enzymes can be detected using appropriate enzyme stains (see Appendix 2). In many cases, at least some activity can be regained after SDS-PAGE once the SDS has been removed so that enzyme detection is not necessarily limited to only native gel electrophoresis. The SDS can be removed from electroblotted proteins simply by washing the blot membrane. Alternatively, it may be better to remove SDS from the original gel and incubate under conditions which will encourage renaturation *before* electroblotting (see Section 10.2.4). Proteins with other forms of biological activity may also be detectable, particularly those which bind to known ligands such as DNA or RNA, hormone receptors, heparin, calmodulin, collagen, etc. (see refs 125, 127, 166 for reviews). Almost any ligand can be used to detect its specific binding protein(s). A good example (167) is the binding of α-bungarotoxin to the α-subunit of the acetylcholine receptor (*Figure 20*). Even proteins with cell-attachment promoting activity can be detected in a procedure known as 'cell blotting'. The electroblots are incubated with whole cells and these proteins become visualized as bands of attached cells (e.g. ref. 168).

Ligand blotting is considered in detail in a companion volume of this series (139).

7.10.4 Multiple probing of a single electroblot

It is quite possible to probe an electroblot, record the result and then elute the probe while retaining the immobilized sample proteins ready for examination with a second, different detection system (e.g. ref. 169). The procedure for stripping off the first probe depends on the nature of the probe. For example, lectins can be removed by washing the filter with an appropriate saccharide while antibodies will usually elute at low pH. More severe washing procedures include $2-5\%$ SDS or 8 M urea (125). Unfortunately, these procedures will also tend to strip the sample proteins from nitrocellulose membranes and so electroblots prepared with DBM paper or DPT paper should ideally be used. Some nylon membranes such as Zetaprobe may also prove successful in multiple probing if the stripping procedure is not too harsh.

Another approach is to prepare multiple blots from the same gel, either by trapping excess protein that passes through the first nitrocellulose membrane using additional nitrocellulose sheets or by changing the blotting membrane at intervals during the electrolytic transfer. However, it is very often difficult to obtain multiple blots which are identical so this method is not in widespread use.

8. Recovery of separated proteins

Several types of specialist apparatus are commercially available for the large-scale purification of proteins by polyacrylamide gel electrophoresis. However, for two

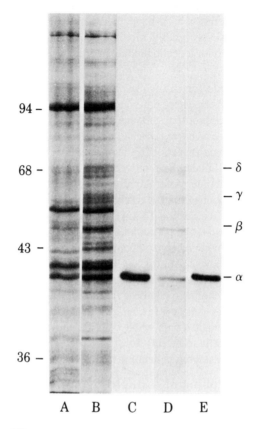

Figure 20. Binding of ^{125}I-labelled α-bungarotoxin to blots of crude *Torpedo* electric organ membranes and purified acetylcholine receptor (AcChoR). Crude *Torpedo* electric organ membranes (50 μg of protein; lanes A, B, and C) or purified AcChoR (5 μg; lanes D and E) were solubilized at room temperature in LiDodSO$_4$-containing sample buffer and run at room temperature (lane A) or at 4°C (lanes B, C, D, and E) on a 10% polyacrylamide gel. The gels either were stained with Coomassie Brilliant Blue (lanes A, B, and D) or were transferred to a Zetabind filter and incubated with 2 nM (2 × 10^6 c.p.m.) of ^{125}I-labelled α-bungarotoxin for 12 h at room temperature. After being washed in cold phosphate-buffered saline, the transfers were autoradiographed (lanes C and E). The position of molecular mass standards (in kd) and AcChoR subunits are indicated. (Reproduced from ref. 167 with permission.)

reasons, this approach is now rarely used for preparative work. First, because the resolution obtainable using the large gel columns associated with such apparatus is usually quite low, bands that migrate close to each other but which are clearly distinct by analytical gel electrophoresis are unlikely to be separated. Second, the amounts of protein which can be purified directly from analytical-scale gels are often sufficient for the further analyses planned. The protocol is therefore to locate the protein band of interest in a rod or slab gel and then recover it, usually by gel excision and elution of the protein. Because of their dimensions, standard size rod gels (Section 5.1.1) can be used to prepare only relatively small amounts of individual proteins ($1-50$ µg); above this level the gels become grossly overloaded and resolution declines rapidly. Therefore a much more common format is to use slab gels. Instead of forming individual sample wells, the gel mixture is either overlayered with buffer to form a continuous flat meniscus or a sample comb with only a single 'tooth' spanning the width of the slab gel may be used to form a single sample well. After polymerization, the sample is loaded onto the gel and forms a continuous layer along the entire upper edge of the gel. Up to 2.0 ml of sample containing several milligrams of protein can be loaded using a discontinuous buffer system in the usual slab gel format. However, the sample capacity of the gel can be increased still further by using thicker side-spacers to give a thicker gel. In practice, the sample load is limited by two factors. First, as sample load increases, so does band width. Therefore the maximum load possible will depend on the separation distance of the desired protein from neighbouring bands. Second, as the total protein load increases, a point is reached when severe vertical streaking down the slab gel occurs, sometimes together with band distortion. Despite these limitations, loading of up to 50 mg total protein per slab gel is not uncommon.

8.1 Localization of protein bands

Once electrophoresis is complete the protein bands must be localized prior to excision. In general the yield of proteins from fixed, stained bands is much lower than from unfixed gels and so several methods have been devised for protein localization in the latter.

(a) Proteins may be localized by slicing the gel into segments (Section 7.5.2), eluting each into a suitable buffer by diffusion (Section 8.2.1) and then assaying for the protein of interest. For native proteins the assay is based upon the biological (e.g. enzymic) activity of the protein. Denatured proteins of interest separated by SDS-PAGE can be located by eluting into 0.01 M NH_4HCO_3, 0.05% SDS at 37°C overnight, followed by lyophilization and then running a portion of each eluate in separate tracks of an analytical slab gel which is then stained and destained in the usual way.

(b) The most popular method is to cut two longitudinal strips from the sides of the slab gel and a narrow longitudinal strip from the centre, and to stain and destain these with Coomassie Blue or other appropriate stain whilst keeping the rest of the gel on a glass plate covered with cling film in a refrigerator. Following this, the stained side strips are lined up along the edges of the unstained gel (but not touching it) and used as guides to cut out bands of interest from the unstained gel. Using rod gels instead, one rod gel may be stained and destained and the protein R_f value used as a guide for cutting out bands from other identical but unstained rods. However, the method is usually applied to slab gels rather than rods because of the more uniform mobility achievable in this format. The method works well using slab gels provided the bands to be recovered are well separated from potential contaminants. The only problem that may occur is if the gel strips shrink somewhat during staining and destaining, thus making accurate alignment with the unstained segments difficult. This may be particularly problematical with concentration gradient gels (Section 9.3) which shrink un-evenly. To overcome this, the stained segments can be allowed to re-swell in aqueous buffers until they regain their original dimensions. Alternatively, before cutting the gel initially, a series of small holes can be punched at fixed distances along the anticipated cut sites. The guide strips are then cut from the main gel by cutting through the holes. After staining and destaining, the appropriate part of the guide strips can be accurately aligned with the unstained gel segments by matching the position of the 'half-circle' perforations (171).

(c) Protein bands in gels lacking SDS can be detected by immersing the whole gel in a suitable fluorescent reagent (see Section 7.4). The method is superior to the guide-strip staining method (b) for bands which are not separated by more than a few millimetres.

(d) An increasingly popular and extremely sensitive method for localizing polypeptides after fractionation by SDS-PAGE is to first mix the bulk sample with an aliquot of sample polypeptides which have been labelled with a fluorescent molecule (Section 7.4). During electrophoresis the labelled polypeptides co-migrate with their unmodified counterparts and serve as markers for these by monitoring the gel after electrophoresis with a UV lamp (172, 173). Since fluorescent labelling alters the charge of the polypeptides, the method is usually only useful for electrophoresis in the presence of SDS which masks the small charge differences created by the treatment. The added advantage of the method is that one can also easily monitor the progress of polypeptide recovery from the gels during electrophoretic or diffusion elution (Section 8.2) simply by view-ing the gel under UV light.

(e) Protein bands can also be localized in unstained gels by virtue of the phos-phorescence of tyrosine and tryptophan residues following UV irradiation (e.g. ref. 174) but the method has been little used.

(f) Also little used at present are a number of other methods for localizing polypeptides after SDS-PAGE without the use of dyes, for example, SDS−polypeptide complexes can be visualized by precipitation either by chill-ing the gel at $0-4°C$, or by incubation with potassium ions or cationic surfactant. References to these and other 'direct detection' methods are given in Appendix 2.

(g) Some enzymes can be localized *in situ* by incubating the gel in an appropriate reaction mixture (Section 7.9.1, Appendix 2). The enzyme activity usually produces some change in colour or fluorescence which leads to its localization. Since only the specific enzyme is visualized, the excised bands may well be con-taminated with other proteins, especially since the coloured or fluorescent reaction products tend to diffuse to yield broad zones.

(h) Protein bands may also be visualized directly by staining with Coomassie Blue. Since the staining conditions will inactivate native proteins, this is only suitable for proteins which are not required in an active form for later analysis. The staining and destaining period should be as brief as possible to allow visualization of the bands but to minimize stain and fixative penetration into the gel. The stained gel is rinsed with water to remove excess fixative and then the relevant protein bands excised. This method of protein localization and elution forms the basis of the Cleveland method of peptide mapping (Chapter 5).

(i) Finally, rather than detect the protein of interest in the gel itself, the gel can be electroblotted onto nitrocellulose (Section 7.10) and the protein then visualized on the blot. Since protein is bound to the nitrocellulose by only non-covalent forces, the section of filter bearing the desired band can be cut out and the protein eluted using detergent (e.g. 1% Triton X-100, 0.5% SDS etc.), denaturing agents (e.g. 8 M urea) or acid (pH 2−3). Naturally the protein binding capacity of nitrocellulose (80−100 μg/cm^2) limits the yield but the method is useful for small-scale preparative work (see also Section 8.3).

8.2 Elution of proteins

If the aim of gel fractionation has been to purify sufficient protein for antibody preparation there is usually no need to separate protein from the polyacrylamide since the latter does not interfere with antibody production. The gel segments are simply homogenized with Freund's adjuvant and injected directly (see Appendix 6). Antibodies raised against SDS−polypeptide complexes often precipitate the native protein as well as the denatured polypeptide.

For other purposes it is usually necessary to isolate the protein free of the gel matrix. Theoretically the simplest method would be to solubilize the gel (Section 7.5.2). However, not only are these solubilizing agents potentially harmful to proteins but also only the gel crosslinkages are labile, leaving the protein contaminated with long chain polyacrylamide molecules. Unfortunately no suitable methodology presently exists to remove this contaminant easily and quantitatively.

The two usual methods of protein recovery are elution by diffusion and by electrophoresis. The main advantages of electrophoretic elution are its speed and the fact that the protein is usually eluted in a much smaller volume than by diffusion. Diffusion elution is simpler but protein recoveries, especially in the absence of SDS, may be poor.

8.2.1 Elution by diffusion

The gel slices are macerated by chopping finely with a scalpel or homogenization. This increases the gel surface area so that, when a suitable buffer is added, elution can occur more readily. Unfortunately there is a conflicting need to keep the volume as small as possible to minimize recovery problems after elution whilst using a large enough volume of buffer to obtain a good yield. Elution with about three gel volumes of buffer overnight with mixing, for example by putting the sample in a stoppered, siliconized tube fixed to a vertical rotating disc, and then a repeated extraction the next day, is a reasonable protocol. The process can be speeded up using higher temperatures if the protein is stable under these conditions. Thus SDS−polypeptide complexes can be eluted overnight at 37°C into buffer containing 0.05% SDS. The degree of protein recovery by diffusion elution is extremely variable with native proteins but can be as good as 70% if the elution buffer contains 0.05−0.1% SDS. In all cases gel fragments are easily removed by centrifugation.

Once gels have been stained, the proteins have been fixed and denatured. However, using SDS-containing buffers it is sometimes possible to obtain quite good recovery from such gels (e.g. ref. 175). The dye can be removed by gel filtration on Sephadex G-25 or by a solvent extraction procedure (175).

In many cases, such as sequence analysis (Sections 10.2.1, 10.2.3) or antibody preparation (Appendix 6, Section 10.2.5) the SDS may not need to be removed from protein preparations. If SDS does need to be removed but denaturation of the protein is acceptable, the sample should be lyophilized and then dissolved in buffer at an alkaline pH. Protein is then precipitated with 10% TCA and SDS is removed by washing the precipitate with ethanol or acetone. Other methods have been reviewed by Furth (176). On the other hand, it may be desirable to remove the SDS to allow at least partial renaturation of the protein so that it can be examined for biological activity. Procedures for achieving this are described in Section 10.2.4.

8.2.2 Electrophoretic elution

A number of devices have been designed for electrophoretic elution of proteins from polyacrylamide gel slices, most of which rely on electrophoresis of proteins out of the gel into a chamber which makes electrical contact with the reservoir buffer via a dialysis membrane. Proteins are eluted and then retained in the elution chamber. The apparatus ranges from those requiring elaborate workshop facilities for their manufacture to simple adaptations of standard rod gel methodology (e.g. refs. 172, 177, 178). Some apparatus is commercially available. The method can be applied to stained or unstained gels and can be used with buffer alone or buffer containing urea and/or SDS. Yields are usually higher if SDS is included. Protein recoveries can exceed 90% (e.g. ref. 178).

Fluorescent pre-staining of the sample proteins before electrophoresis (Section 7.4) is useful because it allows the researcher to follow the progress of elution of the desired protein (172).

An even simpler approach which is useful for multiple samples from SDS-PAGE gels is to adapt a common method used to elute DNA from agarose gel slices. Each set of gel slices is sealed in a small dialysis bag with just enough buffer (0.1% SDS, 0.1 M sodium phosphate pH 7.4) to cover them and then placed on the platform of a horizontal gel electrophoresis chamber (179). The electrophoresis tank is filled with enough of the same buffer to just cover the dialysis bags and then elution is carried out (100 mA, 20−200 V) for a minimum of 2−8 h. The buffer recovered from each bag contains the eluted protein and can be freed of polyacrylamide debris by centrifugation or filtration (179).

Although this method works well when isolating reasonable quantities of proteins, problems can occur with low microgram amounts because protein can be lost by absorption to the large area of dialysis membrane. Hunkapiller *et al.* (180) have described an elution cell that minimizes the amount of dialysis membrane used and so provides a reliable method for recovery of even a few tenths of a microgram of most proteins even if the gels have been stained with Coomassie Blue.

8.3 The recovery of proteins from electroblots

Electroblotting (Section 7.10) is an extremely simple way to purify reasonable amounts of individual proteins for a number of uses, for example raising specific antibodies. Naturally if this is the aim, the filter matrix used must be one which does not bind the proteins covalently. Most work has been done using nitrocellulose. The protein band of interest is first cut from the filter and then eluted with a suitable reagent. Detergents have been used for elution but Parek *et al.* (170) found that the highest recovery of protein was achieved with 20% acetonitrile or pyridine.

9. Modifications to the basic techniques

9.1 Molecular mass analysis of oligopeptides

Oligopeptides with molecular masses below about 15 000 are not well resolved by SDS-PAGE using the Laemmli buffer system even using 15% polyacrylamide gels (*Figure 6*). However, Swank and Munkres (181) showed that the separation of oligopeptides can be considerably improved by using 12.5% polyacrylamide gels prepared with a high ratio of bisacrylamide crosslinker ($C = 10\%$) and the inclusion of 8 M urea in a continuous 0.1 M Tris–phosphate, pH 6.8 buffer system (*Protocol 16*). These conditions yield a good linear relationship between oligopeptide molecular mass and mobility over the molecular mass range of 2500–17 000 (*Figure 21a*). As with the separation of large proteins (Section 4.3) the use of discontinuous

---

Protocol 16. Urea – SDS-PAGE for the separation of oligopeptides[a]

Acrylamide–bisacrylamide: 12.5 g of acrylamide and 1.25 g of bisacrylamide are dissolved in and made to 50 ml with water, then filtered through a Whatman No. 1 filter. Stable at 4°C.

Gel buffer stock: 1% SDS, 1.0 M H_3PO_4 adjusted to pH 5.0 with Tris base.

Reservoir buffer stock: 0.1% SDS, 0.1 M H_3PO_4 adjusted to pH 6.8 with Tris base.

Sample buffer: 1% SDS, 8 M urea, 1% 2-mercaptoethanol, 0.01 M H_3PO_4 adjusted to pH 6.8 with Tris base.

1. Prepare the gel mixture by mixing:
   - acrylamide–bisacrylamide
   - gel buffer stock
   - urea
   - 6% ammonium persulphate (freshly prepared)
   - water to 30 ml final volume.

Protocol 16. *continued*

2. Degas the solution for 1 min, add 20 μl of TEMED and pour the gel without delay. For rod gels, this is the only gel that needs to be poured. Slab gels are preferable to rod gels for comparison between samples but the highly-crosslinked resolving gel used in this method is sufficiently brittle that sample-well divisions often break during preparation. Therefore for slab gels polymerize a low concentration gel (3.75%) on top of the 12.5% gel, but with the same buffer composition, and form the sample wells in this 3.75% polyacrylamide gel.

3. Dissolve the oligopeptides in sample buffer and add Bromophenol Blue to 0.002% (w/v) final concentration. Next, heat the samples to denature the oligopeptides then cool to room temperature prior to use. In the original study the heating step was 60°C for 10 min but recent analyses have used 100°C for 3 min. Molecular mass marker peptides can be prepared by cyanogen bromide cleavage but are also available commercially.

4. Electrophorese at $6-8$ V cm^{-1} for rod gels until the Bromophenol Blue nears the end of the gel. Slab gels can be run at 100 V overnight.

5. After electrophoresis, stain the gels in 0.25% Coomassie Blue R-250 in methanol:water:glacial acetic acid (5:5:1). Destaining is by diffusion using 12.5% isopropanol, 10% acetic acid; electrophoretic destaining is not recommended since some small peptides migrate under these conditions.

[a] Based on the method given in ref. 181.

---

buffers gives improved resolution. Thus Anderson *et al.* (182) recommended the discontinuous system given in *Protocol 17*. Its fractionation range is shown in *Figure 21(b)*. This discontinuous buffer system has been used successfully to study both native small polypeptides (183) and cyanogen bromide polypeptide digests (184).

As a further refinement, Hashimoto *et al.* (185) introduced the use of concentration gradient gel electrophoresis in the presence of SDS and urea for the separation of small polypeptides (see also Section 9.3.3). Most recently, Schagger and Von Jagow (186) have described an approach which does not rely upon the use of urea. Glycine, the tracking ion in the Laemmli buffer system (16) is replaced with Tricine which, at the pH values which occur during electrophoresis, migrates much faster than glycine in the stacking gel. This has the effect of reducing the upper molecular mass range of proteins which can be stacked in the stacking gel but with the benefit that small SDS−polypeptide complexes separate from SDS and so are resolved.

An example of the superior resolution of this Tricine SDS-discontinuous buffer system for the resolution of small polypeptides compared to the Laemmli buffer system is shown in *Figure 22* compared in each case using a $10\%T\,3\%C$ resolving gel and a $4\%T\,3\%C$ stacking gel. The $10\%T\,3\%C$ resolving gel is an ideal choice

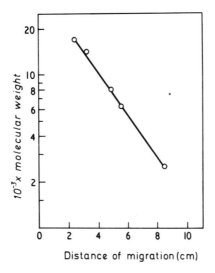

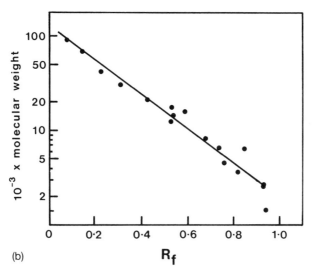

(b)

Figure 21. (a) Mobility of oligopeptides in SDS-urea polyacrylamide gel electrophoresis according to the Swank and Munkres technique (184). Electrophoresis was carried out in a slab gel format as described in *Protocol 16*. The oligopeptides, in order of decreasing molecular mass, were horse heart myoglobin (M_r 16 950), horse heart myoglobin cyanogen bromide fragments I + II (14 404), I (8159), II (6214), and III (2512). The molecular masses of horse heart myoglobin and its cyanogen bromide cleavage fragments are calculated from the sequence data given in ref. 39. (b) Calibration curve of molecular mass against R_f (mobility relative to the tracking dye) for oligopeptides separated by the discontinuous buffer system of Anderson *et al.* (182) described in *Protocol 17*. The identities of the 18 proteins and oligopeptides used, together with their molecular masses, are given in the original reference. (Redrawn from ref. 182 with permission.)

Protocol 17. The Anderson discontinuous urea – SDS-PAGE system for separation of oligopeptides[a]

1. Prepare the following stock solutions:

   Resolving gel acrylamide stock (36%*T*, 5%*C*): Prepare by dissolving 34.2 g of acrylamide, 1.8 g of bisacrylamide, in a total volume of 100 ml of water.

   Stacking gel acrylamide stock (6.25%*T*, 20%*C*): Prepare by dissolving 5.0 g of acrylamide, 1.25 g of bisacrylamide in a total volume of 100 ml of water.

   Buffer solution A: 1 M Tris, 0.2% SDS (pH 7.8 with conc. H_2SO_4).

   Buffer solution B: 0.074 M Tris, 0.1% SDS (pH 7.8 with conc. H_2SO_4).

2. Prepare the resolving gel (8%*T*, 5%*C*) by mixing:
   - 5.33 ml of resolving gel acrylamide stock
   - 4.8 ml of buffer solution A
   - 11.52 g of urea
   - 0.5 ml of freshly prepared 2.4% ammonium persulphate[b]
   - Add water to 23.5 ml

   After the urea has dissolved, degas the solution. Then add 12 μl of TEMED, mix and pour the resolving gel. Overlayer with *n*-butanol.

3. After polymerization, remove the *n*-butanol and rinse the gel surface with water and then with a 1:5 dilution of buffer solution A.

4. Prepare the stacking gel (3.125%*T*, 20%*C*) by mixing:
   - 2.5 ml of stacking gel acrylamide stock
   - 1.0 ml of buffer solution A
   - 0.1 ml freshly prepared 2.4% ammonium persulphate[b]
   - 1.4 ml of water

   Degas the solution, add 5 μl of TEMED and pour the stacking gel.

5. Prepare the samples in 0.139 M Tris, 0.5% SDS, 20% sucrose (pH 7.8 with glacial acetic acid). The most appropriate dye to include is Pyronin Y (2 μl of 2% Pyronin Y per sample).

6. Just before electrophoresis, rinse out the sample wells with water and then add 50 μl of 0.139 M Tris, 0.5% SDS (pH 7.8 with glacial acetic acid) or sufficient volume to give a height of 0.5 cm of liquid. This is sample buffer lacking the sucrose and is needed to ensure that there is sufficient acetate for stacking during electrophoresis.

7. Load the samples by underlayering under this buffer. Then fill each well with buffer solution B.

8. Fill the buffer reservoir (cathodic) with buffer solution B and the lower buffer reservoir (anodic) with buffer solution A diluted 5-fold.

Protocol 17. *continued*

9. Electrophorese at 20 W constant power until the tracking dye reaches within 1 cm of the bottom of the gel (about 4−5 h). The gel should be cooled during electrophoresis given the relatively high power being used.

a From Anderson *et al.* (182).
b Prepared daily.

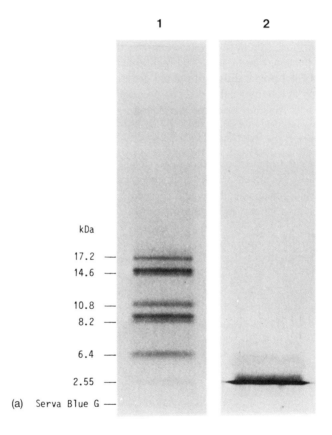

Figure 22. Comparison of the resolving power of the Tricine SDS – discontinuous buffer system of Schagger and Von Jagow (186) with the Laemmli buffer system. Electrophoresis of a commercial set of proteins (Pharmacia) spanning the molecular weight range 2550 – 17 200 using a 10%*T*, 3%*C* gel run under either (lane 1) the tricine buffer system or (lane 2) the Laemmli buffer system (16). The molecular masses of the proteins used are indicated in the left-hand side of the figure. Serva Blue G (Serva) was used as a tracking dye. (Reproduced from ref. 186 with permission.)

for initial examination of a complex sample in which the presence of small polypeptides is suspected since it spans the molecular mass range up to 100 000 yet allows the researcher to detect polypeptides down to a molecular mass of 1000. However, the resolution below 5000 M_r is poor. For fractionations in this size range, the authors recommend the use of a 16.5%T 3%C resolving gel overlaid with a 10%T 3%C 'spacer' gel and a 4%T 3%C stacking gel (fractionation range 1000−70 000 M_r). In addition, 16.5%T 6%C gels are described which have their best resolution between 5000 and 20 000 M_r. The resolution below 5000 M_r is increased by urea. A comparison of the resolving power of these gel types is shown in *Figure 23*. This new buffer system has considerable potential. Apart from its analytical application, the separation of proteins down to 1000 M_r in low (10%) acrylamide concentration and relatively low crosslinkage is ideal for preparative work. In addition, the omission of glycine and urea may aid subsequent protein sequencing of eluted polypeptides. Furthermore, the buffer system tolerates the application of high amounts of protein and proteins can be loaded in high ionic strength media (e.g. 2 M NaCl). Full details of gel preparation and use are given in ref. 186. Other procedures for the separation of low molecular mass polypeptides in the absence of urea are given in refs 187 and 188, but these use acrylamide gradient gels.

9.2 Separation of special classes of proteins

Although the electrophoretic methods described earlier in this chapter will separate the vast majority of cellular proteins, particular classes of proteins tend to be insoluble under the usual buffer conditions and so need special electrophoretic systems for their fractionation. Some of these are described below.

9.2.1 Histones

Histones are small, highly basic proteins. Five major species can be resolved in most cases; H1, H2A, H2B, H3, and H4, and a number of variants of these can occur via charge modification through acetylation or phosphorylation, as well as ADP-ribosylation. The major problem associated with electrophoretic analysis of these proteins is their insolubility in the absence of denaturing agents. Using SDS as denaturant, these proteins can be analysed by SDS-PAGE, under which conditions fractionation occurs mainly on the basis of size; therefore SDS-PAGE is useful for histone size fractionation, but is not able to analyse histone charge variants. A combined charge and size fractionation via polyacrylamide gel electrophoresis requires the use of denaturants able to prevent histone aggregation whilst preserving the charge differences between protein species. The usual method of achieving this has been the use of an acetic acid−urea buffer system (189) but even here it is often difficult to completely resolve histones H2A, H2B, and H3 or to identify minor modified forms. The resolution of these histones can be improved markedly by including a non-ionic detergent (usually Triton X-100 or Triton DF-16) in the acid−urea buffer system (190, 191). These are the main gel systems used for histone fractionation and are described in more detail below. Finally, some histone modifications are acid-labile (e.g. some phosphorylated species) and cannot be

B. David Hames

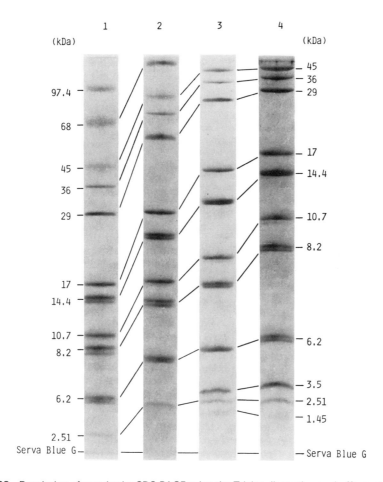

Figure 23. Resolution of proteins by SDS-PAGE using the Tricine discontinuous buffer system of Schagger and Von Jagow (186). Electrolysis of proteins on (1) a 10%*T*, 3%*C* gel (2) a 16.5%*T*, 3%*C* gel (3) a 16.5%*T*, 6%*C* gel (4) a 16.5%*T*, 6%*C* gel plus 6 M urea. In each case the gels were electrophoresed using the tricine buffer system (186). The molecular masses of the proteins used are indicated on each side of the gel lanes. Serva Blue G (Serva) was used as tracking dye. (Reproduced from ref. 186 with permission.)

examined in the acid–urea system but can be fractionated by polyacrylamide gel electrophoresis using a neutral pH buffer system (192). Hardison and Chalkley (192) have reviewed many of these methods for polyacrylamide gel electrophoresis of histones.

Very effective two-dimensional gel systems also exist for the fractionation of histones using a combination of SDS-PAGE acid–urea gels and Triton–acid–urea gels (see Chapter 3). By definition these have higher resolution than any of these fractionation methods in a one-dimensional format.

109

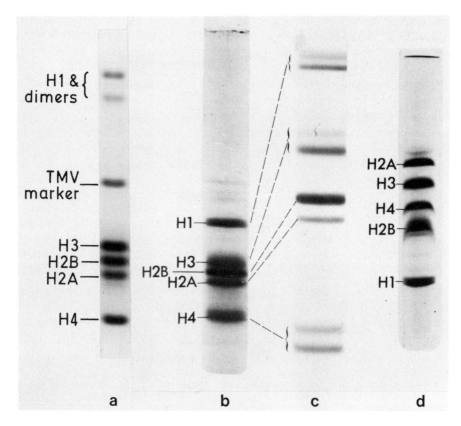

Figure 24. Separation of histones by polyacrylamide gel electrophoresis. (a) SDS-PAGE (slab gel track reproduced from ref. 193 with permission); (b) 2.5 M urea – 0.9 M acetic acid (9 cm rod gel); (c) 2.5 M urea – 1.0 M acetic acid (25 cm rod gel); (d) 2.5 M urea – 0.9 M acetic acid – 0.4% Triton X-100 (9 cm rod gel). The data in (b), (c), and (d) are reproduced from ref. 192 with permission.

i. SDS-PAGE of histones

A number of protocols exist for SDS-PAGE of histones which differ in the proportion of bisacrylamide used in the polyacrylamide gel and the exact buffer conditions, for example the presence or absence of urea as well as SDS. Thomas and Kornberg (193) used the basic system of Laemmli (16) (*Table 2*) but with three modifications: the concentration of Tris buffer in the resolving gel (18% polyacrylamide) is increased to 0.75 M, the ratio of acrylamide:bisacrylamide is increased to 30:0.15, and the electrode buffer comprises 0.05 M Tris, 0.38 M glycine, 0.1% SDS, pH 8.3. All other details of reagent and gel preparation are as described in Section 6. Slab gels are 1.5-mm thick and either 15 cm or 30 cm long. Samples used for electrophoresis can be either purified histones or intact chromatin. Electrophoresis is at 30 mA for about 6 h (15-cm-long gels) or 4 W for 24 h (30-cm-long gels) until the Bromophenol Blue tracking dye almost reaches the gel bottom. Staining with Coomassie Blue and destaining is as described in Section 7.2.1. An example of the separation achieved is shown in *Figure 24*.

Although histones separate mainly on the basis of size when analysed by SDS-PAGE, they migrate slower than would be expected on the basis of their known molecular masses (194). This is presumably due to a reduced overall-negative charge as a result of the high proportion of basic amino acids in histones. The practical consequence of this phenomenon is that the molecular masses of histones will be overestimated if these are calculated solely by comparison of histone mobility with that of standard, non-basic proteins in SDS-PAGE.

ii. Acid – urea gels

The banding pattern of histones separated by charge and size fractionation in an acetic acid – urea gel system depends on both the pH and the urea concentration but a useful system for routine analysis is 0.9 M acetic acid, 2.5 M urea (pH 2.7) in 15% polyacrylamide (189). No stacking gel is used. In this system the five major histones can be separated on a 9-cm resolving gel (*Figure 24*) whereas some minor bands resulting from covalent modifications or sequence heterogeneity are visible only after electrophoresis on much longer (25 cm) gels (*Figure 24*). Details of the acid – urea gel system are given in *Protocol 18*.

One of the disadvantages of the acetic acid – urea technique is that these solvents do not fully dissociate histones from DNA so that these proteins must be first purified free of DNA prior to electrophoresis. Alternatively, this problem appears to be over-come by including the cationic detergent CTAB in the sample buffer (195).

Protocol 18. Acetic acid – urea buffer system for histone separation[a]

Solution A: acrylamide – bisacrylamide; dissolve 60.0 g of acrylamide and 0.4 g of bisacrylamide and adjust to 100 ml with water. Deionize this solution by stirring for 30 min with 3 g of Amberlite MB-1 mixed-bed resin and filter it prior to use. Stable at 4°C.

Solution B: 4% (v/v) TEMED, 43.2% (v/v) glacial acetic acid in water. Stable at 4°C.

Solution C: 40 mg of ammonium persulphate in 20 ml of deionized 4 M urea. Make fresh prior to use.

1. Mix 1 vol. of solution B with 5 vol. of solution C and deaerate. Also deaerate 2 vol. of solution A but in a separate vacuum flask.

2. Combine the deaerated solutions, pour the gels, and overlayer them with cold 0.9 M acetic acid. Leave the gels for an additional 30–60 min after polymer-ization for optimal resolution of histones.

3. After filling the buffer reservoir with 0.9 M acetic acid, pre-electrophorese the gels at 130 V until constant current is obtained. This pre-electrophoresis is essential for good histone fractionation.

Protocol 18. *continued*

4. Load the histone sample, free of DNA and dissolved in 2.5 M urea, 0.9 M acetic acid at 1 mg ml^{-1}. Electrophorese from anode (+) to cathode (−) at 130 V for 9-cm-long rod gels. This takes about 3−4 h at room temperature. It is useful to include methyl green as tracking dye since the blue component of the dye migrates just ahead of the fastest migrating histone (H4) in this system and so gives an indication of the progress of the electrophoretic run. For very long (25 cm) gels, run at 200 V constant voltage at 4°C for about 48 h.

5. After electrophoresis, stain the gels with 0.1% Coomassie Blue R-250 in methanol:water:acetic acid (5:5:1) and destain electrophoretically or by diffusion in 10% acetic acid, 25% methanol.

[a] Based on the methods given in refs. 189 and 192.

---

iii. Triton − acid − urea gels

Addition of a non-ionic detergent, usually Triton X-100, to the acid−urea gel system results in the formation of micelles between the detergent and the hydrophobic regions of the histones. This results in complexes of histone and detergent and consequently a reduction in electrophoretic mobility of the individual histones proportional to their hydrophobicity (*Figure 24*).

The detergent binding is reduced by urea so that the electrophoretic patterns obtained vary markedly depending on the exact concentrations of urea and detergent. These concentrations are therefore chosen based upon which histone components need to be maximally resolved. Since the aim is for separation to be achieved largely on the basis of differences in hydrophobicity, polyacrylamide gels with a very low degree of crosslinking are used. As well as the original continuous buffer system (190, 191) a discontinuous buffer system has been described which reportedly gives superior resolution (196). The latter method (*Protocol 19*) is also more convenient in that no pre-electrophoresis step is needed and crude acid extracts can be loaded directly onto the gel for analysis. Using Triton−acid−urea gels it is possible to

---

Protocol 19. Discontinuous Triton − acid − urea system for histone separations[a]

1. For 30 ml resolving gel (15% acrylamide, 0.1% bisacrylamide, 8 M urea, 8 mM Triton X-100) mix:
   - 7.5 ml of 60% acrylamide
   - 1.2 ml of 2.5% bisacrylamide
   - 1.8 ml of glacial acetic acid
   - 0.15 ml of TEMED
   - 0.09 ml of conc. NH$_4$OH

Protocol 19. *continued*

- 14.4 g of urea
- Water to 27.4 ml

Dissolve the urea with slight warming if necessary
Then add 0.6 ml of 25% Triton X-100 and degas the solution.

2. Add 2 ml of 0.004% riboflavin solution to the resolving gel mixture and mix. Pour the resolving slab gel and overlayer with water. Photopolymerize using a light box containing four 15 W fluorescent lamps.

3. For every 20 ml of stacking gel[b], mix:
   - 1.1 ml of 60% acrylamide
   - 1.3 ml of 2.5% bisacrylamide
   - 1.2 ml of glacial acetic acid
   - 0.1 ml of TEMED
   - 0.06 ml of conc. NH_4OH
   - 9.6 g of urea
   - Water to 18.7 ml

   After the urea has dissolved, degas the solution.

4. Add 1.3 ml of 0.004% riboflavin solution to the stacking gel mixture. Pour off the water overlayer in the resolving gel, pour on the stacking gel and insert the sample comb. Photopolymerize as before; see step 2.

5. The reservoir buffer for both the anodic and cathodic reservoirs is 0.1 M glycine, 1 M acetic acid. Prepare this fresh from a 2 M glycine stock solution and glacial acetic acid.

6. Electrophorese samples with methylene blue as the tracking dye at 5−10 mA constant current overnight at room temperature.

[a] From Bonner *et al.* (196) with permission.
[b] No Triton X-100 is needed in the stacking gel.

---

resolve histones which differ in the substitution of only a single, neutral, amino acid when this occurs in a hydrophobic region and results in a change in detergent binding (190). The main disadvantage of the Triton−acid−urea system is artefactual histone heterogeneity due to the oxidation of methionine residues; oxidized histones have altered Triton-binding characteristics. Therefore special care must be taken to avoid this problem. Full details of the method using the continuous and discontinuous buffer systems are given by Zweidler (191) and Bonner *et al.* (196) respectively.

9.2.2 Nuclear non-histone proteins

These proteins have a pronounced tendency to aggregate and if one attempts electrophoretic separations using non-dissociating conditions many of the proteins remain as an aggregate at the top of the gel. Usually the inclusion of neither SDS nor urea alone is sufficient to ensure complete disaggregation of the proteins, rather both must be present in the gel and the sample solution. In practice it has been found that optimal resolution can be obtained using discontinuous SDS-PAGE as described in *Table 2* with the modification that urea is added to the gel to a final concentration of 4 M by the addition of urea to the appropriate solution (197). For most complex mixtures a 12−15% polyacrylamide gel has proved to be suitable.

When preparing samples for electrophoresis, all procedures involving precipitation should be avoided. Samples should ideally be concentrated by lyophilization after dialysis against 0.1% (w/v) SDS. Alternatively, concentration of the proteins in SDS or urea solutions can be carried out using polyethylene glycol or Sephadex G-200 (Section 6.6.1). The protein sample is applied to the gel in 1% (w/v) SDS, 5% (v/v) 2-mercaptoethanol in 8 M urea. Usually, because of the possible problem of cyanate ions in the urea, it is advisable to add 5 mM Tris−HCl (pH 6.8). Similarly, heating of the sample should be avoided if at all possible since the heating of samples greatly enhances the rate of formation of cyanate ions. The presence of 8 M urea is sufficient to increase the density of the sample to ensure that there is no difficulty in applying the sample to the gel without the addition of sucrose or glycerol.

The gels can be stained with Coomassie Blue as described elsewhere (Section 7.2.1). After destaining, most non-histone fractions are revealed to be extremely complex mixtures. For this reason two-dimensional gel electrophoresis gives much better estimates of the complexity of such samples and has now been accepted as the preferred analytical method (see Chapter 3).

Nucleosomes can also be analysed directly, either in one-dimensional or two-dimensional gels (198).

9.2.3 Ribosomal proteins

Ribosomal proteins are basic proteins which are not readily solubilized in non-dissociating buffers. They can be successfully analysed on a size basis using the SDS−phosphate buffer system with a 10% acrylamide gel (e.g. ref. 199) The SDS-discontinuous buffer system also may be used and gives good resolution especially with a 6−18% linear gradient gel. In each case the ribosomes or ribosomal subunits can be dissolved directly in the SDS-containing sample buffer without removing the RNA.

A combined size and charge separation of ribosomal proteins requires the presence of urea to maintain solubility. The separation can be carried out in a continuous buffer system (e.g. 0.9 M acetic acid, 6.0 M urea; ref. 200) or in the low pH discontinuous system of Reisfeld *et al.* (ref. 19; *Table 4*) modified to include 8 M urea (201). However, although SDS-PAGE is still widely used for one-dimensional electrophoretic analysis of ribosomal proteins, most detailed studies of individual ribosomal proteins which include a charge separation now use two-dimensional polyacrylamide gel electrophoresis (see Chapter 3).

9.2.4 Membrane proteins

Peripheral membrane proteins, which do not intercalate into the hydrophobic domain of the bilayer, usually possess the properties of water soluble proteins when removed from the membrane by pH changes, chelating agents or by low or high salt washes. They can therefore generally be fractionated by polyacrylamide gel electrophoresis using standard methods. In contrast, integral membrane proteins which are deeply integrated into the hydrophobic milieu are more difficult to analyse. The interested reader is referred to a companion volume in this series (202) for a detailed consideration of all aspects of this complex field. Here, only a few observations will be made to indicate the variety of polyacrylamide gel fractionation systems which can be applied to membrane proteins.

SDS, when used in excess, solubilizes all, or almost all, membrane protein components of prokaryotic and eukaryotic cells. Hence SDS-PAGE is now the most widely used method for investigating the complexity of membrane protein mixtures and polypeptide molecular mass determination. The protocols for SDS-PAGE of membranes are the same as described earlier for other proteins, although 8 M urea can be included in the sample buffer to ensure membrane protein denaturation (e.g. ref. 203). In addition, plasma membrane proteins exposed on the cell surface can be identified by labelling their tyrosine residues with ^{125}I using lactoperoxidase (204) followed by SDS-PAGE of whole cell extracts then autoradiography. Several other reagents for radiolabelling cell-surface proteins are also available (205).

The one drawback is that SDS extraction of membranes usually inactivates the solubilized protein although an increasing number of proteins appear to be stable in its presence (e.g. ref. 206). In contrast, extraction of membranes with the non-denaturing detergent Triton X-100 instead of SDS often preserves the protein's biological activity although the disadvantage here is that not all membrane proteins are extractable with this detergent. After extraction, the protein is fractionated using the SDS-discontinuous buffer protocol described earlier (Section 6) but replacing 10% (w/v) SDS in the recipes with 10% (v/v) Triton X-100. Further details of membrane protein preparation and electrophoresis using Triton X-100, plus assays for a number of membrane marker enzymes, are given by Dewald *et al.* (207). Other detergents such as sodium deoxycholate (207), Lubrol (208), and Sarkosyl (208) have also been used in combination with polyacrylamide gel electrophoresis for analysis of membrane proteins. Hjelmeland and Chrambach (209) and Findlay (210) have reviewed strategies for the solubilization of membrane proteins using the large range of detergents now available.

Other electrophoretic systems exist which aim to solubilize all membrane proteins but, unlike SDS-PAGE, allow fractionation on a charge and size basis. A common early method involved solubilization with phenol:acetic acid:urea followed by electrophoresis in acid gels containing 5 M urea and 35% acetic acid (211). Chloral hydrate will also solubilize membrane proteins and allow a charge and size fractionation by subsequent electrophoresis in polyacrylamide gels containing this reagent (212). These methods are little used at the present time since solubilization and/or disaggregation can be poor and chloral hydrate is toxic. One can also carry out 'charge-shift electrophoresis' of membrane proteins in polyacrylamide gels. The basis of the method is that integral membrane proteins possess hydrophobic domains which anchor them to the membrane lipid whilst peripheral membrane proteins or soluble proteins do not. Thus hydrophilic proteins bind little or no Triton X-100 whereas membrane proteins possessing hydrophobic domains bind large amounts. Integral membrane proteins therefore show altered mobility in mixtures of Triton X-100 and ionic detergent (when they form charged Triton X-100−ionic detergent−protein complexes) compared to Triton X-100 alone, whereas the mobility of hydrophilic proteins is unaffected. The original work of Helenius and Simons (213) utilized agarose gel electrophoresis with the anionic detergent sodium deoxycholate and the cationic detergent cetyltrimethylammonium bromide (CTAB). However, the technique can also be applied to polyacrylamide gel electrophoresis. Thus Bordier *et al.* (214) found that by replacing the SDS in the SDS-discontinuous buffer system with 0.5% Triton X-100 plus 0.25% sodium deoxycholate, the mobility of four membrane protein complexes was greatly increased compared to the situation when Triton X-100 was the only detergent present, showing that the membrane proteins contain hydrophobic domains and so are probably integral membrane proteins. The magnitude of the shift in mobility should be a measure of the size of the hydrophobic domain in the membrane protein.

Finally, recent studies have experimented with the use of N-acroyloyl-morpholine (ACM), a hydrophobic analogue of acrylamide, as a partial or complete substitute for acrylamide in polyacrylamide gels (215, 216). Gels containing various proportions of ACM can be used successfully with a number of organic solvents for the fractionation of hydrophobic proteins.

9.3 Concentration gradient gels

9.3.1 Uses of concentration gradient gels

Polyacrylamide gels which have a gradient of increasing acrylamide concentration (and hence decreasing pore size) with increasing migration distance are now being extensively used instead of single concentration gels, both for analysis of the protein composition of samples and molecular mass estimation using SDS as the dissociating agent. Although step gradients (in which gels of different concentration are layered one upon the other) were used in early work, they can give artefactual multicomponent bands at the interface between two layers. It is now usual to use continuous acrylamide gradients. The usual limits are $3-30\%T$ in linear or concave gradients with the particular range chosen depending upon the size of the proteins to be fractionated. During electrophoresis in gradient gels, proteins migrate until the decreasing pore size impedes further progress (217). Once this 'pore limit' is approached the protein banding pattern does not change appreciably with time although migration does not cease completely. If electrophoresis is unduly prolonged, proteins with molecular masses below about $60\ 000-80\ 000$ eventually migrate off these gels.

One of the main advantages of gradient gel electrophoresis is that the migrating proteins are continually entering areas of gel with decreasing pore size such that the advancing edge of the migrating protein zone is retarded more than the trailing edge, resulting in a marked sharpening of the protein bands. In addition, the gradient in pore size increases the range of molecular masses which can be fractionated

116

simultaneously on one gel. Therefore a gradient slab gel used for SDS-PAGE will not only fractionate a complex protein mixture into sharper bands than is usually possible with a single concentration gel, but also allows the molecular mass estimation of almost all the components irrespective of their size. Non-dissociating buffers have also been used for analysis of native proteins on gradient gels but less frequently (see below).

9.3.2 Molecular mass estimation using gradient gels

i. SDS-PAGE (SDS-denatured proteins)

With uniform concentration polyacrylamide gels, there is a linear relationship between $\log_{10}$ molecular mass and R_f value or distance migrated by the SDS-polypeptide complex (Section 4.7.2). However, with linear concentration gradient gels, the linear relationship is between $\log_{10}$ molecular mass and $\log_{10}$ polyacrylamide concentration ($\%T$) (218−220). This relationship would be expected to hold even in the case of non-linear gradients, but the calculation of $\%T$ is facilitated by using linear gradients. The procedure is to load the unknown sample proteins and a suitable set of calibration proteins onto a single linear gradient slab gel, and then carry out electrophoresis until the tracking dye reaches the last centimetre or so of gel. After staining, the position of the protein bands is measured and $\%T$ calculated for each. For example, a polypeptide which has migrated halfway through a 5−20% polyacrylamide gel will have reached a $\%T$ of 12.5%. A plot of $\log_{10}$ molecular mass versus $\log\%T$ for the calibration proteins then yields a standard curve from which the molecular mass of the sample proteins can be determined. It is essential that molecular mass marker proteins should be included with each gel run and a standard curve drawn for that particular gel.

The range of polypeptide molecular masses over which there is a linear relationship between $\log_{10}$ molecular mass and $\log\%T$ depends on the polyacrylamide gradient conditions chosen. Examples are 7−25%T (C_{Bis} = 1%), M_r approximately 14 300−330 000 (91); 5−20%T (C_{Bis} = 2.6%), M_r approximately 14 300−210 000 (*Figure 25*); 3−30%T (C_{Bis} = 8.4%), M_r approximately 13 000−950 000 (219). Each of these represents a considerable improvement over the very restricted range available for molecular mass determination using any one uniform concentration polyacrylamide gel (Section 4.7.2). Furthermore, there is some evidence (219) that glycoproteins do not behave anomalously during SDS-PAGE using gradient gels, unlike their behaviour in single concentration gels (Section 4.7.2), possibly because the gradient of pore size within gradient gels causes molecular sieving to predominate over the anomalous charge effect caused by reduced binding of SDS by glycoproteins. Gradient gels may therefore become the gel medium of choice for glycoprotein molecular mass determination.

Rothe and Purkhanbaba (221) have also shown that the molecular masses of proteins separated in linear SDS polyacrylamide gradient gels can be estimated from plots of $\log_{10}$ molecular mass (M) versus the square root of the migration distance ($\sqrt{D}$). The straight line produced is described by the equation

$$\log_{10} M = -a\sqrt{D} + b$$

117

where a is the slope and b is the intercept. A straight line is obtained irrespective of the buffer system, the concentration of the crosslinker within the range of $1-8\%C$, and the concentration range of the gradient within $3-30\%T$, at least over a gel length of $8-15$ cm, although the slope and intercept values do vary with these parameters. This relationship can be applied to reduced, SDS–polypeptide complexes, non-reduced SDS–polypeptide complexes, glycoproteins and non-glycoproteins. Furthermore the relationship is time-independent so that electrophoresis can be stopped at the most appropriate time to gain maximal resolution of the protein bands of interest rather than waiting until protein migration ceases.

It is important to note that even gradient gels in the $3-30\%T$ range cannot resolve polypeptides smaller than about 13 000 M_r using the standard SDS buffer conditions. However, small polypeptides down to 1500 M_r can be separated using $10-18\%$ gradient gels if 7 M urea is included in the buffer (185).

Despite the above advantages of gradient gels it is important to note that these cannot match the resolution of two protein components obtainable with gels of a uniform, optimal polyacrylamide concentration. The approach should therefore be to use a gradient gel for SDS-PAGE initially to determine the complexity of a protein mixture and to obtain an estimate of the molecular mass of the components. If the

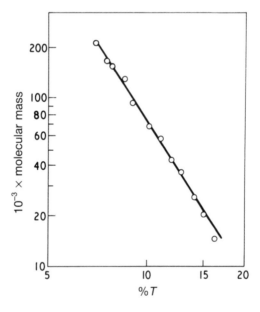

Figure 25. Calibration curve of $\log_{10}$ polypeptide molecular mass versus $\log_{10}$ %T for a 5–20% linear gradient slab gel. The marker polypeptides, in order of decreasing molecular mass, were myosin (M_r 212 000), RNA polymerase β' (165 000) and β (155 000) subunits, β-galactosidase (130 000), phosphorylase a (92 500), bovine serum albumin (68 000), catalase (57 500), ovalbumin (43 000), glyceraldehyde-3-phosphate dehydrogenase (36 000), chymotrypsinogen A (25 750), soybean trypsin inhibitor (20 100), and lysozyme (14 300).

polypeptides are well resolved this is all that may be required. However, if all the proteins of interest fall into a narrow molecular mass range, SDS-PAGE should then be carried out with an appropriate uniform concentration gel to obtain optimal resolution of the components.

ii. Native proteins

As with SDS-PAGE, the fractionation of native proteins in polyacrylamide gradient gels enables a much larger range of protein sizes to be fractionated than gels of uniform concentration, and the protein bands are sharpened by the gel gradient. In gradient gels, native proteins migrate as in uniform concentration gels; that is, according to both their size and charge. Several investigators have used polyacrylamide gradient gels to determine the molecular size of native proteins based on the distance migrated, but other workers find that this is only accurate when the proteins under study are monomers and their homologous oligomers (218) and suggest that the molecular mass of a native protein electrophoresed in a linear polyacrylamide gradient gel is best estimated by measuring the rate of its migration through the gel (219). The interested reader is referred to an article by Rothe and Maurer (222) who have described the determination of the molecular mass of native proteins in gradient gels at some length. However, as with uniform concentration gels, molecular mass estimation for native proteins using gradient gels is little used compared to polypeptide molecular mass estimation using buffers containing SDS with these gels.

9.3.3 Preparation of concentration gradient gels

Both linear and concave gradient gels can easily be produced in either slab or rod formats in the laboratory, although gradient slab gels are far more widely used than the corresponding rod gels. Multiple rod or slab gradient gel formation requires the use of a purpose-built gel forming tower (Section 9.6). However, gradient slab gels are often only required one or two at a time and so, although multiple gels may be made using a tower and stored, it is often convenient to produce them singly. This is easily done provided that a peristaltic pump and a suitable gradient maker are available (see below).

In addition to a gradient in acrylamide concentration, a density gradient of sucrose or glycerol is often included to minimize mixing by convective disturbances caused by the heat evolved during polymerization. Some workers avoid the latter problem by including a gradient of polymerization catalyst to ensure that polymerization occurs first at the top of the gel (low acrylamide concentration) progressing to the bottom. Gradient gels produced to a high standard of reproducibility are now also available commercially (e.g. Pharmacia).

i. Linear gradient slab gels

Linear gradient makers are available commercially or can be constructed in the laboratory workshop from Perspex. The two chambers must be of exactly equal cross-section (or it will not be possible to produce linear gradients) and are joined by an inter-connecting tunnel controlled by a two-way tap. A typical set-up for producing linear gradient slab gels using this apparatus is shown in *Figure 26*. The outlet from the gradient maker is connected via fine-bore Tygon tubing to a peristaltic pump and thence to the glass-plate sandwich which has been previously assembled as described in Section 6.5. The Tygon tubing is cut at an angle of 45° and the cut edge taped in position facing the rear (un-notched) plate at the centre of the assembly (*Figure 26*). According to this procedure, the most concentrated acrylamide solution enters the glass-plate sandwich first and runs down the inside of the un-notched glass plate to reach the gel bottom. As the level of acrylamide mixture rises in the sandwich, the acrylamide concentration steadily decreases. Surprisingly little mixing occurs if this is carried out carefully. Some commercial apparatuses and some homemade devices provide for a gradient of acrylamide to be introduced at the base of a single slab gel holder. In this case, the least concentrated acrylamide mixture enters first.

The exact composition of the acrylamide mixtures used for gradient gel preparation will depend on the concentration range of the gradient and the buffer system used. Most gradient gels are used for SDS-PAGE with the SDS-discontinuous buffer system (*Table 2*). Two acrylamide mixtures, corresponding to the lowest and highest concentrations in the gradient, are prepared according to the details in this table but with the amount of polymerization catalyst reduced, to allow time to pour the gradient before polymerization occurs, and with the highest acrylamide concentration mixture containing 15% (w/v) sucrose (4.5 g per 30 ml gel mixture) to give a stabilizing density gradient. Thus for a 5−20% linear gradient gel (2.6% bisacrylamide) using the SDS-discontinuous buffer system, the mixtures are as shown in *Table 8*.

The gel mixtures are degassed and TEMED (10 μl per 30 ml gel mixture) is then

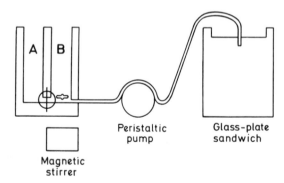

Peristaltic pump

Glass-plate sandwich

Magnetic stirrer

Figure 26. Apparatus for the formation of gradient polyacrylamide slab gels. A and B refer to the reservoir and mixing chambers of the gradient maker, respectively, which is supported about 1 cm above the magnetic stirrer using a clamp stand. Mixing chamber B contains a magnetic follower and is connected to reservoir A by a tunnel controlled by a two-way tap. The Tygon tubing used to connect the gradient maker via the peristaltic pump to the glass-plate sandwich is 0.075 cm or 0.15 cm i.d. Most simply, the Tygon tubing itself is cut at a 45° angle and then taped to the centre of the rear (unnotched) plate with the cut edge facing this plate, the tubing being pushed a centimetre or so between the plates to 'clamp' the cut end in position. If the tubing o.d. is too large to allow this, a suitably cut piece of fine-bore Teflon tubing or syringe needle inserted into the end of the Tygon tubing will suffice instead.

Table 8. Gel mixtures for a 5 – 20% gradient gel

5% acrylamide mixture[a]
      5.0 ml of acrylamide – bisacrylamide (30:0.8)
      3.75 ml of resolving gel buffer stock; 3.0 M Tris – HCl (pH 8.8)
      3.0 ml of 10% SDS
      0.7 ml of 1.5% ammonium persulphate
   20.25 ml of water

20% acrylamide mixture[a]
      20.0 ml of acrylamide – bisacrylamide (30:0.8)
      3.75 ml of resolving gel buffer stock; 3.0 M Tris – HCl (pH 8.8)
      0.3 ml of 10% SDS
      0.7 ml of 1.5% ammonium persulphate
      4.5 g of sucrose (equivalent to 2.5 ml volume)
      2.75 ml of water

[a] Add 10 μl of TEMED to initiate polymerization just before pouring the gel (see the text).

---

added to each. The low concentration mixture is added to reservoir A (*Figure 26*) and the connecting tube between the chambers of the gradient maker opened to fill it and then closed. Any gel solution which flowed into mixing chamber B (*Figure 26*) is returned to reservoir A. An equal volume of high concentration mixture is now added to chamber B. Each volume is calculated to correspond to half the volume of the final resolving gel. For the standard notched plate assembly (*Figure 11*), this requires 15 ml of acrylamide mixture per chamber. Chamber B can be mixed using a magnetic stirrer in which case care must be taken to prevent heat from the magnetic stirrer causing premature polymerization. To avoid this, the gradient maker should be supported (using a clamp) about 1 cm above the stirrer. The chambers of the gradient maker are connected and the peristaltic pump and stirrer turned on. The flow rate of the gel mixture into the glass-plate sandwich should be about 3.0 ml/min. Following the quantitative delivery of the gel mixture into the sandwich, the outlet tubing from the gradient former is dipped into a flask of buffer with the same composition as in the resolving gel (3.75 ml of 3.0 M Tris–HCl, pH 8.8, 0.30 ml of 10% w/v SDS, 25.95 ml of water) and the flow rate reduced to 0.5 ml/min. This serves to overlayer the resolving gel. Immediately the gradient has been poured, the gradient maker is flushed out with water to prevent acrylamide polymerizing in the apparatus. After polymerization of the slab gel, the overlay is removed by tilting the gel and a stacking gel is polymerized in place, complete with sample wells, as described previously (Section 6.5.2).

Some workers prefer to use riboflavin as the polymerization catalyst instead of ammonium persulphate when pouring gradient gels, since it allows more time for manipulations without the fear of premature gel polymerization. In this case, riboflavin at a final concentration of 0.5 mg/ml of gel mixture replaces the ammonium persulphate and the gradient gel is polymerized, after pouring, by exposure to a fluorescent light for about 30 min. Then the stacking gel may be added. In practice, the sucrose gradient is sufficient to stabilize the acrylamide gradient during polymerization and so a gradient of polymerization catalysts to ensure that the slab gel polymerizes at the top first is unnecessary. However, where convective mixing is a problem, the catalyst concentrations should be adjusted so as to cause the low concentration acrylamide mixture to gel in about 25−30 min and the high concentration mixture in about 40−45 min.

The preparation of urea−SDS-polyacrylamide gradient gels used for fractionating small polypeptides (185) is essentially the same as for standard SDS-polyacrylamide gradient gels. The gel and buffer solutions required are given in *Table 9*.

ii. Concave gradient slab gels

Concave gradient polyacrylamide slab gels can be produced using the same gradient maker used for linear gradient gels. For a concave '5−20%' gradient gel of 30 ml total volume, for example, 7.5 ml of 20% acrylamide mixture is placed in the mixing chamber B and this is then stoppered with a rubber bung. The pressure inside this chamber is equalized with atmospheric pressure by momentary insertion of a hypodermic syringe needle. Reservoir A receives 22.5 ml of 5% acrylamide mixture. The slab gel is then poured as for linear gradient gels (see above). During pouring,

---

Table 9. Gradient SDS-PAGE for the separation of low molecular mass polypeptides[a]

Stacking gel
5% acrylamide, 0.13% bisacrylamide
0.067 M Tris−HCl (pH 6.8)
0.1% SDS
0.067% ammonium persulphate
0.067% TEMED

Resolving gel
10−18% linear acrylamide gradient, 0.5−0.9% bisacrylamide
0.45 M Tris−HCl (pH 8.8)
0.1% SDS
7 M urea
0.05% ammonium persulphate
0.05% TEMED
0−10% linear gradient sucrose

Reservoir buffer
0.05 M Tris, 0.38 M glycine (pH 8.5)
0.1% SDS

Sample buffer
10% sucrose
0.0625 M Tris−HCl (pH 6.8)
2% SDS
10 mM DTT (or 1% 2-mercaptoethanol)
0.0025% Bromophenol Blue

Samples are heated for 2 min at 100°C and then electrophoresed at 120 V for about 15 h (slab gel 0.1 cm thick, 14 cm high, 15 cm wide)

[a] From Hashimoto *et al.* (185).

---

the volume of chamber B remains constant and is continuously diluted by the incoming solution from reservoir A. This generates an exponentially decreasing gradient of acrylamide and yields a gel with the concave profile shown in *Figure 27*, ranging from 20% at the gel bottom to about 6% at the gel top. A stacking gel may be polymerized in place if required.

9.3.4 Sample preparation and electrophoresis

For standard SDS-polyacrylamide gradient gels, sample preparation is the same as for uniform concentration gels. Electrophoresis conditions will depend on the buffer system and gel concentrations comprising the gradient, but for the 5−20% linear gradient gel described above using the SDS-discontinuous buffer system, electrophoresis is at 25 mA constant current for about 5 h, or 90 V constant voltage

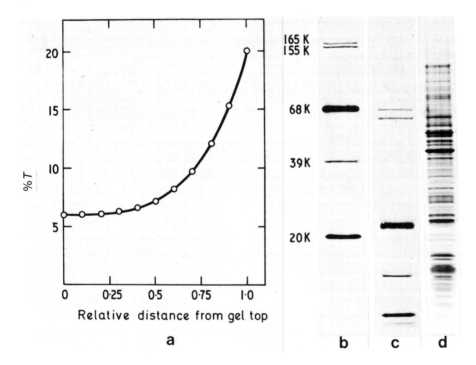

Figure 27. Use of gradient polyacrylamide gels. (a) A profile of a '5 – 20%' concave ('exponential') gradient gel prepared as described in the text. (b) and (c) Polypeptide band distribution after fractionation of RNA polymerase α, β, and β' subunits, bovine serum albumin, and soybean trypsin inhibitor on a linear 5 – 20% and concave '5 – 20%' polyacrylamide gradient slab gel, respectively, using the SDS-discontinuous buffer system. The molecular masses of the polypeptides are shown to the side of the gel tracks. (d) An example of a complex polypeptide mixture analysed using a linear gradient polyacrylamide slab gel; the sample was a subfraction of [35S]-methionine-labelled proteins from *Dictyostelium discoideum* spores analysed on a 6 – 18% linear gradient polyacrylamide slab gel using the SDS-discontinuous buffer system. For other examples of complex polypeptide fractionation using gradient slab gels see *Figures 2* and *14*.

overnight, or until the tracking dye nears the gel bottom. Analysis of gels after electrophoresis is the same as with uniform concentration polyacrylamide gels. The protein bands visible after staining are equally sharp with either linear or concave gradient gels (*Figure 27b* and *27c*, respectively) but the band distribution depends on the type of gradient chosen, concave gradient gels tending to separate polypeptides of high molecular mass better than linear gradient gels but at the expense of reducing band separation between lower molecular mass polypeptides. In the author's laboratory, linear gradient slab gels are preferred for one-dimensional SDS-PAGE of complex mixtures of polypeptides covering a wide range of sizes (e.g. *Figure 27d*) since this gradient profile is better suited than concave gradient gels to polypeptide molecular mass estimation.

Sample preparation for the urea−SDS-polyacrylamide gradient gels (*Table 9*) of Hashimoto *et al.* (185) uses the same buffer as for standard discontinuous SDS-PAGE (Section 6.6.1), namely 0.0625 M Tris−HCl (pH 6.8), 2% SDS, 5% 2-mercapto-ethanol (on 10 mM dithiothreitol), 10% sucrose (or glycerol), 0.002% Bromophenol Blue. Electrophoresis is typically at 120 V for at least 15 h.

9.4 Transverse gradient gel electrophoresis

The vast majority of concentration gradient gels are used with only the polyacrylamide concentration, and hence the pore size, varying in the direction of electrophoretic migration as described in Section 9.3. Transverse gradient gel electrophoresis was originally conceived as applying a protein sample to a transverse gel gradient, that is with the slab gel rotated through 90° to its usual orientation (223). At different distances along the gel, left to right, the protein therefore experiences different gel concentrations during migration and migrates in a smooth continuous arc (*Figure 28*). Using this format, Ferguson plots can be constructed from the migration pattern of the protein in a single gel (223, 272) but are reported to be not as accurate as those obtained from a series of uniform concentration gels (224). Nevertheless, even when used qualitatively the method can be very useful to estimate the gel concentration which gives the best separation of complex protein mixtures for subsequent analyses.

Variations of transverse gradient electrophoresis are also being increasingly used where the variable component is not the gel concentration but some other parameter of interest. For example, Creighton (225) used slab gels with a 0−8 M transverse urea gradient to follow denaturation of various proteins as a function of urea concentration (*Figure 28*). Thus urea gradient gel electrophoresis can be used to give information about both the thermodynamics and kinetics of the unfolding process. Feinstein and Mondrianakis (226) also used a transverse urea gradient, this time in the presence of SDS, to examine mitochondrial protein subunit structure. Fernandes *et al.* (227) used both urea and Triton gradients in this buffer system for studies of membrane proteins. The versatility of transverse gradient electrophoresis and the wealth of information gained, at least at the qualitative level, suggests that this approach will become more popular in future. A detailed description of the procedures involved and some applications of the method are given in refs. 228 and 272.

9.5 Micro- and mini-gels

A variety of smaller format gel systems have been developed. The extreme situation is the formation of polyacrylamide gels in capillary tubes which can analyse the amounts of protein found in single cells. Both single concentration and gradient polyacrylamide gels may be produced in the 1 μl to 100 μl volume range using appropriately sized capillaries. Using 5 μl volume capillary gels Neuhoff (229) has detected 1 ng of albumin after staining. Micro-slab gels have also been used for protein analysis but here the highest sensitivity obtained has been 20 ng for a single protein.

9.5.1 Rod gels

The detailed methodology for producing and analysing capillary gels, both uniform concentration and gradient gels, has been described by Neuhoff (229). Basically, for uniform concentration gels, glass capillaries are filled to about two-thirds of their total volume by being dipped in the gel mixture. They are then pressed into a plasticine cushion, about 2 mm thick, to seal the capillary bottom, and overlaid with water using a fine Pyrex glass capillary pipette of smaller diameter than the capillary used to hold the gel. After polymerization, the overlay is removed and the protein solution (0.1 − 1.0 μl of 1 − 3 mg/ml) is applied by capillary pipette, any free space in the capillary then being filled with electrode buffer. After filling, that length of capillary penetrated by the plasticine is snapped off and the gel subjected to electrophoresis. Grossbach (230) avoids the use of plasticine as a gel sealer and by using a micro-manipulator with 50 μl capillary gels is able to polymerize both a resolving gel and a stacking gel in place prior to sample addition. Gradient gels may be produced in capillary tubing using either capillarity (231) or a gradient maker (232).

Despite these methods, capillary rod gels have been little used in recent years, possibly because some experience is needed to obtain good gels (particularly in water overlayering of individual gels) and the resolution of protein components is poor, at least in the lower volume capillary gels, compared to the standard rod gel. Furthermore, quantitation of stained protein bands can only be achieved using a microdensitometer rather than the more generally available scanning spectro-photometer. However, Condeelis (233) has described the use of intermediate-size gels made in 100 μl and 250 μl volume Drummond microcaps which can analyse proteins in the nanogram range (*Figure 29*) and yet are almost as easy to prepare and handle as standard size rod gels. I shall refer to these as mini-rod gels. Any of the continuous and discontinuous gel buffer systems already described in this chapter could be used but Condeelis (233) found no increase in resolution when us-ing a discontinuous buffer system with gels of this size. Details of the protein loading capacity and sensitivity of these mini gels are given in *Table 10*. Perhaps the most useful will be the 250 μl size which can be used to detect 2 ng of a single protein when stained with Coomassie Blue R-250 and yet protein bands can be quantitated using a scanning spectrophotometer as used with standard-size gels.

9.5.2 Slab gels

A number of micro-slab polyacrylamide gel apparatuses have been described, the

One-dimensional polyacrylamide gel electrophoresis

Staphylococcal nuclease
pH 8·0

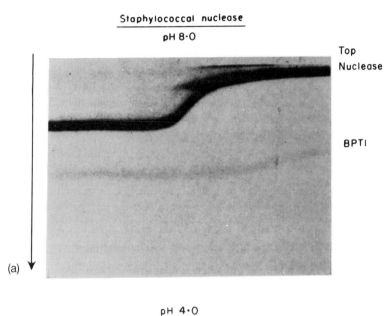

(a)

pH 4·0

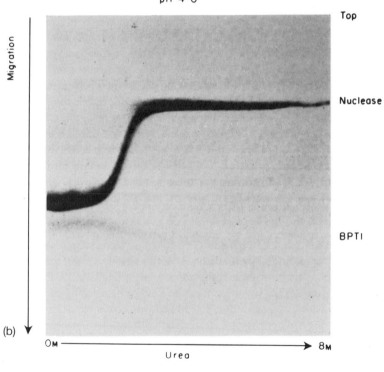

(b)

Migration

0ᴍ ⟶ 8ᴍ

Urea

126

smallest format being glass plates cut from microscope slides e.g. 7.5 cm × 2.5 cm (234), 7.6 cm × 3.8 cm (235). These are able to detect 20 ng of protein in a sharp band stained with Coomassie Blue. However, such small gel sizes cause serious problems of usage. With electrophoretic migration of samples along the long axis of these gels (234) only three samples can be co-electrophoresed, whilst electrophoretic migration along the shorter axis is too restricted for the separation of components comparable to that of the standard gel format. Not surprisingly therefore, micro-slab gels have been little used. Matsudaira and Burgess (236) reported an intermediate-size slab gel format (8.2 cm × 9.2 cm) which enables the analysis of 21 samples with good resolution yet with a sensitivity such that 20 ng of a protein can be detected, although 100 ng is required for a strongly-stained band. The time taken for a complete analysis with such mini-slab gels is only 2 h.

Commercial mini-slab gel apparatus with similar dimensions to that of Matsudaira and Burgess (236) is available either in a format resembling the Studier type of design (e.g. LKB, Hoeffer Scientific) or as a miniaturized version of standard commercial cooled apparatus (e.g. Bio-Rad; *Figure 30*). The main advantages of this small format are considerably shorter electrophoresis and processing times and cost savings because of the reduced use of gel and buffer reagents. The extra apparatus required for applications such as casting multiple gels or electroblotting using mini-slab gels is also available commercially. As the ultimate in integration, Pharmacia market the Phast System^{TM} which allows the loading, electrophoresis, staining, and destaining of pre-packaged gradient mini-slab gels using a single multifunctional piece of equipment—all without the need for aqueous buffers!

9.6 Large numbers of gels

9.6.1 Slab gels

Provided a multiple channel peristaltic pump and a gradient former of sufficient volume are available, it is possible to produce several gradient gels simultaneously. Larger numbers of gradient slab gels can be produced using a purpose-built gel tower built cheaply in the laboratory from Perspex. Essentially it consists of a Perspex box to hold the slab gel assemblies, with the base of the box tapering down to form a funnel. The gel mixture enters the apparatus at the base of the funnel, which serves to decelerate the flow of liquid and to expand the horizontal cross-section of the liquid equal to that of the bottom of the gel holders. Details of gel tower construction and use are given elsewhere (217, 236, 237). The greatest practical problem in using a gel tower to produce gradient slab gels is to avoid convection currents during the

Figure 28. Transverse urea gradient electrophoresis of Staphylococcal nuclease using linear gradients of 0 – 8 M urea (a) at pH 8.0 and (b) at pH 4.0. Bovine pancreatic trypsin inhibitor (BPTI) was added to each sample as internal standard. Urea is seen to induce unfolding of Staphylococcal nuclease at very low urea concentrations at pH 4.0 but higher urea concentrations are needed to unfold the protein at pH 8.0. In each case, the protein bands are continuous indicating rapid equilibration of the folded and unfolded forms of the protein but the presence of 'spurs' may reflect structural or conformational heterogeneity of the protein used. (From ref. 225, with permission.)

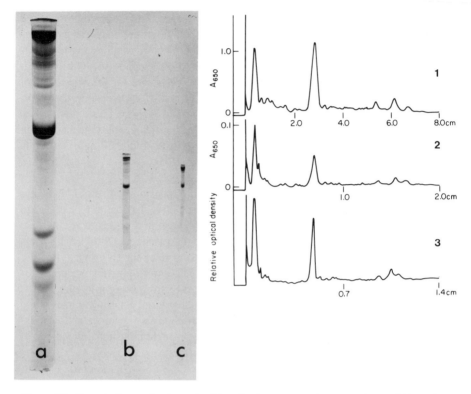

Figure 29. Use of micro-rod polyacrylamide gels. An actomyosin extract from rabbit skeletal muscle was electrophoresed on (**a**) a standard 0.5 cm diameter rod gel loaded with 20 μg total protein, (**b**) a 0.15 cm micro-rod gel loaded with 0.5 μg total protein, (**c**) a 0.07 cm micro-rod gel loaded with 0.1 μg total protein. The gels were electrophoresed using the SDS-discontinuous buffer system and then stained with Coomassie Blue. **1**, **2**, and **3** show densitometer scans of three gels (**a**), (**b**), and (**c**), respectively. (Reproduced from ref. 233 with permission.)

Table 10. Properties of mini-rod gels[a]

Gel format	Maximum loading of protein (μg)	Minimum detectable per band (μg)	Sample volume (μl)
Gel in 250 μl Drummond microcap	2.0	2×10^{-3}	0.25 – 2.0 (0.5 optimal)
Gel in 100 μl Drummond microcap	0.5	3×10^{-4}	0.1 – 0.5 (0.2 optimal)

[a] Data from ref. 233 with permission.

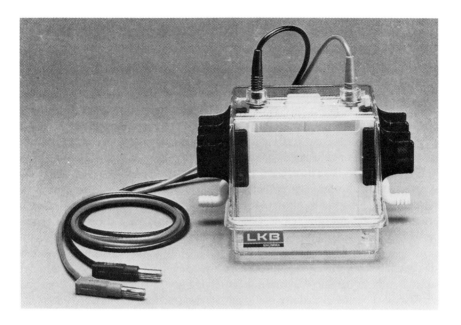

Figure 30. A typical mini-slab gel apparatus. The model shown is the Midget vertical slab gel unit from Pharmacia-LKB (LKB 2050-001). It can run two small (10 cm × 8 cm) vertical gels simultaneously using a notched plate system with facilities for cooling. Run times as short as 45 min are possible. Several other designs are also commercially available including a mini-version of the Protean gel apparatus (*Figure 12*) from Bio-Rad Laboratories, the mini-Protean II. This incorporates a buffer apparatus in which the gel is immersed during electrophoresis, thus providing some cooling and reducing band distortion.

exothermic polymerization reaction of such a large volume of acrylamide. For this reason, it is essential that the polymerization catalyst concentrations are arranged so as to cause polymerization to occur first at the top and then to proceed down the gel.

Several major suppliers of electrophoresis equipment (e.g. Hoefer Scientific, Bio-Rad, LKB) also market apparatus for casting multiple slab gels but, in each case, the apparatus is obviously suited to the preparation of slab gels of the size used by that supplier's lab gel electrophoretic apparatus.

Standard commercial cooled vertical slab gel apparatus (e.g. see *Figure 12*) can typically run up to two or four slab gels simultaneously but modified apparatus is also available for running larger numbers of gels at the same time (e.g. Hoefer Scientific).

9.6.2 Rod gels

In addition to producing gradient slab gels, a gel tower can be used to prepare large numbers of uniform concentration slab gels or rod gels. However, apparatus designed specifically to cast multiple rod gels (either gradient gels or uniform concentration gels) is also commercially available or can be readily constructed in most laboratory workshops (e.g. ref. 238). Some apparatuses are capable of preparing several hundred gradient rod gels simultaneously (e.g. ref. 239).

9.7 Agarose/acrylamide composite gels

Polyacrylamide gels cannot be prepared at a total monomer concentration less than about 2.2%. Above this concentration, but below about 3%, gelation does occur but the gels are very difficult to handle, being almost viscous at the lower concentrations. For the vast majority of protein electrophoretic analyses it is unlikely that the investigator will need to use acrylamide gels at this low concentration. However, such gels have been used, with agarose added to 0.5% to give mechanical strength, for fractionation of complex ribonucleoproteins such as spliceosomes, ribosomes, and polyribosomes (240) and some viruses. Particles as large as polyribosomes comprising eight ribosomes enter the gel with mobilities that approximate to the $\log_{10}$ of the particle weight. Full details of gel preparation and use are given by Dahlberg and Grabowski (240). Moulin *et al.* (241) have described a discontinuous buffer system with a 2% polyacrylamide/0.7% agarose stacking gel and a 3% polyacrylamide/0.7% agarose resolving gel. Gradient agarose/acrylamide gels have also been successfully used both for the fractionation of large complexes of native proteins (e.g. 242) and denatured proteins (243).

9.8 Affinity electrophoresis

The principle of affinity electrophoresis is that sample proteins are electrophoresed in a gel containing immobilized ligands (the 'affinity gel') which interact with the protein(s) of interest so retarding the protein(s) compared to their mobility in a control gel. The control gel can either lack immobilized ligands or contain ligands incapable of interacting with the protein(s) of interest. Although the procedure can be applied to a broad range of protein studies, it has yet to be widely used and so will not be discussed at length here. The interested reader is referred to a full discussion elsewhere (244).

10. Homogeneity and identity

10.1 Homogeneity of separated proteins

The detection of only one protein band by zone electrophoresis in polyacrylamide gels under a single set of experimental parameters does not prove homogeneity. Since the use of non-dissociating buffer conditions in polyacrylamide gel electrophoresis causes fractionation on the basis of both size and charge, it is possible to have two proteins which differ in each of these properties but which, under the electrophoresis conditions chosen, migrate at the same rate to yield a single band. Therefore, native proteins should always be analysed by electrophoresis at several pH values and gel concentrations before homogeneity or identity of protein samples is claimed. Identity testing by Ferguson plot analysis is covered in ref. 222. Since SDS-PAGE fractionates only on the basis of molecular size, here again the number of protein bands observed does not necessarily equate with the number of distinct polypeptides in the sample mixture; any of the bands may comprise multiple components of identical molecular mass. Because of these problems, an increasingly common method to check on protein homogeneity is to use two-dimensional polyacrylamide gel electrophoresis (see Chapter 3) where proteins can be separated on the basis of charge in the first dimension followed by a size fractionation according to SDS-PAGE in the second. A single spot after two-dimensional polyacrylamide gel electrophoresis is a good indication of homogeneity, although not absolute proof. One excellent test of homogeneity is N-terminal amino acid sequence analysis (see Section 10.2.3). If multiple sequences are obtained, this points to a mixture of protein components or a single type of protein which exists in several modified (or cleaved) forms. On the other hand, an unambiguous single N-terminal sequence is compelling evidence that the sample contains mainly, or wholly, one specific polypeptide.

10.2 Further characterization of separated proteins

Assuming that the protein band of interest resolved by one-dimensional polyacrylamide gel electrophoresis represents a single pure protein, information about the protein structure and its identity or non-identity with other proteins can be obtained by a number of methods.

10.2.1 Analysis of amino acid composition

The amino acid composition of stained protein bands is readily determined without elution from the gel by acid hydrolysis of only $5-10$ nmol of protein and then loading onto an amino acid analyser (245, 246). Unfortunately, the large amount of ammonia produced by hydrolysis of the polyacrylamide can prevent the determination of basic amino acids and is detrimental to the equipment. Andrews (3) reduced this problem by repeated lyophilization of the acid hydrolysate in 5% Na_2CO_3 before loading onto the analyser. The alternative approach is to elute the protein (Section 8.2) before hydrolysis in which case a gel slice containing only 5 μg of protein may be enough for amino acid analysis. One of the problems still encountered is that polyacrylamide gel often contains impurities which become increasingly important as the sensitivity of amino acid composition studies increases. As would be expected, the degree of interference depends on the volume of gel eluted. Since the background contamination is fairly reproducible, some correction for this can be made by analysis of an equivalent volume of blank gel slices. Currently, electroelution followed by dialysis of about 50 pmol of a protein should allow good analysis of its amino acid composition using modern HPLC methods and detection systems (247).

10.2.2 Peptide mapping

Polypeptides separated by polyacrylamide gel electrophoresis can also be characterized by peptide mapping. Given the small amount of polypeptide usually available, the sensitivity of peptide detection can be increased by elution of polypeptide from the gel followed by radioiodination prior to tryptic digestion (248, 249). An even simpler method has been described by Elder *et al.* (250) whereby polypeptides separated by SDS-PAGE are both radioiodinated and treated with trypsin whilst still in the gel. The tryptic peptides are then easily eluted from the gel and analysed by two-dimensional chromatography followed by autoradiography. Entire multi-component systems can be analysed using only microgram amounts of total protein. The method works on fixed and stained protein bands and even on dried gels after years of storage, and requires only a few days' work. If a protein mixture is too complex for adequate resolution of proteins it is best analysed by two-dimensional polyacrylamide gel electrophoresis (Chapter 3) and then the Elder method of mapping applied. The only disadvantage of this peptide mapping method methodology is that the protein analysed must contain amino acid residues susceptible to some form of radioiodination (see Appendix 3). Another rapid method of peptide mapping has been reported by Cleveland *et al.* (251). It involves partial proteolytic cleavage of proteins separated by polyacrylamide gel electrophoresis, without elution of protein, followed by a second SDS-PAGE separation during which the peptide products separate on a size basis. In principle, the peptide mapping approach need not be limited to peptides generated by proteases; any reliable method of producing peptides by cleavage at specific sites can be used. For example, acid hydrolysis of Asp−Pro bonds, hydroxylamine cleavage of Asn−Gly bonds, *N*-chlorosuccinamide cleavage at tryptophan residues, and cyanogen bromide cleavage at methionine residues have all been used for peptide mapping. However, it is important to avoid methods which generate large numbers of very small peptides since these may be difficult to resolve or even lost from the gel during processing. Peptide mapping in the presence (and absence) of SDS is described in detail in Chapter 5.

10.2.3 Amino acid sequencing

N-terminal amino acid sequences of proteins separated by polyacrylamide gel electrophoresis can be obtained by several protocols all of which rely on the sequential removal of N-terminal amino acids following reaction with phenylisothiocyanate (PITC), namely the Edman degradation. Originally Wiener *et al.* (252) described the elution of proteins into buffer containing SDS followed by a modified dansyl−Edman degradation. Although relatively simple to perform, this technique permits only small numbers of residues to be identified. A considerable improvement is achieved by automated sequencing. The protein is usually electroeluted from the gel and covalently attached to activated glass supports (253) or blotted onto synthetic membranes (254). Following sequencing, the amino acid derivatives are identified by HPLC with sensitivities extending to below 1 pmol. Proteins may also be labelled *in vivo* prior to polyacrylamide gel electrophoresis and sequencing (255). Although theoretically capable of very high sensitivity, this approach is only applicable to the restricted number of proteins where *in vivo* labelling to high specific activity is feasible.

Procedures for isolating microgram quantities of protein from polyacrylamide gels for sequence analysis, even after staining, are given in companion volumes in this series (247, 256).

10.2.4 Renaturation of biological activity

As described earlier, the biological activity of native proteins can be examined and compared after gel electrophoresis in non-dissociating buffers either *in situ* or after elution from the gel (Section 7.9). Increasingly, this is also possible with denatured proteins after separation by SDS-PAGE. The key step is removal of all SDS from the protein whereupon many proteins renature to regain at least some of their biological activity. Thus the use of SDS does not necessarily lead to irreversible denaturation. Many factors affect the ability of proteins to renature after SDS denaturation.

- The purity of the SDS used in the original fractionation is important. Commercial preparations of SDS can vary in their contamination with longer chain alkyl sulphates and may also contain small proportions of other lipophiles such as dodecyl alcohol (DDA), 1-dodecene, didodecyl ether (DDE), and didodecyl sulphate (DDS) which bind more tightly to proteins than SDS, thus inhibiting the renaturation process.
- Several proteases are fairly resistant to denaturation and hence can degrade sample proteins sufficiently during sample preparation to impair biological activity during subsequent renaturation unless steps are taken to avoid this.
- Irrespective of the fractionation and renaturation conditions, renaturation is most effective for proteins where disulphide bonds are not required for activity and which are not composed of heterologous subunits.
- One of the most important aspects is the complete elimination, as far as possible, of all bound and unbound SDS prior to renaturation. Ideally the removal of SDS should be arranged to occur slowly so that the denatured proteins have sufficient time to renature into their correct native conformation (257) but the time required varies depending on the protein.
- Cofactors and co-enzymes which will have been removed during sample preparation and electrophoresis should be included in the renaturation buffer. The presence of high concentrations of substrates, sodium chloride, glycerol, and a thiol-reducing agent such as dithiothreitol also assists the renaturation of many enzymes (270). However, the concentration of thiol reagents must be kept at a low level if the enzyme detection system is based upon the use of phenazine methosulphate (PMS) and tetrazolium salts (259).
- For hydrophobic proteins, the removal of SDS must often be counterbalanced by its replacement with a non-ionic or zwitterionic detergent to maintain protein solubility. Proteins can be renatured after SDS-PAGE either in the gel, after electroelution and immobilization onto a blotting membrane or after elution into free solution. The renaturation of SDS solubilized membrane proteins has been discussed by Hjerten (206).

i. Renaturation in situ

This type of renaturation is clearly possible only for monomeric proteins and those composed of identical subunits since for heteromeric proteins the different-sized subunit polypeptide chains will be located at different places in the gel. Most of the SDS can be removed by fixing and washing with isopropanol:acetic acid:water (5:2:13 by vol.) and then with aqueous buffer (e.g. ref. 260) or by washing with buffered 25% isopropanol (e.g. ref. 261). Alternatively the gel can be simply washed repeatedly in buffer alone or supplemented with glycerol, substrate, thiol reagent, high salt, and/or non-ionic detergents (e.g. 270). SDS has also been removed by continued electrophoresis of the polypeptides in the presence of urea. The two-dimensional system of Manrow and Dottin (258) is particularly noteworthy. The protein mixture is fractionated by SDS-PAGE in the first dimension in a rod gel, and then by electrofocusing in the presence of urea and Nonidet P-40 in a slab gel in the second dimension. The urea and non-ionic detergent in the second dimension gel remove SDS from the polypeptides so that, when the slab gel is then equilibrated with buffer to remove the urea, the polypeptides renature and can be located by assaying for biological activity. Thus, the method combines the high resolution associated with two-dimensional gel electrophoresis with the ability to identify the separated polypeptides on a functional basis.

ii. Renaturation after electroblotting

This will probably become the method of choice for renaturation where the original band pattern achieved by the electrophoretic separation needs to be preserved, at least in the initial stages of study. After electroblotting (Section 7.10), SDS can be readily removed from the immobilized proteins by washing the membrane or by a brief additional electrophoresis in buffer lacking SDS. The proteins can then be assayed for biological activity either while still immobilized or after elution.

iii. Renaturation after elution into free solution

The immediate problem after elution of SDS−polypeptide complexes from SDS-PAGE gels is the complete removal of the SDS. Dialysis alone is usually not sufficiently effective. Weber and Kuter (262) used urea to displace the SDS followed by removal of the SDS using Dowex resin. The urea can then itself be removed by dialysis which allows the protein to renature slowly. An alternative protocol, which is better than the use of urea for the renaturation of at least some proteins, has been described by Hager and Burgess (257). The protein band is visualised in the gel after SDS-PAGE by KCl 'staining', then the band is cut out, crushed, and the polypeptide is eluted by diffusion in a buffer containing 0.1% SDS. The polypeptide is concentrated (and most of the SDS removed) by acetone precipitation. Renaturation of the polypeptide occurs after the precipitate is dissolved in guanidine hydrochloride and then diluted. The percentage of original activity recovered depends on the renaturation protocol chosen and on the particular protein under investigation.

Other procedures for the removal of SDS are ion-pair extraction (263) and the use of ion-retardation resins (264). Using ion-pair extraction, the protein recovery is generally 80% or better even in the sub-microgram range and the removal of SDS is complete.

10.2.5 Immunological characterization

Many detailed characterizations of individual proteins have only been possible by a combination of immunoprecipitation, using antibodies raised against the protein, and SDS-PAGE. Either polyclonal or monoclonal antibodies can be used, the only requirement being that they are sufficiently specific for the protein(s) under study. Antibodies can be raised against proteins separated by gel electrophoresis either after eluting the protein or by macerating entire gel slices and injecting these (Appendix 6). If SDS-PAGE has been used to obtain the polypeptide, it is usually not necessary to remove the SDS prior to injection and the resulting antibodies frequently precipitate both native and denatured forms of the protein from a complex mixture either by direct immunoprecipitation using carrier antigen or indirect immunoprecipitation using, say, monospecific rabbit IgG followed by goat anti-rabbit IgG. Contaminating proteins are then removed by repeated washing with detergents such as Triton X-100 or centrifugation through sucrose-containing detergent.

In practice, the use of a soluble antibody to precipitate antigen–antibody complexes has declined with the introduction of solid-phase immunoadsorbents since these are easier to handle and wash than conventional second antibody complexes. One class of immunoadsorbents is based on *S.aureus* coat protein A which has high affinity for the Fc portion of some immunoglobulins including IgG. Addition of killed *S.aureus* Cowan strain II cells or protein A-Sepharose binds the immune complexes which are then pelleted and washed (265, 266). Use of protein A Sepharose requires fewer washing steps than the killed bacteria and gives lower backgrounds. More recent techniques include direct immunoprecipitation using antibody covalently linked to Sepharose (e.g. ref. 267) or indirect immunoprecipitation with the second antibody linked to Sepharose (e.g. ref. 268).

After purification by any of these methods, immune complexes are solubilized by heating in SDS-containing sample buffer and subjected to SDS-PAGE. Polypeptides specifically precipitated by the antibody are identified by electrophoresis of a control sample exposed to pre-immune serum in a parallel track. Further identification of polypeptide antigens after electrophoresis is possible either by molecular mass estimation relative to standard polypeptides or by comigration with purified antigens. If the polypeptide under study has a molecular mass which corresponds to that of the antibody light chain (M_r approx. 23 500) or heavy chain (M_r approx. 50 000), comigration in SDS-PAGE can be avoided by omitting the thiol reagent during sample dissociation or by use of papain-treated antibody respectively.

The technique works best when the protein antigen to be precipitated exceeds about 0.1% of the total protein although lower concentrations of specific proteins have been successfully analysed. Among the more common uses for this methodology are as follows.

- The relatedness of two proteins from different sources can be analysed by testing for immunological cross-reaction using antibody raised against one of the proteins followed by SDS-PAGE.

• Immune precipitation followed by SDS-PAGE is a common method of following *in vivo* or *in vitro* synthesis of specific proteins (e.g. refs. 265, 266). The protein is labelled using a radioactive amino acid (usually [³H]leucine or [³⁵S]methionine), immunoprecipitated, and fractionated by SDS-PAGE. After electrophoresis, the specific protein band is detected and quantitated by fluorography, autoradiography or a gel slicing and scintillation counting method (see Section 7.5). One big advantage of the combined immunoprecipitation/SDS-PAGE approach over immunoprecipitation alone is that SDS-PAGE separates specifically-precipitated protein from non-specific protein contaminants which may be carried through the immunoprecipitation procedure, thus reducing background counts considerably.

• The technique can also be used to isolate microgram amounts of immunologically related proteins for further analysis. For example, Platt *et al.* (269) used antibody raised against *E.coli lac* repressor to isolate a mutant protein which was compared to the wild-type protein by N-terminal amino acid sequence determination.

Once antibodies to specific proteins are available, they can also be used directly to probe for identical or cross-reacting proteins in complex mixtures of native or denatured proteins by gel electrophoresis followed by immunoblotting. This technique and its wide variety of applications are discussed in more detail in Section 7.10.3.

11. Artefacts and troubleshooting

(a) Lack of polymerization is usually due to incorrect concentrations of the prepared reagents, the omission of a reagent from the gel mixture, impurities in the reagents, or the use of an old ammonium persulphate stock solution. The simplest remedy is usually to discard solutions and prepare a fresh batch using pure reagents. High concentrations of thiol reagents will also inhibit polymerization.

(b) Polymerization should occur according to the time periods specified in this chapter, usually $10-30$ min for single concentration gels at room temperature. Too fast or too slow polymerization can most easily be corrected by varying the concentrations of the polymerization catalysts.

(c) Cracking of the gel during polymerization (usually only high concentration gels) is often due to excessive heat production by the polymerization reaction itself and is remedied by using cooled solutions. For rod gels, siliconizing the tubes may also help. This problem is usually not encountered at most gel concentrations when rod or slab gels are made individually. Gel cracking during electrophoresis occurs due to excessive current input overheating the gel, and is remedied by using less current over a longer period of time.

(d) Provided that the ionic strength of non-dissociating buffers is high enough to prevent aggregation of native proteins, insoluble material in the sample usually represents denatured protein and must be removed by centrifugation prior to electrophoresis. Insolubility in SDS-containing buffers indicates too little SDS, too little reducing agent, or two low a pH, especially after TCA precipitation of proteins. Alternatively, for some proteins, it may be advisable to add urea in addition to SDS to ensure solubilization.

(e) Failure of the sample to form a layer at the well bottom when applied to slab gels indicates either the accidental omission of sucrose or glycerol from the sample buffer or the use of a sample comb where the teeth do not form a snug fit with the glass plates, allowing gel to polymerize between the teeth and glass plates which then interferes with sample loading. The remedy for the latter problem is to use a better fitting comb, but in the short term, excess gel can be removed from the wells using a syringe needle connected to a water aspirator.

(f) Detachment of slab gels from the glass plates during gel electrophoresis usually indicates inadequately cleaned plates. Low concentration gels sometimes detach from rod gel tubes even though these are clean; the problem is overcome by attaching a piece of nylon mesh to the bottom of the tube (Section 6.4).

(g) Sometimes the top of a rod gel collapses inward upon itself and away from the tube wall during electrophoresis, causing deformation of the gel and allowing sample to migrate between the gel and the tube wall. A similar phenomenon is observed with slab gel tracks where the base of the sample well appears to be dragged downwards in the direction of electrophoresis. Both these effects can be caused by trapping of high molecular mass, highly charged species at the gel surface and is particularly common with high concentrations of nucleic acid in the sample (low concentrations are usually tolerable).

(h) Poor staining after SDS-PAGE is often cured by increasing the volume of staining solution to dilute out the SDS present. Uneven staining with any buffer system indicates that more time should be allocated to the staining step to allow the dye to penetrate fully. If stained bands are being lost on destaining, reduce the destaining time or use a better fixative.

(i) A metallic sheen on gels after staining with Coomassie Blue R-250 usually indicates that solvent has been allowed to evaporate causing dye to dry on the gel at that point. Slight films of Coomassie Blue sometimes observed on the gel surface after destaining can be removed by a quick rinse in 50% methanol or by gently swabbing the gel surface with methanol-soaked tissue paper.

(j) Blue blotches near gel borders are usually fingerprints caused by gel handling without gloves! Silver staining, because of its considerably higher sensitivity, is even more prone to user-induced artefacts of this kind.

(k) Protein bands observed in all tracks of a slab gel or all rod gels may indicate contamination of the sample buffer. Contaminated reservoir buffer produces a continuous stained region from the gel origin to near the buffer front, even in tracks which have not been loaded with sample. The same protein bands observed in several neighbouring lanes of a slab gel indicates that sample from one well has contaminated adjacent wells, usually by overflowing. Reduce the volume of sample loaded.

(l) A high background of protein staining along individual rod gels or slab gel tracks with indistinct protein bands usually indicates extensive sample proteolysis [see (r) below for remedy]. However, this phenomenon has also been observed with the SDS-discontinuous system when using certain impure grades of SDS; analysis of the same samples with purified SDS gave sharp bands.

(m) Distorted bands may occur for one of several reasons. Common reasons are insoluble material or bubbles in the gel or inconsistent pore size throughout the gel. In future, filter the gel reagents before use and ensure that the gel mixture is well mixed and degassed before pouring the gel. If an uncooled gel apparatus is being used, some distortion may occur because of uneven heating of the gel. Either use a cooled apparatus or reduce the current at which electrophoresis is performed. Variations in staining density along the width of a stained band often indicates an uneven gel surface, resulting in sample accumulating at the low points prior to electrophoresis. Uneven gel surfaces can be caused by insufficient care or experience in overlayering the gel or as a result of vibration during gel polymerization.

(n) The inability of a substantial portion of the protein to enter the resolving gel may cause a heavily stained band at the gel origin. Provided that the gel concentration is sufficiently low that one would have expected the proteins to be able to enter the gel matrix, the problem may be caused by aggregated protein in the sample prior to electrophoresis [see (d) above] or, in the case of non-dissociating discontinuous buffer systems, precipitation of the proteins due to the formation of highly concentrated zones during electrophoresis in the stacking gel. If the latter is correct, one would be advised to use less concentrated samples with a continuous buffer system.

(o) Protein streaking along individual rod gels or slab gel tracks is often accompanied by protein at the gel origin and indicates protein precipitation followed by dissolution of the precipitates during electrophoresis. For some remedies see (d) and (n) above. However, overloading of the gel is also a major cause of vertical streaking in which case the amount of sample loaded should clearly be decreased.

(p) Cracking of single concentration gels during drying under vacuum will occur if the vacuum is accidentally released before the gel is properly dry, or if thick gel slices or slabs (>1.5 mm) are being used, or if the gel has been allowed to swell appreciably before drying.

(q) Artefactual blackening of X-ray film during fluorography may be caused by inadequate removal of DMSO. With future gels, ensure sufficient soaking in water before drying the gel.

(r) Irreproducibility of the protein band pattern is usually caused by problems of sample preparation rather than by polyacrylamide gel electrophoresis. Reduction in the staining intensity or complete loss of individual components may be indicative of proteolysis, as is the appearance of previously unobserved fast migrating bands. Working at low temperature and the use of protease inhibitors during sample preparation may eliminate the problem. If the problem occurs

B. David Hames

with SDS-PAGE, check that the sample is being heated to at least 90°C for 2 min during dissociation.

(s) A high silver-staining background is often caused by acrylic acid contamination in the acrylamide and/or bisacrylamide. The highest quality reagents, including the purest (deionized) water, should always be used.

Acknowledgements

Thanks are due to D. Rickwood for critical reading of this manuscript, to J. B. C. Findlay for welcome advice on the sections concerned with membrane protein analysis and protein sequencing and to Ms Paula Duncan and Mrs S. Gray for typing from my less than ideal handwriting.

References

1. Gordon, A. H. (1975). Electrophoresis of Proteins in Polyacrylamide and Starch Gels. *Laboratory Techniques in Biochemistry and Molecular Biology,* (ed. T. S. Work and E. Work), Vol. 1, Part 1. North-Holland, Amsterdam.
2. Smith, I. (1975). In *Chromatographic and Electrophoretic Techniques,* (ed. I. Smith), Vol. 2, *Zone Electrophoresis,* p. 153. William Heinemann Medical Books Ltd, London.
3. Andrews, A. T. (1986). *Electrophoresis: Theory, Techniques and Biochemical and Clinical Applications.* 2nd edn. Oxford University Press, New York.
4. Fawcett, J. S. and Morris, C. J. O. R. (1966). *Separation Studies,* **1**, 9.
5. Campbell, W. P., Wrigley, C. W., and Margolis, J. (1983). *Anal. Biochem.,* **129**, 31.
6. Righetti, P. G., Brost, B. C. W., and Snyder, R. S. (1981). *J. Biochem. Biophys. Methods,* **4**, 347.
7. Mócz, G. and Bálint, M. (1984). *Anal. Biochem.,* **143**, 283.
8. Shapiro, A. L., Vinuela, E., and Maizel, J. V. (1967). *Biochem. Biophys. Res. Commun.,* **28**, 815.
9. Weber, K. and Osborn, M. (1969). *J. Biol. Chem.,* **244**, 4406.
10. Zweidler, A. and Cohen, L. H. (1972). *Fed. Proc. Fed. Am. Soc. Exp. Biol.,* **31**, 926.
11. Alfageme, C. R., Zweidler, A., Mahowald, A., and Cohen, L. H. (1974). *J. Biol. Chem.,* **249**, 3729.
12. Bonner, W. M., West, M. H. P., and Stedman, J. D. (1980). *Eur. J. Biochem.,* **109**, 17.
13. Rovera, G., Magarian, C., and Borun, T. W. (1978). *Anal. Biochem.,* **85**, 506.
14. Ornstein, L. (1964). *Ann. NY Acad. Sci.,* **121**, 321.
15. Davis, B. J. (1964). *Ann. NY Acad. Sci.,* **121**, 404.
16. Laemmli, U. K. (1970). *Nature,* **277**, 680.
17. Neville, D. M. (1971). *J. Biol. Chem.,* **246**, 6328.
18. Jovin, T. M. (1973). *Biochemistry,* **12**, 871, 890; *Ann. NY Acad. Sci.,* **209**, 477; for buffer details see *Multiphasic Buffer Systems Output,* Public Board Numbers 196090, 258309−259312. National Technical Information Service, Springfield, VA, USA.
19. Reisfeld, R. A., Lewis, V. J., and Williams, D. E. (1962). *Nature,* **195**, 281.
20. Williams, D. E. and Reisfeld, R. A. (1964). *Ann. NY Acad. Sci.,* **121**, 373.
21. McLellan, T. (1982). *Anal. Biochem.,* **126**, 94.
22. Choules, G. L. and Zimm, B. H. (1965). *Anal. Biochem.,* **13**, 336.

23. Jordan, E. M. and Raymond, S. (1969). *Anal. Biochem.*, **27**, 205.
24. Ferguson, K. A. (1964). *Metabolism*, **13**, 21.
25. Rodbard, D., Chrambach, A., and Weis, G. H. (1974). In *Electrophoresis and Isoelectric Focusing in Polyacrylamide Gel,* (ed. R. C. Allen and H. R. Maurer), p. 62. Walter de Gruyter, Berlin.
26. Hedrick, J. L. and Smith, A. J. (1968). *Arch. Biochem. Biophys.*, **126**, 155.
27. Rodbard, D. and Chrambach, A. (1974) In *Electrophoresis and Isoelectric Focusing in Polyacrylamide Gel,* (ed. R. C. Allen, and H. R. Maurer), p. 28. Walter de Gruyter, Berlin.
28. Segrest, J. P. and Jackson, R. L. (1972). In *Methods in Enzymology*, (ed. V. Ginsburg), Vol. 28B, p. 54. Academic Press, New York.
29. Allore, R. J. and Barber, B. H. (1984). *Anal. Biochem.*, **137**, 523.
30. Studier, F. W. (1973). *J. Mol. Biol.*, **79**, 237.
31. Swaney, J. B., Vande Wonde, G. F., and Bachrach, H. L. (1974). *Anal. Biochem.*, **58**, 337.
32. Margulies, M. M. and Tiffany, H. L. (1984). *Anal. Biochem.*, **136**, 309.
33. Petropakis, H. J., Angelmeir, A. F., and Montgomery, M. W. (1972). *Anal. Biochem.*, **46**, 594.
34. Gelfi, C. and Righetti, P. G. (1981). *Electrophoresis*, **2**, 220.
35. Lambin, P. C. (1978). *Anal. Biochem.*, **85**, 114.
36. Payne, J. W. (1973). *Biochem. J.*, **135**, 867.
37. Carpenter, F. H. and Harrington, K. T. (1972). *J. Biol. Chem.*, **247**, 5580.
38. Dautrevaux, M., Boulanger, Y., Han, K., and Biserte, G. (1969). *Eur. J. Biochem.*, **11**, 267.
39. Banker, G. A. and Cotman, C. W. (1972). *J. Biol. Chem.*, **247**, 5856.
40. Frank, R. N. and Rodbard, D. (1975). *Arch. Biochem. Biophys.*, **171**, 1.
41. Rice, R. H. and Means, G. E. (1971). *J. Biol. Chem.*, **246**, 831.
42. Ghadge, G. D., Bodhe, A. M., Modak, S. R., and Vartak, H. G. (1983). *Anal. Biochem.*, **128**, 469.
43. Dhamankar, V. S., Choudhury, M. D., and Vartak, H. G. (1986). *Anal. Biochem.*, **157**, 289.
44. Lee, C., Levin, A., and Branton, D. (1987). *Anal. Biochem.*, **166**, 308.
45. Leblance, E. A. and Cochrane, B. J. (1987). *Anal. Biochem.*, **161**, 172.
46. Wilson, C. M. (1983). In *Methods in Enzymology,* (ed. C. H. W. Hirs and S. N. Timasheff), Vol. 91, p. 236. Academic Press, New York.
47. Righetti, P. G. (1983). In *Laboratory Techniques in Biochemistry and Molecular Biology*, (ed. T. S. Work and E. Work), Vol. 11, p. 148. North-Holland, Amsterdam.
48. Diezel, W., Kopperschläger, G., and Hofmann, E. (1972). *Anal. Biochem.*, **48**, 617.
49. Chrambach, A., Reisfeld, R. A., Wyckoff, M., and Zaccari, J. (1967). *Anal. Biochem.*, **20**, 150.
50. Fenner, C., Traut, R. R., Mason, D. T., and Wilkman-Coffelt, J. (1975). *Anal. Biochem.*, **63**, 595.
51. Wong, P., Barbeau, A., and Roses, A. D. (1985). *Anal. Biochem.*, **150**, 288.
52. Kohler, P. O., Bridson, W. E., and Chrambach, A. (1971). *J. Clin. Endocrinol.*, **32**, 70.
53. Maizel, J. V. (1971). In *Methods in Virology,* (ed. K. Maramorosch and H. Koprowski), Vol. 5, p. 179. Academic Press, New York and London.
54. Blakesley, R. W. and Boezi, J. A. (1977). *Anal. Biochem.*, **82**, 580.
55. Switzer, R. C., Merril, C. R., and Shifrin, S. (1979). *Anal. Biochem.*, **98**, 231.

56. Dunn, M. J. and Burghes, A. H. M. (1983). *Electrophoresis*, **4**, 173.
57. Ohsawa, K. and Ebata, N. (1983). *Anal. Biochem.*, **135**, 409.
58. Oakley, B. R., Kirsch, D. R., and Morris, N. R. (1980). *Anal. Biochem.*, **105**, 361.
59. Sammons, D. W., Adams, L. D., and Nishizawa, E. E. (1981). *Electrophoresis*, **2**, 135.
60. Morrissey, J. H. (1981). *Anal. Biochem.*, **117**, 307.
61. Confavreux, C., Gianazza, E., Chazot, G., Lasne, Y., and Arnaud, P. (1982). *Electrophoresis*, **3**, 206.
62. Merril, C. R. (1986). In *Electrophoresis '86*, (ed. M. J. Dunn), p. 273. VCH Verlagsgesellschaft mbH.
63. Wray, W., Bonlikas, T., Wray, V. P., and Hancock, R. (1981). *Anal. Biochem.*, **118**, 197.
64. Ansorge, W. (1985). *J. Biochem. Biophys. Methods*, **11**, 13.
65. Merril, C. R., Harrington, M., and Alley, V. (1984). *Electrophoresis*, **5**, 289.
66. Goldman, D., Merril, C. R., and Ebert, M. H. (1980). *Clin. Chem.*, **26**, 1317.
67. Dunbar, B. S. (1987). *Two Dimensional Gel Electrophoresis and Immunological Techniques*. Plenum Press, New York.
68. Berson, G. (1983). *Anal. Biochem.*, **134**, 230.
69. Merril, C. R., Goldman, D., and van Keuran, M. C. (1982). *Electrophoresis*, **3**, 17.
70. Hallinan, F. U. (1983). *Electrophoresis*, **4**, 265.
71. Tsutsui, K., Kurosaki, T., Tsutsui, K., Nagai, H. and Oda, T. (1985). *Anal. Biochem.*, **146**, 111.
72. Dzandu, J. K., Deh, M. E. and Wise, G. E. (1985) *Biochem. Biophys. Res. Commun.*, **126**, 50.
73. Ragland, W. L., Pace, J. L., and Kemper, D. L. (1974) *Anal. Biochem.*, **59**, 24.
74. Eng, P. R. and Parker, C. O. (1974) *Anal. Biochem.*, **59**, 323.
75. Barger, B. O., White, F. C., Pace, J. C., Kemper, D. L., and Ragland, W. L. (1976) *Anal. Biochem.*, **70**, 327.
76. Ragland, W. C., Benton, T. L., Pace, J. L., Beach, F. G., and Wade, A. E. (1978) In *Electrophoresis '78* (ed. N. Catsimpoolas), Vol. 2, p. 217. Elsevier/North-Holland Publishing Co., Amsterdam.
77. Wiederkamm, E., Wallach, D. F. H., and Flückinger, R. (1973). *Anal. Biochem.*, **54**, 102.
78. Hartman, B. K. and Udenfriend, S. (1969). *Anal. Biochem.*, **30**, 391.
79. Harowitz, P. M. and Bowman, S. (1987). *Anal. Biochem.*, **165**, 430.
80. Liebowitz, M. J. and Wang, R. W. (1984). *Anal. Biochem.*, **137**, 161.
81. Laskey, R. A. and Mills, A. D. (1975). *Eur. J. Biochem.*, **56**, 335.
82. Laskey, R. A. and Mills, A. D. (1977). *FEBS Lett.*, **82**, 314.
83. Bonner, W. M. and Laskey, R. A. (1974). *Eur. J. Biochem.*, **46**, 83.
84. Laskey, R. A. (1980). In *Methods in Enzymology,* (ed. L. Grossman and K. Moldave), Vol. 65, p. 363. Academic Press, New York.
85. Harding, C. R. and Scott, I. R. (1983). *Anal. Biochem.*, **129**, 371.
86. Skinner, M. K. and Griswold, M. D. (1983). *Biochem. J.*, **209**, 281.
87. Chamberlain, J. P. (1979). *Anal. Biochem.*, **98**, 132.
88. Roberts, P. L. (1985). *Anal. Biochem.*, **147**, 521.
89. Bonner, W. M. (1983). In *Methods in Enzymology*, (ed. S. Fleischer and B. Fleischer), Vol. 96, p. 215. Academic Press, New York.
90. Rigby, P. J. W. (1981). *Amersham Research News* No. 13.
91. Laskey, R. A. (1981). *Amersham Research News* No. 23.

92. Vachon, D. Y. (1981). *J. Investigative Radiology*, **16**, 221.
93. McConkey, E. H. (1979). *Anal. Biochem.*, **96**, 39.
94. Walton, K. E., Stryer, D., and Gruenstein, E. (1979). *J. Biol. Chem.*, **254**, 795.
95. Cooper, P. C. and Burgess, A. W. (1982). *Anal. Biochem.*, **126**, 301.
96. Kronenberg, L. H. (1979). *Anal. Biochem.*, **93**, 189.
97. Davidson, J. B. and Case, A. (1982). *Science*, **215**, 1398.
98. Burbeck, S. (1983). *Electrophoresis*, **4**, 127.
99. Peterson, J., Tipton, H. W., and Chrambach, A. (1972). *Anal. Biochem.*, **62**, 274.
100. Goodman, D. and Matzura, H. (1971). *Anal. Biochem.*, **62**, 274.
101. Paus, P. N. (1971). *Anal. Biochem.*, **42**, 372.
102. Hansen, J. N., Pheiffer, B. H., and Boehnert, J. A. (1980). *Anal. Biochem.*, **105**, 192.
103. Faulkner, R. D., Carraway, R., and Bhatnagar, Y. M. (1982). *Biochim. Biophys. Acta*, **708**, 245.
104. Anderson, L. E. and McClure, W. O. (1973). *Anal. Biochem.*, **51**, 173.
105. Spath, P. J. and Koblet, H. (1979). *Anal. Biochem.*, **93**, 275.
106. Gelfi, C. and Righetti, P. G. (1981). *Electrophoresis*, **2**, 213.
107. Meredith, C. and Johnson, M. K. (1987). *Anal. Biochem.*, **162**, 403.
108. O'Connell, P. B. H. and Brady, C. J. (1976). *Anal. Biochem.*, **76**, 63.
109. Basch, R. S. (1968). *Anal. Biochem.*, **26**, 184.
110. Ames, G. F. (1974). *J. Biol. Chem.*, **249**, 634.
111. Aloyo, V. J. (1979). *Anal. Biochem.*, **99**, 161.
112. Martin, A. F., Prior, G., and Zak, R. (1976). *Anal. Biochem.*, **72**, 577.
113. Airhart, J., Kelley, J., Brayden, J. E., and Low, R. B. (1979). *Anal. Biochem.*, **96**, 45.
114. Dubray, G. and Bezard, G. (1982). *Anal. Biochem.*, **119**, 325.
115. Auerbach, S. and Pederson, T. (1975). *Biochem. Biophys. Res. Commun.*, **63**, 149.
116. Olden, K. and Yamada, K. M. (1977). *Anal. Biochem.*, **78**, 483.
117. Burridge, K. (1978). In *Methods in Enzymology*, (ed. V. Ginsburg), Vol. 50, p. 54. Academic Press, New York.
118. Cutting, J. A. and Roth T. F. (1984). *Methods in Enzymology*, **104**, 451.
119. Debruyne, I. (1983). *Anal. Biochem.*, **133**, 110.
120. Showe, M. K., Isobe, E., and Onorato, L. (1976). *J. Mol. Biol.*, **107**, 55.
121. Saltzgaber-Müller, J. and Schatz, G. (1978). *J. Biol. Chem.*, **252**, 305.
122. Scheele, G., Pash, J., and Bieger, W. (1980). *Anal. Biochem.*, **112**, 303.
123. Sri Venugopal, K. S. and Adiga, P. R. (1980) *Anal. Biochem.*, **101**, 215.
124. Bertazonni, V., Scovassi, A. I., Mezzina, M., Sarasin, A., Franchi, E., and Izzo, R. (1986). *Trends Genet.*, **2**, 67.
125. Gershoni, J. M. and Palade, G. E. (1983). *Anal. Biochem.*, **131**, 1.
126. Renart, J. and Sandoval, I. V. (1984). In *Methods in Enzymology*, (ed. W. B. Jakoby), Vol. 104, p. 455. Academic Press, New York.
127. Towbin, H. and Gordon, J. (1984). *J. Immunol. Methods*, **72**, 313.
128. Lin, W. and Kasamatsu, H. (1983). *Anal. Biochem.*, **128**, 302.
129. Gershoni, J. M. and Palade, G. E. (1982). *Anal. Biochem.*, **124**, 396.
130. Alwine, J. C., Kemp, D. J., and Stark, G. R. (1977). *Proc. Natl. Acad. Sci. USA*, **74**, 5350.
131. Resier, J. and Wardale, J. (1981) *Eur. J. Biochem.*, **114**, 569.
132. Clarke, L., Hitzemann, R., and Carbon, J. (1979) In *Methods in Enzymology*, (ed. R. Wu), Vol. 68, p. 436. Academic Press, New York.
133. Towbin, H., Staehelin, T., and Gordon, J. (1979). *Proc. Natl. Acad. Sci. USA*, **76**, 4350.

B. David Hames

134. Burnette, W. N. (1981). *Anal. Biochem.*, **112**, 195.
135. Nielsen, P. J., Manchester, K. L., Towbin, H., Gordon, J., and Thomas, G. (1982). *J. Biol. Chem.*, **257**, 12316.
136. Gibson, W. (1981). *Anal. Biochem.*, **118**, 1.
137. Gershoni, J. M., Davis, F. E., and Palade, G. E. (1985). *Anal. Biochem.*, **144**, 32.
138. Waterborg, J. H. and Harrington, R. E. (1987). *Anal. Biochem.*, **162**, 430.
139. Soutar, A. K. and Wade, D. P. (1989). In *Protein Function: A Practical Approach*, (ed. T. E. Creighton), p. 55. IRL Press Ltd, Oxford and Washington DC.
140. Hancock, K. and Tsang, V. C. W. (1983). *Anal. Biochem.*, **133**, 157.
141. Yuen, K. C. C., Johnson, T. K., Dennell, R. E., and Consigli, R. A. (1982). *Anal. Biochem.*, **126**, 398.
142. Rohringer, R. and Holden, D. W. (1985). *Anal. Biochem.*, **144**, 118.
143. Moeremans, M., Daneels, G., De Raeymaeker, M., and De May, J. (1986). In *Electrophoresis '86*, (ed. M. J. Dunn), p. 328. VCH Verlagsgesellschaft mbH.
144. Wotjkowiak, Z., Briggs, R. C., and Hnilica, L. S. (1983). *Anal. Biochem.*, **129**, 486.
145. Kittler, J. M., Meisler, N. T., Viceps-Madore, D., Cidlowski, J. A,. and Thomassi, J. W. (1984). *Anal. Biochem.*, **137**, 210.
146. Bio-Radiations (1985) from Bio-Rad Laboratories Ltd; No. 56EG.
147. Della-Penna, D., Christofferson, R. E., and Bennet, A. B. (1986). *Anal. Biochem.*, **152**, 329.
148. Dunn, S. D. (1986). *Anal. Biochem.*, **157**, 144.
149. Johnson, D. A., Gausch, J. W., Sportsman, J. R., and Elder, J. H. (1984). *Gene Anal. Techn.*, **1**, 3.
150. Batteiger, B., Newhall, W. J., and Jones, R. B. (1982). *J. Immunol. Methods*, **55**, 297.
151. Towbin, H., Ramjone, H. P., Kuster, H., Liverani, D., and Gordon, J. (1982). *J. Biol. Chem.*, **257**, 12709.
152. Perides, G., Plagen, V., and Traub, P. (1986). *Anal. Biochem.*, **152**, 94.
153. The, T. H. and Feltkamp, T. E. W. (1970). *Immunology*, **18**, 865.
154. Weigele, M., De Bernardo, S., Leimgruber, W., Cleeland, R., and Grunberg, E. (1973). *Biochem. Biophys. Res. Commun.*, **54**, 899.
155. Blake, M. S., Johnston, K. H., Russel-Jones, G. J., and Gotschlich, E. C. (1984). *Anal. Biochem.*, **136**, 175.
156. Hsu, Y. (1984). *Anal. Biochem.*, **142**, 221.
157. Surek, B. and Latzko, E. (1984). *Biochem. Biophys. Res. Commun.*, **121**, 284.
158. Brada, D. and Roth, J. (1984). *Anal. Biochem.*, **142**, 79.
159. Jackson, P. and Thompson, R. J. (1984). *Electrophoresis*, **5**, 35.
160. Gershoni, J. M., Bayer, E. A., and Wilcheck, M. (1985). *Anal. Biochem.*, **146**, 59.
161. Keren, Z., Berke, G., and Gershoni, J. M. (1986). *Anal. Biochem.*, **155**, 182.
162. Moroi, M. and Jung, S. M. (1984). *Biochim. Biophys. Acta*, **798**, 295.
163. Glass, W. F., Briggs, R. C., and Hnilica, L. S. (1981). *Anal. Biochem.*, **115**, 219.
164. Clegg, J. C. S. (1982). *Anal. Biochem.*, **127**, 389.
165. Bradbury, W. C., Mills, S. D., Preston, M. A., Barton, L. J., and Penner, J. L. (1984). *Anal. Biochem.*, **137**, 129.
166. Gershoni, J. M. (1985). *Trends Biochem.*, **10**, 103.
167. Gershoni, J. M., Hawrot, E., and Lentz, T. L. (1983). *Proc. Natl. Acad. Sci. USA*, **80**, 4973.
168. Melcher, V., Eidels, L., and Uhr, J. (1975). *Nature*, **258**, 434.
169. Geysen, J., De Loof, A., and Vandesande, F. (1984). *Electrophoresis*, **5**, 129.

170. Parekh, B. S., Mehta, H. B., West, M. D., and Motelaro, R. C. (1985). *Anal. Biochem.*, **148**, 87.
171. Graesslin, D., Weise, H. C., and Rick, M. (1976). *Anal. Biochem.*, **71**, 492.
172. Stephens, R. E. (1975). *Anal. Biochem.*, **65**, 369.
173. Schetters, H. and McLeod, B. (1979). *Anal. Biochem.*, **98**, 329.
174. Liebowitz, M. J. and Wang, R. W. (1984). *Anal. Biochem.*, **137**, 161.
175. Bray, D. and Brownlee, S. M. (1973). *Anal. Biochem.*, **55**, 213.
176. Furth, A. J. (1980). *Anal. Biochem.*, **109**, 207.
177. Spiker, S. and Isenberg, I. (1983). In *Methods in Enzymology*, (ed. C. H. W. Hirs and S. N. Timasheff), Vol. 91, p. 214. Academic Press, New York.
178. Ihara, S., Suzuki, H., and Kawakami, M. (1987). *Anal. Biochem.*, **166**, 349.
179. McDonald, G., Fawell, S., Pappin, D., and Higgins, S. (1986). *Trends Genet.*, **2**, 35.
180. Hunkapillar, M. W., Lujan, E., Ostrander, F., and Hood, L. E. (1983). In *Methods in Enzymology*, (ed. C. H. W. Hirs and S. N. Timasheff), Vol. 91, p. 227. Academic Press, New York.
181. Swank, R. W. and Munkres, K. D. (1971). *Anal. Biochem.*, **39**, 462.
182. Anderson, B. L., Berry, R. W., and Telser, A. (1983). *Anal. Biochem.*, **132**, 365.
183. Kadenach, B. and Merle, P. (1981). *FEBS Lett.*, **135**, 1.
184. Kyle, J. and Rodrigues, H. (1983). *Anal. Biochem.*, **133**, 515.
185. Hashimoto, F., Horigome, T., Kanbayashi, M., Yoshida, K., and Sugano, H. (1983). *Anal. Biochem.*, **129**, 192.
186. Schagger, H. and Von Jagow, G. (1987). *Anal. Biochem.*, **166**, 368.
187. De Wald, D. B., Adams, L. D., and Pearson, J. D. (1986) *Anal. Biochem.*, **154**, 502.
188. Fling, S. P. and Gregerson, D. S. (1986). *Anal. Biochem.*, **155**, 83.
189. Panyim, S. and Chalkley, R. (1969). *Arch. Biochem. Biophys.*, **130**, 337.
190. Franklin, S. G. and Zweidler, A. (1977). *Nature*, **266**, 273.
191. Zweidler, A. (1978). In *Methods in Cell Biology*, (ed, G. Stein, J. Stein, and L. J. Kleinsmith), Vol. 27, p. 223. Academic Press, New York.
192. Hardison, R. and Chalkley, R. (1978). In *Methods in Cell Biology*, (ed. G. Stein, J. Stein, and L. J. Kleinsmith), Vol. 17, p. 235. Academic Press, New York.
193. Thomas, J. O. and Kornberg, R. D. (1975). *Proc. Natl. Acad. Sci. USA*, **72**, 2626.
194. Panyin, S. and Chalkley, R. (1971). *J. Biol. Chem.*, **246**, 7557.
195. Shmatchenko, V. V. and Varshavskey, A. J. (1978). *Anal. Biochem.*, **85**, 42.
196. Bonner, W. M., West, M. H. P., and Stedman, J. D. (1980). *Eur. J. Biochem.*, **109**, 17.
197. MacGillivray, A. J., Cameron, A., Krauze, R. T., Rickwood, D., and Paul, J. (1972). *Biochim. Biophys. Acta*, **277**, 384.
198. Nicholas, R. H. (1990). In *Gel Electrophoresis of Nucleic Acids: A Practical Approach*, 2nd edn., (ed. D. Rickwood and B. D. Hames). Oxford University Press, Oxford.
199. Bickle, T. A. and Traut, R. R. (1974). In *Methods in Enzymology* (ed. K. Moldave and L. Grossman), Vol. 30, p. 545. Academic Press, New York.
200. Kanda, F., Ochiai, H., and Iwabuchi, M. (1974). *Eur. J. Biochem.*, **44**, 469.
201. Traub, P., Mizushima, S., Lowry, C. V., and Nomura, M. (1971). In *Methods in Enzymology*, (ed. K. Moldave and L. Grossman), Vol. 20, p. 391. Academic Press, New York.
202. Findlay, J. B. C. and Evans, W. H. (1987). *Biological Membranes: A Practical Approach*. IRL Press Ltd, Oxford.
203. Siu, C. H., Lerner, R. A., and Loomis, W. F. (1977). *J. Mol. Biol.*, **116**, 469.
204. Hubbard, A. L. and Cohn, Z. A. (1972). *J. Cell Biol.*, **55**, 390.

B. David Hames

205. Gahmberg, C. G. (1977). In *Dynamic Aspects of Cell Surface Organization, Cell Surface Reviews,* (ed. G. Poste and G. L. Nicolson), Vol. 3, p. 371. Elsevier/North-Holland Publishing Co., Amsterdam.
206. Hjerten, S. (1983). *Biochim. Biophys. Acta,* **736**, 130.
207. Dewald, B., Dulaney, J. T., and Touster, O. (1974). In *Methods in Enzymology,* (ed. S. Fleischer and L. Packer), Vol. 32, p. 82. Academic Press, New York.
208. Newby, A. C. and Chrambach, A. (1979). *Biochem. J.,* **177**, 623.
209. Hjelmeland, L. M. and Chrambach, A. (1984). In *Methods in Enzymology,* (ed. W. B. Jacoby), Vol. 104, p. 305. Academic Press, New York.
210. Findlay, J. B. C. (1990). In *Protein Purification Applications: A Practical Approach,* (ed. E. Harris and S. Angal). Oxford University Press, Oxford.
211. Zahler, W. L. (1974). In *Methods in Enzymology,* (ed. S. Fleischer and L. Packer), Vol. 32, p. 70. Academic Press, New York.
212. Ballou, B. and Smithies, O. (1977). *Anal. Biochem.,* **80**, 616.
213. Helenius, A. and Simons, K. (1977). *Proc. Natl. Acad. Sci. USA,* **74**, 529.
214. Bordier, C., Loomis, W. F., Elder, J., and Lerner, R. (1978). *J. Biol. Chem.,* **253**, 5133.
215. Vecchio, G., Righetti, P. G., Zanoni, M., Artoni, G., and Gainazza, E. (1984). *Anal. Biochem.,* **137**, 410.
216. Artoni, G., Gianazza, E., Zanoni, M., Gelfi, G., Tanzi, M. C., Barozzi, C., Ferruti, P., and Righetti, P. G. (1984) *Anal. Biochem.,* **137**, 420.
217. Margolis, J. and Kenrick, K. G. (1967). *Nature,* **214**, 1334.
218. Lambin, P. C. (1978). *Anal. Biochem.,* **85**, 114.
219. Lambin, F. and Fine, J. M. (1979). *Anal. Biochem.,* **98**, 160.
220. Poduslo, J. F. and Rodbard, D. (1980). *Anal. Biochem.,* **101**, 394.
221. Rothe, G. M. and Purkhanbaba, H. (1982). *Electrophoresis,* **3**, 43.
222. Rothe, G. M. and Maurer, W. D. (1986). In *Gel Electrophoresis of Proteins* (ed. M. T Dunn), p. 37. Wright, Bristol.
223. Margolis, J. and Kenrick, K. G. (1968). *Anal. Biochem.,* **25**, 347.
224. Kapadia, G., Chrambach, A., and Rodbard, D. (1984). In *Electrophoresis and Isoelectric Focusing in Polyacrylamide Gel,* (ed. R. C. Allen and H. R. Maurer), p. 115. Walter de Gruyter, Berlin.
225. Creighton, T. E. (1979). *J. Mol. Biol.,* **129**, 235.
226. Feinstein, D. L. and Mondrianakis, E. N. (1984). *Anal. Biochem.,* **136**, 362.
227. Fernandes, P. B., Nardi, R. V., and Franklin, S. G. (1978). *Anal. Biochem.,* **91**, 101.
228. Goldenberg, D. P. (1989). In *Protein Structure: A Practical Approach,* (ed. T. E Creighton), p. 225. Oxford University Press, Oxford.
229. Neuhoff, V. (1973). In *Micromethods in Molecular Biology,* (ed. V. Neuhoff), p. 1. Springer Verlag, New York.
230. Grossbach, U. (1974). In *Electrophoresis and Isoelectric Focusing in Polyacrylamide Gel,* (ed. R. C. Allen and H. R. Maurer), p. 207. Walter de Gruyter, Berlin.
231. Ruche, R. (1974). In *Electrophoresis and Isoelectric Focusing in Polyacrylamide Gel,* (ed. R. C. Allen and H. R. Maurer) p. 215. Walter de Gruyter, Berlin.
232. Dames, W. and Maurer, H. R. (1974). In *Electrophoresis and Isoelectric Focusing in Polyacrylamide Gel,* (ed. R. C. and H. R. Maurer) p. 221. Walter de Gruyter, Berlin.
233. Condeelis, J. S. (1977). *Anal. Biochem.,* **77**, 195.
234. Maurer, H. R. and Dati, F. A. (1972). *Anal. Biochem.,* **46**, 19.
235. Amos, W. B. (1976). *Anal. Biochem.,* **70**, 612.
236. Matsudaira, P. T. and Burgess, D. R. (1978). *Anal. Biochem.,* **87**, 386.

237. Leaback, D. H. (1975). In *Chromatographic and Electrophoretic Techniques,* Vol. 2. *Zone Electrophoresis*, (ed. I. Smith), p. 250. W. Heinemann Medical Books Ltd, London.
238. Alligood, J. P. (1983). *Anal. Biochem.*, **134**, 284.
239. Backhaus, R., Schwippert, N., and Wolf, G. (1984). *Anal. Biochem.*, **136**, 298.
240. Dahlberg, A. E. and Grabowski, P. J. (1990). In *Gel Electrophoresis of Nucleic Acids: A Practical Approach* (2nd edn), (ed. D. Rickwood and B. D. Hames), Oxford University Press, Oxford.
241. Moulin, J., Fuchart, J. C., Dewailly, P., and Sezille, G. (1979). *Clin. Chem. Acta*, **91**, 159.
242. Lee, L. T., Lefevre, M., Wong, L., Roheim, P. S., and Thomson, J. J. (1987). *Anal. Biochem.*, **162**, 420.
243. Warren, D. F., Naughton, M. A., and Fink, L. M. (1982). *Anal. Biochem.*, **121**, 331.
244. Horejsi, V. (1984). In *Methods in Enzymology*, (ed. W. B. Jacoby), Vol. 104, p. 275. Academic Press, New York.
245. Kyte, J. (1971). *J. Biol. Chem.*, **246**, 4157.
246. Huston, L. L. (1971). *Anal. Biochem.*, **44**, 81.
247. Findlay, J. B. C. and Geisow, M. (1989). *Protein Sequencing: A Practical Approach.* IRL Press Ltd, Oxford.
248. Bray, D. and Brownlee, S. M. (1973). *Anal. Biochem.*, **55**, 213.
249. Raison, R. L. and Marchalonis, J. J. (1977). *Biochemistry*, **16**, 2036.
250. Elder, J. H., Pickett, R. A., Hampton, J., and Lerner, R. A. (1977). *J. Biol. Chem.*, **252**, 6510.
251. Cleveland, D. W., Fischer, S. G., Kirshner, M. C., and Laemmli, U. K. (1977). *J. Biol. Chem.*, **252**, 1102.
252. Wiener, A. M., Platt, T., and Weber, K. (1972). *J. Biol. Chem.*, **247**, 3242.
253. Findlay, J. B. C. (1987). In *Biological Membranes: A Practical Approach* (ed. J. B. C. Findlay and W. H. Evans), p. 179. Oxford University Press, Oxford.
254. Matsudaira, P. (1987). *J. Biol. Chem.*, **262**, 10035.
255. Ballou, B. T., McKean, D. J., Freedlender, E. F., and Smithies, O. (1973). *Proc. Natl. Acad. Sci. USA*, **73**, 4487.
256. Harris, E. and Angal, S. (1989). *Protein Purification: A Practical Approach.* Oxford University Press, Oxford.
257. Hager, D. A. and Burgess, R. R. (1980). *Anal. Biochem.*, **109**, 76.
258. Manrow, R. E. and Dottin, R. P. (1980). *Proc. Natl. Acad. Sci. USA*, **77**, 730.
259. Dottin, R. P., Manrow, R. E., Fishel, B. R., Aukerman, S. L., and Culleton, J. L. (1979). In *Methods in Enzymology*, (ed. R. Wu), Vol. 68, p. 513. Academic Press, New York.
260. Dupuis, G. and Doucet, J. P. (1981). *Biochim. Biophys. Acta*, **669**, 171.
261. Blank, A., Silber, J. R., Thelen, M. P., and Dekker, C. A. (1983). *Anal. Biochem.*, **135**, 423.
262. Weber, K. and Kuter, D. J. (1971). *J. Biol. Chem.*, **246**, 4505.
263. Konigsberg, W. H. and Henderson, H. (1983). In *Methods in Enzymology*, (ed. C. H. W. Hirs and S. N. Timasheff), Vol. 91, p. 254. Academic Press, New York.
264. Vinogradov, S. N. and Kapp, O. H. (1983). In *Methods in Enzymology*, (ed. C. H. W. Hirs and S. N. Timasheff), Vol. 91, p. 259. Academic Press, New York.
265. Anderson, D. J. and Blobel, G. (1983). In *Methods in Enzymology*, (ed. S. Fleischer and B. Fleischer), Vol. 96, p. 111. Academic Press, New York.

266. Sastre, L., Kishimoto, T. K., Gee, C., Roberts, T., and Springer, K. A. (1986). *J. Immunol.*, **137**, 1060.
267. Kurzinger, K. and Springer, T. A. (1982). *J. Biol. Chem.*, **257**, 12412.
268. Sanchez-Madrid, F., Simon, P., Thompson, S., and Springer, T. A. (1983). *J. Exp. Med.*, **158**, 586.
269. Platt, J., Weber, K., Ganem, D., and Muller, J. H. (1972). *Proc. Natl. Acad. Sci. USA*, **69**, 897.
270. Manrow, R. E. and Dottin, R. P. (1982). *Anal. Biochem.*, **120**, 181.
271. Hofemann, M., Kittsteiner-Eberle, R., and Schrier, P. (1983). *Anal. Biochem.*, **128**, 217.
272. Goldenberg, D. P. and Creighton, T. E. (1984). *Anal. Biochem.*, **138**, 1.
273. Zaccharias, R. J., Zell, T. E., Morrison, J. H., and Woodlock, J. J. (1969). *Anal. Biochem.*, **31**, 148.
274. Eckhardt, A. E., Hayes, C. E., and Goldstein, I. E. (1976). *Anal. Biochem.*, **73**, 192.
275. Rauchsen, D. (1979). *Anal. Biochem.*, **99**, 474.
276. Satoh, K. and Busch, H. (1981). *Cell Biol. Int. Rep.*, **5**, 857.
277. Green, M. R., Pastewka, J. V., and Peacock, A. C. (1973). *Anal. Biochem.*, **56**, 43.
278. See, Y. P. and Jackowski, G. (1988). In *Protein Structure: A Practical Approach,* (ed. T. E. Creighton), p. 1. Oxford University Press, Oxford.
279. Kaufmann, E., Geisler, N., and Weber, K. (1984). *FEBS Lett.*, **170**, 81.
280. Seeburg, P. M., Colby, W.W ., Capon, D. J., Goedel, D. V., and Levinson, A. D. (1984). *Nature*, **312**, 71.
281. Fasano, O., Aldrich, R., Tamanoi, F., Taparowsky, E., Furth, M., and Wigler, M. (1985). *Proc. Natl. Acad. Sci. USA*, **81**, 4008.
282. Caillet-Boudin, M. L. and Lemay, P. (1986). *Electrophoresis*, **7**, 309.

2

Isoelectric focusing

PIER GIORGIO RIGHETTI, ELISABETTA GIANAZZA,
CECILIA GELFI and MARCELLA CHIARI

1. Introduction

All fractionations which rely on differential rates of migration of sample molecules (separative transports) for example along the axis of a chromatographic column or along the electric field lines in electrophoresis, generally lead to concentration bands or zones of sample molecules which are essentially always out of equilibrium. The narrower the band or zone the steeper the concentration gradients and the greater the tendency of these gradients to dissipate spontaneously. This dissipative transport is thermodynamically driven; it relates to the tendency of entropy to break down all gradients, to maximize dilution and, during this process, to mix thoroughly all components (1). Most frequently, entropy exerts its effects via diffusion which causes molecules to move down concentration gradients and so produces band broadening and component intermixing (*Figure 1*).

The process of isoelectric focusing (IEF) in carrier ampholytes (2) and in immobilized pH gradients (IPG) (3) provides an additional force which counteracts the diffusion of carrier ampholytes and so maximizes the ratio of separative transport to dissipative transport. This substantially increases the resolution of the fractionation method. The sample focuses towards its isoelectric point (pI) driven by the voltage gradient and by the slope of the pH gradient along the separation axis (*Figure 2*). The separation can be optimized by using thin or ultrathin matrices (0.5 mm or less in thickness) and by applying very low sample loads (as permitted by high sensitivity detection techniques, such as silver and gold staining, radioactive labelling, immunoprecipitation followed by amplification with peroxidase- or alkaline phosphatase-linked secondary antibodies).

2. Conventional IEF in amphoteric buffers

2.1 General considerations

2.1.1 The standard method

IEF is an electrophoretic technique in which amphoteric compounds are fractionated according to their isoelectric points (pIs) along a continuous pH gradient (4). Con-

149

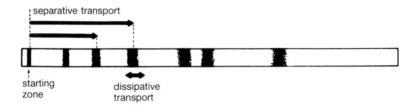

Figure 1. Illustration of separative and dissipative transport in a zone electrophoresis system. (By permission from Giddings; see ref. 1.)

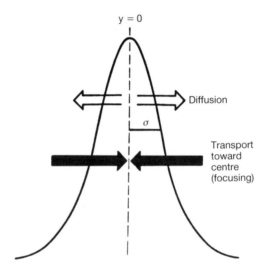

Figure 2. Illustration of the forces acting on a condensed zone in IEF. The focused zone is represented as a symmetrical Gaussian peak about its focusing point (pI; $y = 0$). Migration of sample towards the pI position is driven by the voltage gradient and by the slope of the pH gradient. σ is the standard deviation of the peak. (By courtesy of Dr O.Vesterberg.)

trary to zone electrophoresis, where the constant (buffered) pH of the separation medium establishes a constant charge density at the surface of the molecule and causes it to migrate with constant mobility (in the absence of molecular sieving), the surface charge of an amphoteric compound in IEF keeps changing, and decreasing, according to its titration curve, as it moves along a pH gradient until it reaches its equilibrium position, that is, the region where the pH matches its pI. There, its mobility equals zero and the molecule comes to a halt.

The pH gradient is created, and maintained, by the passage of an electric current through a solution of amphoteric compounds which have closely spaced pIs, encompassing a given pH range. Their electrophoretic transport causes the carrier ampholytes to stack according to their pIs, and a pH gradient, increasing from anode to cathode, is established. At the beginning of the run, the medium has a uniform pH, which equals the average pI of the carrier ampholytes (*Figure 3a*). Thus most ampholytes have a net charge and a net mobility. The most acidic carrier ampholyte moves toward the anode, where it concentrates in a zone whose pH equals its pI, while the more basic carrier ampholytes are driven toward the cathode. A less acidic ampholyte migrates adjacent and just cathodal to the previous one and so on, until all the components of the system reach a steady-state (*Figure 3b*). After this stacking process is completed, some carrier ampholytes still enter zones of higher or lower pH, by diffusion, where they are no longer in isoelectric equilibrium. But as soon as they enter these zones, the carrier ampholytes become charged and the applied

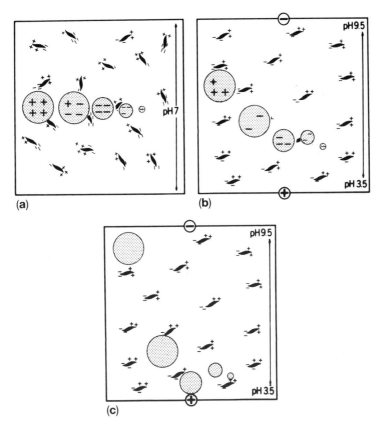

Figure 3. The principle of IEF in carrier ampholyte buffers. (a) Sample proteins are placed on a gel containing the carrier ampholyte buffers. The gel has the average pH of the mixture of carrier ampholytes. (b) When the field is applied, the small ampholytes migrate to their pIs. Due to their good buffering capacity, they form a continuous pH gradient in the gel. The larger sample proteins begin a slower migration directed by the initial pH environment. (c) Within the established pH gradient, each protein molecule migrates to its pI position, where it bears a zero net charge (courtesy of LKB Produkter AB).

voltage forces them back to their equilibrium position. This pendulum movement, diffusion versus electrophoresis, is the primary cause of the residual current observed under isoelectric steady-state conditions. Finally, as time progresses, the sample protein molecules also reach their isoelectric point (*Figure 3c*).

2.1.2 Applications and limitations

The technique only applies to amphoteric compounds and more precisely to good ampholytes with a steep titration curve around their pI, a necessary condition for any compound to focus as a narrow band. This is very seldom a problem with proteins but it may be so for short peptides, that need to contain at least one acidic, or basic, amino acid residue, in addition to the $-NH_2$ and $-COOH$ termini. Peptides which have only these terminal charges are isoelectric over the entire range from approximately pH 4 to pH 8 and so do not focus. Another limitation with short peptides is encountered at the level of the detection methods: carrier ampholytes are reactive to most peptide stains. This problem may be circumvented by using specific stains, when appropriate (5, 6) or by resorting to IPGs which do not give background reactivity to ninhydrin and other common stains for primary amino groups (e.g. dansyl chloride, fluorescamine) (7).

In practice, notwithstanding the availability of carrier ampholytes covering the pH 2.5−11 range, the practical limit of carrier ampholyte-IEF (CA-IEF) is in the pH 3.5−10 interval. Since most protein pIs cluster between pH 4 and 6 (8), this may pose a major problem only for specific applications.

When a restrictive support like polyacrylamide is used, a size limit is also imposed for sample proteins. This can be defined as the size of the largest molecules which retain an acceptable mobility through the gel. A conservative evaluation sets an upper molecular mass limit of about 750 000 when using standard techniques. The molecular form in which the proteins are separated depends mainly upon the presence of additives, such as urea and/or detergents. Moreover, supramolecular aggregates or complexes with charged ligands can be focused only if their K_d is lower than 1 μM and if the complex is stable at a pH equal to its pI (9). An aggregate with a higher K_d is easily split by the pulling force of the current.

2.1.3 Specific advantages

- IEF is an equilibrium technique; therefore the results do not depend (within reasonable limits) upon the mode of sample application, the total protein load or the time of operation.

- An intrinsic physicochemical parameter of the protein (its pI) can be measured.

- IEF requires only a limited number of chemicals, is completed within a few hours, is less sensitive than most other techniques to the skill (or lack of it) of the operator.

- IEF allows excellent resolution of proteins whose pIs differ by only 0.02 pH units (with immobilized pH gradients, up to ∼0.001 pH units); the protein bands are very sharp due to the focusing effect.

2.1.4 Carrier ampholytes

Table 1 lists the general properties of carrier ampholytes, that is, the amphoteric buffers used to generate and stabilize the pH gradient in IEF. The fundamental and performance properties listed in this table are usually required for a well-behaved IEF system, whereas the 'phenomena' properties are in fact the drawbacks or failures inherent to the technique. For instance, the 'plateau effect' or 'cathodic drift' is a slow decay of the pH gradient with time, whereby, upon prolonged focusing at high voltages, the pH gradient with the focused proteins drifts towards the cathode and is eventually lost in the cathodic compartment. There seems to be no remedy to this problem (except for abandoning CA-IEF in favour of the IPG technique), since there are complex physicochemical causes underlying it, including a strong electro-osmotic flow generated by the covalently bound negative charges of the matrix (carboxyls and sulphates in both polyacrylamide and agarose) (10). In addition, it appears that basic carrier ampholytes may bind to hydrophobic proteins, such as membrane proteins, by hydrophobic interaction. This cannot be prevented during electrophoresis, whereas ionic carrier ampholyte−protein complexes are easily split by the voltage gradient (11).

In chemical terms, carrier ampholytes are oligoamino, oligocarboxylic acids, available from different suppliers under different trade names (Ampholine from LKB Produkter AB, Pharmalyte from Pharmacia Fine Chemicals, Biolyte from BioRad, Servalyte from Serva GmbH, Resolyte from BDH). There are two basic synthetic approaches: Vesterberg's approach, which involves reacting different oligoamines (tetra-, penta- and hexa-amines) with acrylic acid (12); and the Pharmacia synthetic process, which involves the co-polymerization of amines, amino acids and dipeptides with epichlorohydrin (4).

The wide-range synthetic mixtures (pH 3−10) contain hundreds, possibly thousands, of different amphoteric chemicals with pIs evenly distributed along the

---

Table 1. Properties of carrier ampholytes

Fundamental 'classical' properties
1. Buffering ion has a mobility of zero at pI.
2. Good conductance.
3. Good buffering capacity.

Performance properties
1. Good solubility.
2. No influence on detection systems.
3. No influence on sample.
4. Separable from sample.

'Phenomena' properties
1. 'Plateau' effect (i.e. drift of the pH gradient).
2. Chemical change in sample.
3. Complex formation.

---

pH scale. Since they are obtained by different synthetic approaches, carrier ampholytes from different manufacturers are bound to have somewhat different pIs. Thus, if higher resolution is needed, particularly for two-dimensional maps of complex samples, we suggest using blends of the different commercially available carrier ampholytes. A useful blend is 50% Pharmalyte, 30% Ampholine, and 20% Biolyte (by vol.).

Carrier ampholytes from any source should have an average molecular mass of about 750 (size interval 600−900, the higher molecular mass species are usually the more acidic carrier ampholyte species). Thus carrier ampholytes should be readily separable (unless they are hydrophobically complexed to proteins) from macromolecules by gel filtration (13). Dialysis is not recommended due to the tendency of carrier ampholytes to aggregate. Salting out of proteins with ammonium sulphate seems to eliminate completely any contaminating carrier ampholytes.

A further complication arises from the chelating effect of acidic carrier ampholytes, especially towards Cu^{2+} ions, which may inactivate some metallo-enzymes (14). In addition, focused carrier ampholytes represent a medium of very low ionic strength (<1 mEq/litre at the steady-state) (15). Since the isoelectric state involves a minimum of solvation, and thus of solubility, for the protein macro-ion, there is a tendency for some proteins (e.g. globulins) to precipitate during the IEF run near their pI position. This is a severe problem in preparative runs. In analytical procedures it can be minimized by reducing the total amount of sample applied.

2.2 Equipment

2.2.1 Electrophoretic apparatus

Three major items of apparatus are required: an electrophoretic chamber, a power supply and a thermostatic unit.

Electrophoretic chamber
The optimal configuration of the electrophoretic chamber is for the lid to contain movable platinum wires (e.g. in the LKB Ultrophor 2217 and Multiphor 2, in the Pharmacia FBE3000 or in the Bio-Rad chambers models 1045 and 1415). This allows the researcher to use gels of various sizes and to apply high field strengths across just a portion of the separation path. A typical chamber is shown in *Figure 4.*

Power supply
The most suitable power supplies for IEF are those with automatic constant power operation and with voltage maxima at 2−2500 V. The minimal requirements for good resolution are a limiting voltage of 1000 V and a reliable ammeter with a full scale not exceeding 50 mA. Lower field strengths cause the protein bands to spread (resolution is proportional to the square root of the voltage gradient, $\sqrt{E}$). The ammeter monitors the conductivity and so allows periodic manual adjustment of the electrophoretic conditions to keep the power delivered as close as possible to a constant value.

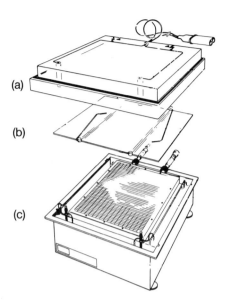

Figure 4. Drawing of the LKB Multiphor II chamber. (a) Cover lid, (b) cover plate with movable platinum electrodes, (c) base-chamber with ceramic cooling block for supporting the gel slab. (Courtesy of LKB Produkter AB.)

Thermostatic unit
Efficient cooling is extremely important for IEF because it allows high field strengths to be applied without overheating. Tap water circulation is adequate for 8 M urea gels but not acceptable for gels lacking urea. Siting the electrophoretic apparatus in a cold room may be beneficial to prevent water condensation around the unit in very humid climates, but it is inadequate as a substitute for coolant circulation.

2.2.2 Polymerization cassette
The polymerization cassette is the chamber that is used to form the gel for IEF. It is assembled from the following elements: a gel supporting plate, a spacer, a cover (moulding) plate and some clamps.

Gel supporting plate
A plain glass plate is sufficient to support the gel when detection of the separated proteins does not require processing through several solutions (e.g. when the sandwich technique for zymograms or immunoblotting are to be applied) or when the poly-acrylamide matrix is sturdy (gels >1 mm thick, >5%T). However, for thin, soft gels a permanent support is required. Glass coated with γ-methacryl-oxypropyl-trimethoxy-silane is the most reliable reactive substratum and is the most suitable for autoradiographic procedures (16) (see *Protocol 1* for the procedure). It is also

155

Protocol 1. Use of silanizing solutions

Binding silane (γ-methacryloxy-propyl-trimethoxy-silane)[a]
Two alternative procedures are available.
Either
1. Add 4 ml of silane to 1 litre of distilled water adjusted to pH 3.5 with acetic acid.
2. Leave the plates in this solution for 30 min.
3. Rinse with distilled water and dry in air.

or
1. Dip the plates for 30 sec in a 0.2% solution of silane in anhydrous acetone.
2. Thoroughly evaporate the solvent using a hair drier.
3. Rinse with ethanol if required.

In either case, store the silanized plates away from untreated glass.

Repel-Silane (dimethyl-dichloro-silane)[b]
1. Swab the glass plates with a wad impregnated with a 2% (w/v) solution of silane in 1,1,1-trichloroethane.
2. Dry the plates in a stream of air and rinse with distilled water.

[a] Prepared by Union Carbide, available through Pharmacia, Serva, or LKB.
[b] From suppliers such as Merck, Serva, etc.

---

the cheapest of such supports: dried gels can be removed with a blade and then a brush with some scrubbing powder. Unreacted silane may be hydrolysed by keeping the plates in Clorox for a few days. This step is unnecessary, however, if they have to go through successive cycles of silanization. The glass plates used as a support should not be thicker than 1.0−1.2 mm. On the other hand, thin plastic sheets designed to bind polyacrylamide gel firmly (e.g. Gel Bond PAG by Marine Colloids, PAG foils by LKB, Gel Fix by Serva) are more practical if the records of a large number of experiments have to be filed, or when different parts of the gel need to be processed independently (e.g. the first step of a two-dimensional separation or a comparison between different stains). The plastic sheet is applied to a supporting glass plate and the gel then cast onto this. The binding of the polyacrylamide matrix to these substrata, however, is not always stable and so care should be taken in using them, especially for detergent-containing gels and when using aqueous staining solutions. For good adherence, the best procedure is to cast 'empty' gels (i.e. poly-acrylamide gels lacking carrier ampholytes), wash and dry them, and allow them to reswell in the solvent of choice (see Section 3.2.2).

Spacer

U-gaskets of any thickness, between 0.2 and 5 mm can be cut from rubber sheets (para−, silicone−, nitrile−rubber). For thin gels, a few layers of Parafilm (each about 120 μm thick) can be stacked and cut with a razor blade (17). The width of such U-gaskets should be about 4 mm. In addition, cover plates with a permanent frame are commercially available (# 2117-901 from LKB). It is the same for a plastic tray with two lateral ridges for horizonal polymerization (Bio-Rad # 170-4261 and 170-4258; see *Figure 6a*). A similar device may be home-made using 250-μm-thick Dymo tape strips to form the permanent spacer frame. Mylar foil strips or self-adhesive tape may be used as spacers for 50−100-μm-thick gels. Rubber- or tape-gaskets should never be left to soak in detergent (which they absorb) but just rinsed and dried promptly.

Cover plate

Clean glass, glass coated with dimethyl-dichloro-silane (Repel Silane; *Protocol 1*) or a thick Perspex sheet are all suitable materials for the cover plate. If you wish to mould sample application pockets into the gel slab during preparation, attach Dymo tape pieces to the plate, or glue small Perspex blocks to the plate with drops of chloroform. Perspex should never be exposed to high temperatures unevenly (e.g. by being rinsed in running hot water) because it bends even if cut in thick slabs.

Clamps

Clamps of adequate size and strength should be chosen for any gel thickness. Insufficient pressure may result in leakage of the polymerizing solution. The pressure of the clamps must be applied on the gasket, never inside it.

2.3 The polyacrylamide gel matrix

2.3.1 Reagents

Stocks of dry chemicals (acrylamide, bisacrylamide, ammonium persulphate) may be kept at room temperature, provided they are protected from moisture by being stored in air-tight containers. Very large stocks are better sealed into plastic bags, together with Drierite (Merck), and stored in a freezer. TEMED stocks should also ideally be kept in a freezer, in an air-tight bottle or better, under nitrogen. Avoid contaminating acrylamide solutions with heavy metals, which can initiate its polymerization.

Acrylamide and bisacrylamide for IEF must be of the highest purity to avoid poor polymerization and strong electro-osmosis resulting from acrylic acid (18). Monomer solutions can be purified from contaminating acrylic acid by treatment with ion-exchange resin (see Chapter 7). Bisacrylamide is more hydrophobic and more difficult to dissolve than acrylamide, so start by stirring it in a small amount of lukewarm distilled water (the solution process is endothermic), then add acrylamide and water as required. *Table 2* gives the composition and general storage conditions for

Table 2. Stock solutions for polyacrylamide gel preparation

Monomer solution[a,b]

30%T, 2.5%C:	Mix 29.25 g of acrylamide and 0.75 g of bisacrylamide. Add water to 100 ml.
30%T, 3%C:	Mix 29.10 g of acrylamide and 0.90 g of bisacrylamide. Add water to 100 ml.
30%T, 4%C:	Mix 28.80 g of acrylamide and 1.20 g of bisacrylamide. Add water to 100 ml.

Initiator solution

40% (w/v) ammonium persulphate	This reagent is stable at 4°C for no more than 1 week.

Catalyst

TEMED	This reagent is used undiluted. It is stable at 4°C for several months.

[a] Monomer solutions can be stored at 4°C for ~1 month. Gels of ≥10%C (*Table 6*) require two monomer stock solutions, 30% acrylamide and 2% bisacrylamide.
[b] %T = g monomers per 100 ml; %C = g crosslinker per 100 g monomers.

monomers and catalyst solutions, while *Table 3* lists the most commonly used additives in IEF.

2.3.2 Gel formulations

In order to allow all the sample components to reach their equilibrium position at essentially the same rate, and the experiment to be terminated before the pH gradient decay process adversely affects the quality of the separation, it is best to choose a non-restrictive anti-convective support. There are virtually no theoretical but only practical lower limits for the gel concentration (the minimum being about 2.2%T, 2%C). Large pore sizes can be obtained both by decreasing %T and by either decreasing or increasing %C from the critical value of 5%. Although the pore size of polyacrylamide can be enlarged enormously by increasing the percentage of crosslinker, two undesirable effects also occur in parallel, namely increasing gel turbidity and susceptibility to syneresis (16, 21, 22). Thus, in practice, the upper limit for a useful highly-crosslinked gel appears to be about 30%C. In this respect, N,N'-(1,2-dihydroxyethylene)bisacrylamide (DHEBA), with its superior hydrophilic properties, appears superior to bisacrylamide. In contrast, N,N'-diallyltartardiamide (DATD) inhibits the polymerization process and so gives porous gels just by reducing the actual %T of the matrix. Because unpolymerized acryloyl monomers may react with amino and sulphydryl groups on proteins and, once absorbed through the skin, act as neurotoxins, the use of DATD should be avoided altogether. *Table 4* gives the upper size of proteins which will focus easily in different %T gels. Because of these limitations, agarose is more suitable than polyacrylamide for the separation of very large proteins (M_r ~1 000 000), or supramolecular complexes. *Table 5*

Table 3. Common additives for IEF

Additive	Purpose	Concentration	Limitations
Sucrose, Glycerol	To improve the mechanical properties of low %T gels and to reduce water transport and drift.	5 – 20%	The increased viscosity slightly slows the focusing process.
Glycine, Taurine	To increase the dielectric constant of the medium. This increases the solubility of some proteins (e.g. globulins) and reduces ionic interactions.	0.1 – 0.5 M	Glycine is zwitterionic between pH 4 and 8, taurine between 3 and 7. Their presence somewhat slows the focusing process and shifts the resulting gradient.
Urea[a]	Disaggregation of supramolecular complexes.	2 – 4 M	Unstable in solution especially at alkaline pH.
	Solubilization of water-insoluble proteins, denaturation of hydrophilic proteins.	6 – 8 M	Urea is soluble at ≥ 10°C; accelerates polyacrylamide polymerization, so reduce the amount of TEMED added.
Non-ionic and zwitterionic detergents[b]	Solubilization of amphiphilic proteins.	0.1 – 1%	To be added to the polymerizing solutions just before the catalysts to avoid foaming; they interfere with polyacrylamide binding to reactive substrata; are precipitated by TCA and require a specific staining protocol.

[a] See ref. 19.
[b] See ref. 20.

Table 4. Choice of polyacrylamide gel concentration

Protein M_r (upper limit for a given %T)	Gel composition
15 000	%T = 7 %C = 5
75 000	%T = 6 %C = 4
150 000	%T = 5 %C = 3
500 000	%T = 4 %C = 2.5

gives details of preparation for 4−7.5% polyacrylamide gels containing 2−8 M urea for total gel volumes ranging from 4 to 30 ml, together with the recommended amounts of catalysts. *Table 6* gives details of preparation for highly-crosslinked gels.

2.3.3 Choice of carrier ampholytes

A simple way to extend and stabilize the extremes of a wide (pH 3−10) gradient is to add acidic and basic (natural) amino acids. Thus lysine, arginine, aspartic acid and glutamic acid are prepared as individual 20 mM stock solutions containing 0.004% sodium azide and stored at 0−4°C. They are added in volumes sufficient to give 2−5 mM final concentration. To cover ranges spanning between 3 and 5 pH units, several narrow cuts of carrier ampholyte need to be blended, with the proviso that the resulting slope of the gradient will be (over each segment of the pH interval) inversely proportional to the amount of ampholytes isoelectric in that region.

Shallow pH gradients are often used to increase the resolution of sample components. However, longer focusing times and more diffuse bands will result unless the gels are electrophoresed at higher field strengths. Shallow pH gradients (shallower than the commercial 2-pH unit cuts) can be obtained in different ways:

- By subfractionating the relevant commercial carrier ampholyte blend—this can be done by focusing the carrier ampholytes at high concentration either in a continuous-flow apparatus (23) or in a sucrose density gradient, or in a Sephadex slurry.
- By allowing trace amounts of acrylic acid to induce a controlled cathodic drift during prolonged runs—this is effective in the acidic pH region, but it is rather difficult to obtain reproducible results from run to run (24).
- By preparing gels containing different concentrations of carrier ampholytes in adjacent strips (25) or with different thickness along the separation path (26).
- By adding specific amphoteric compounds (spacers) at high concentration (27).

In the last case, two kinds of ampholytes may be used for locally flattening the pH gradient: 'good' and 'poor'. Good carrier ampholytes, those with a small $pI - pK_1$ (i.e. possessing good conductivity and buffering capacity at the pI) are able to focus in narrow zones. Low concentrations (5−50 mM) are sufficient to induce

Table 5. Recipes for IEF gels (4 – 7.5%T and 2 – 8 M urea)

Gel vol. (ml)	30%T monomer solution (ml)								2% Carrier ampholytes[a] (ml)		Urea (g)				TEMED (µl)	40% ammonium persulphate[b] (µl)
	4%T	4.5%T	5%T	5.5%T	6%T	6.5%T	7%T	7.5%T	A	B	2 M	4 M	6 M	8 M		
30	4.00	4.50	5.00	5.50	6.00	6.50	7.00	7.50	1.50	1.88	3.60	7.20	10.40	14.40	9.0	30
25	2.34	3.75	4.17	4.58	5.00	5.42	5.83	6.25	1.25	1.56	3.00	6.00	9.00	12.00	7.5	25
20	2.66	3.00	3.34	3.66	4.00	4.33	4.67	5.00	1.00	1.25	2.40	4.80	7.20	9.60	6.0	20
15	2.00	2.27	2.50	2.75	3.00	3.25	3.50	3.75	0.75	0.94	1.80	3.60	5.40	7.20	4.5	15
10	1.33	1.50	1.66	1.83	2.00	2.16	2.33	2.50	0.50	0.63	1.20	2.40	3.60	4.80	3.0	10
8	1.06	1.21	1.33	1.46	1.60	1.73	1.86	2.00	0.40	0.50	0.96	1.92	2.88	3.84	2.4	8
7	0.93	1.05	1.17	1.28	1.40	1.51	1.63	1.75	0.35	0.44	0.84	1.68	2.52	3.36	2.1	7
6	0.80	0.90	1.00	1.10	1.20	1.30	1.40	1.50	0.30	0.38	0.72	1.44	2.16	2.88	1.8	6
5	0.66	0.75	0.83	0.91	1.00	1.08	1.16	1.25	0.25	0.31	0.60	1.20	1.80	2.40	1.5	5
4	0.53	0.60	0.66	0.73	0.80	0.86	0.93	1.00	0.20	0.25	0.48	0.96	1.44	1.92	1.2	4

[a] A, for 40% solution (Ampholine, Servalyte, Resolite); B, for Pharmalyte.
[b] To be added after degassing the solution and just before pouring it into the gel mould.

Table 6. Recipes for highly crosslinked gels (per 1%*T* and per millilitre volume)

%*C*	Volume of 30% acrylamide (μl per ml gel)	Volume of 2% bisacrylamide (μl per ml gel)
10	30.0	50
15	28.3	75
20	26.6	100
25	25.0	125
30	23.3	150

a pronounced flattening of the pH curve around their pIs. A list of these carrier ampholytes is given in *Table 7* (2). Poor carrier ampholytes, on the other hand, form broad plateaux in the region of their pI, and should be used at high concentrations (0.2–1.0 M). Their presence usually slows down the focusing process. Some of them are listed in *Table 8*.

A note of caution: in an IEF system, acids and bases are distributed according to their dissociation curve, in a pattern that may be defined as protonation (or deprotonation) stacking. If large amounts of these compounds originate from the samples (in the form of buffers), the limits of the pH gradient shift from the expected values. For example, 2-mercaptoethanol, as added to denatured samples for two-dimensional PAGE analysis, lowers the alkaline end from pH 10 to about pH 7.5 (28). This effect, however, may sometimes be usefully exploited. For example, high levels of TEMED in the gel mixture appear to stabilize alkaline pH gradients (29).

2.4 Gel preparation and electrophoresis

Protocol 2 outlines the series of steps required for an IEF run. The key steps are described in more detail below.

Protocol 2. CA-IEF flow sheet

1. Assemble the gel mould.
2. Mix all components of the polymerizing mixture, except ammonium persulphate.
3. Degas the mixture for a few minutes, and re-equilibrate (if possible) with nitrogen.
4. Add the required amount of ammonium persulphate stock solution and mix.
5. Transfer the mixture to the gel mould and overlay it with water.
6. Leave the mixture to polymerize (at least 1 h at room temperature or 30 min at 37°C).

Protocol 2. *(continued)*

 7. Open the gel mould and blot any moisture from the gel edges and surface.

 8. Lay the gel on the cooling block of the electrophoretic chamber.

 9. Apply the electrodic strips.

 10. Pre-run the gel, if appropriate.

 11. Apply the samples.

 12. Run the gel.

 13. Measure the pH gradient.

 14. Reveal the protein bands using a suitable detection procedure.

2.4.1 Assembling the gel mould

Assembly of the gel mould is shown in *Figure 5a−c*. *Figure 5a* shows the preparation of the slot former which will give 20 sample application slots in the final gel. In the method shown, the slot former is prepared by gluing a strip of embossing tape onto the cover plate and cutting rectangular tabs, with the dimensions shown, using a scalpel. A rubber U-gasket covering three edges of the cover plate is also glued to the plate. In *Figure 5b*, a sheet of Gel Bond PAG film is applied to the supporting glass plate via a thin film of water. When using reactive polyester foils as the gel backing, use glycerol rather than water. Avoid leaving airpockets behind the gel backing sheet since this will produce gels of uneven thickness which will create distortions in the pH gradient. Also ensure that the backing sheet is cut flush with the glass support since any overhang easily bends. Finally, the gel mould is assembled using clamps (*Figure 5c*). Assembly is usually made easier by wetting and blotting the rubber gasket just before use. Note that the cover plate has three V-shaped indentations on one edge which allow insertion of a pipette or syringe tip into the narrow gap between the two plates of the mould to facilitate pouring the gel.

2.4.2 Filling the mould

One of four methods may be chosen:

i. By gravity

This procedure uses a vertical cassette with a rubber gasket U-frame glued to the cover plate (*Figure 5*). It may be used for a range of thicknesses of gel slab. For thick gels (down to 0.7 mm), the gel mixture may be directly poured into the mould (tilted at an angle of 30°) from the mouth of the vacuum flask if the support and cover plates are arranged to be of slightly different lengths or slightly offset. For thinner gels, the solution may be transferred with a Pasteur pipette (with its tip resting against one limb of the gasket) or with a syringe. Irrespective of the gel thickness, filling the mould is made considerably easier if the cover plate has V-indentations along its free edge as shown in *Figure 5a*. The gel mixture can then be transferred

Table 7. Good carrier ampholytes acting as spacers

Carrier ampholyte	pI	Carrier ampholyte	pI	Carrier ampholyte	pI
Aspartic acid	2.77	p-Aminobenzoic acid	3.62	Lysyl–glutamic acid	6.10
Glutathione	2.82	Glycyl–aspartic acid	3.63	Histidyl–glycine	6.81
Aspartyl–tyrosine	2.85	m-Aminobenzoic acid	3.93	Histidyl–histidine	7.30
o-Aminophenylarsonic acid	3.00	Diiodotyrosine	4.29	Histidine	7.47
Aspartyl–aspartic acid	3.04	Cystinyl–diglycine	4.74	L-Methylhistidine	7.67
p-Aminophenylarsonic acid	3.15	α-Hydroxyasparagine	4.74	Carnosine	8.17
Picolinic acid	3.16	α-Aspartyl–histidine	4.92	α,β-Diaminopropionic acid	8.20
Glutamic acid	3.22	β-Aspartyl–histidine	4.94	Anserine	8.27
β-Hydroxyglutamic acid	3.29	Cysteinyl–cysteine	4.96	Tyrosyl–arginine	8.38–8.68
Aspartyl–glycine	3.31	Pentaglycine	5.32	L-Ornithine	9.70
Isonicotinic acid	3.35	Tetraglycine	5.40	Lysine	9.74
Nicotinic acid	3.44	Triglycine	5.59	Lysyl–lysine	10.04
Anthranilic acid	3.51	Tyrosyl–tyrosine	5.60	Arginine	10.76
		Isoglutamine	5.85		

Table 8. Poor carrier ampholytes acting as spacers

Carrier ampholyte	pK$_1$	pK$_2$
Carrier ampholytes with pIs 7 – 8		
β-Alanine	3.55	10.24
γ-Aminobutyric acid	4.03	10.56
δ-Aminovaleric acid	4.26	10.77
ε-Aminocaproic acid	4.42	11.66
'Good' buffers with acidic pIs[a]		
Mes	1.3	6.1
Pipes	1.3	6.8
Aces	1.3	6.8
Bes	1.3	7.1
Mops	1.3	7.2
Tes	1.3	7.5
Hepes	1.3	7.5
Epps	1.3	8.0
Taps	1.3	8.4

[a] For all these 'Good' buffers, the pK$_1$ is approx. 1.3 (cysteic acid).

simply by using a pipette or a syringe with its tip resting on one of these indentations (*Figure 5d*).

Avoid filling the mould too fast, which will create turbulence and trap air bubbles. If an air bubble appears, stop pouring the solution and try to remove the bubble by tilting and knocking the mould. If this manoeuvre is unsuccessful, displace the bubble with a 1 cm wide strip of polyester foil.

ii. By capillarity

This method is mainly used for casting reasonably thin gels (0.5 – 1.0 mm). It requires a horizontal sandwich with two lateral spacers (see *Figure 6a*). The solution is fed either from a pipette or from a syringe fitted with a short piece of fine-bore tubing. It is essential that the solution flows evenly across the whole width of the mould during casting. If an air bubble appears, do not stop pumping in the solution or you will produce more bubbles. Remove all of the air bubbles at the end, using a strip of polyester foil. A level table is not mandatory but the mould should be left lying flat until the gel is completely polymerized.

iii. The flap technique (30)

This procedure is again mainly used for preparing thin gels. A 20 – 50% excess of gel mixture is poured along one edge of the cover (with spacers on both sides) (*Figure 6b*) and the support plate is slowly lowered on it (*Figure 6c*). Air bubbles can be avoided by using clean plates. If bubbles do get trapped, they can also be removed by lifting and lowering the cover plate once more. Since this method may lead to spilled unpolymerized acrylamide solution, take precautions for its containment (wear gloves and use absorbent towels to mop up the excess solution).

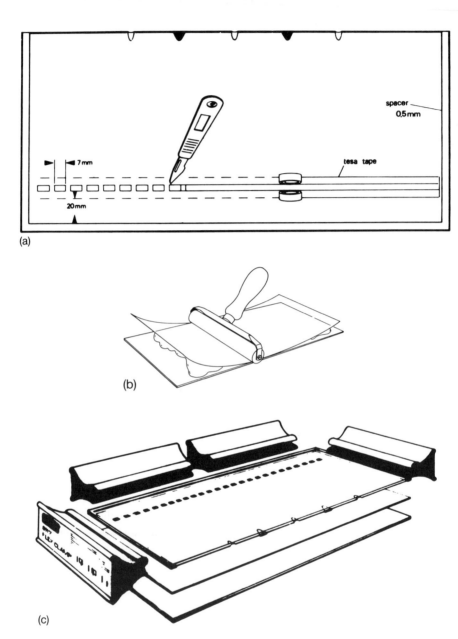

Figure 5. Preparation of the gel cassette. (a) Preparation of the slot former: onto the cover plate (bearing the rubber gasket U-frame) is glued a strip of tesa tape out of which rectangular tabs are cut with a scalpel. (b) Application of the Gel Bond PAG film to the supporting glass plate. (c) Assembling the gel cassette. (d) Pouring the gelling solution in the vertically-standing cassette using a pipette. (Courtesy of LKB Produkter AB.)

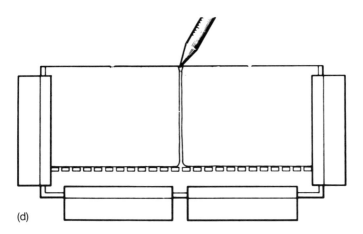

(d)

Figure 5 (cont.)

iv. Wedge gels

This variation (31) is illustrated in *Figure 7*. A gel of a linearly varying thickness (50–300 μm) from anode to cathode (or vice versa, depending on the electrode polarity) is produced by using spacers of different thicknesses at the two extremities of the glass cover. The electrode polarity is chosen so that the protein zones focus in the thin region of the gel, while the sample is applied onto the thick region of the gel. There are two reasons for such an arrangement; in the thick gel region, pH gradient distortions due to salts and buffers present in the sample are minimized while at steady-state the protein bands, by focusing in the thin gel region, are subjected to higher voltage gradients, which produce band sharpening.

In all cases, use clean plates and avoid grease: it hampers polymerization. If there are leakages, search for chipped gaskets or plates, and discard them. If the mould looks intact, check the pressure of the clamps; all clamps must rest on the gasket.

2.4.3 Gel polymerization

Protocol 3. Polymerization of the gel for IEF

1. Mix all the components of the gel formulation (from *Table 5*, except TEMED, ammonium persulphate and detergents, when used) in a cylinder, add distilled water to the required volume and transfer the mixture to a vacuum flask.

2. Degas the solution using suction from a water pump. The operation should be continued as long as gas bubbles develop from the liquid. Manually swirl the mixture or use a magnetic stirrer during degassing. The use of a mechanical vacuum pump is desirable for the preparation of very soft gels but is unnecessarily cumbersome for urea gels (urea would crystallize). At the beginning of the degassing step the solution should be at, or above, room

Protocol 3. *continued*

temperature, to decrease oxygen solubility. Its cooling during degassing is then useful in slowing down the onset of polymerization.

3. If possible, it is beneficial to re-equilibrate the degassed solution against nitrogen instead of air (even better with argon, which is denser than air). From this step on, the processing of the gel mixture should be as prompt as possible.

4. Add detergents, if required. If they are viscous liquids, prepare a stock solution beforehand (e.g. 30%) but do not allow detergent to take up more than 5% of the total volume. Mix briefly with a magnetic stirrer.

5. Add TEMED and then ammonium persulphate, immediately mix by swirling, and then transfer the mixture to the gel mould. Carefully overlayer with water or butanol.

6. Leave to polymerize for at least 1 h at room temperature (overnight for highly crosslinked gels) or 30 min at 37°C. Never tilt the mould to check whether the gel has polymerized; if it has not when you tilt it, then the top never will polymerize. Instead, the differential refractive index between liquid (at the top and usually around the gasket) and gel phase is an effective, and safe, index of polymerization and is shown by the appearance of a distinct line.

7. The gels may be stored in their moulds for a couple of days in a refrigerator. For longer storage (up to 2 weeks for neutral and acidic pH ranges), it is better to disassemble the mould and (after covering the gels with Parafilm and wrapping with Saran-Wrap) store them in a moist box.

8. Before opening the mould, the gel should be allowed to cool at room temperature, if polymerized at a higher temperature. Laying one face of the mould onto the cooling block of the electrophoresis unit may facilitate opening the cassette. Carefully remove the overlay by blotting and remove the clamps.

    (a) If the gel is cast on a plastic foil, remove its glass support, then carefully peel the gel from the cover plate.

    (b) If the gel is polymerized on silanized glass, simply lever the two plates apart with a spatula or a blade.

    (c) If the gel is not bound to its support, place the mould on the bench, with the plate to be removed uppermost and one side protruding a few centimetres from the edge of the table. Insert a spatula at one corner and force against the upper plate. Turn the spatula gently until a few air bubbles form between the gel and the plate, then use the spatula as a lever to open the mould. Wipe any liquid from the gel surface with a moistened tissue but be careful; keep moving the swab to avoid it sticking to the surface. If you get bubbles on both sides of the gel, try at the next corner of the mould. If the gel separates from both glass plates and folds up, you may still be able to salvage it, provided the gel thickness is at least 400 μm and it is at least 5%T.

Protocol 3. *continued*

Using a microsyringe, force a small volume of water below the gel, then carefully make the gel lie flat again by manoeuvring it with a gloved hand. Remove any remaining air bubbles using the needle of the microsyringe. Cover the gel with Parafilm, and gently roll it flat with a rubber roller (take care not to damage it with excess pressure). Carefully remove any residual liquid by blotting.

As a variant to the above procedures, one can polymerize 'empty' gels (i.e. devoid of carrier ampholytes), wash and dry them and reswell them in the appropriate carrier ampholyte solution. This is a direct application of IPG technology (3). While it has

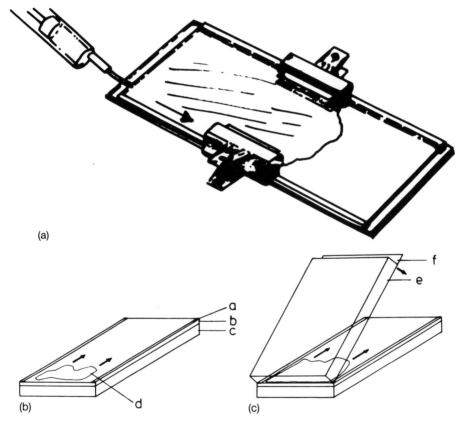

Figure 6. Casting of thin gel plates. (a) By capillary filling of a horizontally placed cassette; (b and c) by the 'flap' technique. Components shown are a, spacer strips; b, silanized glass plate or polyester film; c, glass base plate; d, polymerization mixture; e, glass cover plate; f, cover film. Diagram (a), by courtesy of LKB; (b) and (c), by permission from Radola (see ref. 30).

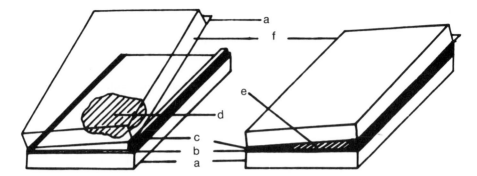

Figure 7. Casting of wedge-shaped gels. a, Supporting glass slab; b, silanized polyester foil; c, spacers of variable thickness (50 and 300 μm); d, gel mixture; e, polymerized gel; f, glass cover; and g, plastic cover foil. (By permission from Pflug; see ref. 31.)

been stated that the carrier ampholyte equilibrium process takes one week (32), we suggest a better and much faster approach, as commonly utilized for IPG gels. After polymerization (as above, but in the absence of carrier ampholytes) the gel is washed three times in 500 ml of distilled water each time to remove catalysts and unreacted monomers. The washed gel is equilibrated (20 min with shaking) in 1.5% glycerol and finally dried onto Gel Bond PAG foil. It is essential that the gel does not bend. Therefore, before drying, the foil should be made to adhere to a supporting glass plate, taking care not to trap air bubbles, and fastening it in position with clean, rust-proof clamps. Drying must be at room temperature, in front of a fan. Finally, the dried gel is mounted back in the polymerization cassette and allowed to reswell in the appropriate carrier ampholyte solution (this process takes at most a couple of hours, not a week!) (refer to *Figure 13* for more details) (46).

In ultrathin gels pH gradients are sensitive to the presence of salts, including TEMED and ammonium persulphate. Moreover, unreacted monomers are toxic and noxious to proteins. The preparation, washing and re-equilibration of 'empty' gels removes these components from the gel and so avoids these problems.

2.4.4 Sample loading and electrophoresis

Protocol 4. Electrophoretic procedure

1. Set the cooling unit at $2-4°C$ for normal gels, at $8-10°C$ for 6 M urea and at $10-12°C$ for 8 M urea gels. To enable rapid heat transfer, pour a few millilitres of a non-conductive liquid (distilled water, 1% non-ionic detergent or light kerosene) onto the cooling block of the electrophoretic chamber. Form a continuous liquid layer between the gel support and the apparatus as you gently lower the plate into place. Avoid trapping air bubbles and splashing water onto or around the gel. Should this happen, remove all liquid by careful blotting. When the gel is narrower than the cooling block, apply the plate on its middle.

Protocol 4. *continued*

   If this is not possible, or if the electrode lid is too heavy to be supported by
   just a strip of gel, insert a wedge (e.g. several layers of Parafilm) between
   the electrodes and cooling plate.

2. Cut electrode strips (e.g. from Whatman No. 17 filter paper, ~5 mm wide)
   about 3 mm shorter than the gel width. Saturate them with electrode solutions
   (see *Table 9*). However, they should not be dripping; blot them on paper towels
   if required. Note that most paper exposed to alkaline solutions becomes swollen
   and fragile.

3. Transfer the strips onto the gel. They must be parallel and aligned with the
   electrodes. The cathodic strip firmly adheres to the gel: do not try to change
   its position once applied. Avoid cross-contaminating the electrode strips
   (including with your fingers). If any electrode solution spills over the gel, blot
   it off immediately, rinse with a few drops of water and blot again. Check that
   the wet electrode strips do not exceed the size of the gel and cut away any
   excess pieces. Be sure to apply the most alkaline solution at the cathode and
   the most acidic at the anode (if you are concerned that you may have misplaced
   them, the colour of the NaOH soaked paper is yellowish but, of course, you
   may also check this with pH paper). If you discover a mistake at this point,
   simply turn your plate around or change the electrode polarity. However, there
   is no remedy after the current has been on for a while.

4. The salt content of the samples should be kept as low as possible. Dialyse them
   against glycine (any suitable concentration) or dissolve them in diluted carrier
   ampholytes. When buffers are required, use low molarity buffers composed
   of weak acids and bases if possible.

5. The samples are best loaded into pre-cast slots. These should not be deeper
   than 50% of the gel thickness. If they are longer than a couple of millimetres
   and it is necessary to pre-electrophorese the gel (see step 6), the pockets should
   be filled with dilute carrier ampholytes. After the pre-electrophoresis (see
   Table 10) remove this solution by blotting. Then apply the samples. The amount
   of sample applied should fill the slots. Try to equalize the volumes and the
   salt content among different samples. After about 30 min of electrophoresis
   at high voltage (*Table 10*), the content of the slots can be removed by blotting
   and new aliquots of the same samples loaded. The procedure can be repeated
   a third time. Alternatively, the samples can be applied to the gel surface,
   absorbed into pieces of filter paper. Different sizes and material of different
   absorbing power are used to accommodate various volumes of liquid (e.g. a
   5 × 10 mm tab of Whatman No. 1 can retain about 5 μl, and Whatman No. 3
   MM about 10 μl). Up to three layers of paper may be stacked. If the exact
   amount of sample loaded has little importance, the simplest procedure is to
   dip the tabs into it, then blot them to remove excess sample. Otherwise, dry
   tabs are aligned on the gel and measured volumes of the samples are applied

Protocol 4. *continued*

with a micropipette. For stacks of paper pieces, feed the solution slowly from one side rather than from the top. Do not allow a pool of sample to drag around its pad, but stop feeding liquid to add an extra tab when required. This method of application of samples is not suitable for samples containing alcohol.

6. Most samples may be applied to the gel near the cathode without pre-electrophoresis. However, pre-running is advisable if the proteins are sensitive to oxidation or unstable at the average pH of the gel before the run. It is not suitable for those proteins with a tendency to aggregate upon concentration, or whose solubility is increased by high ionic strength and dielectric constant, or which are very sensitive to pH extremes. Anodic application should be excluded for high-salt samples and for proteins denatured at acidic pH (e.g. many serum components). Besides this guideline, however, as a rule the optimal conditions for sample loading should be determined experimentally, together with the minimum focusing time, with a pilot run, in which the sample of interest is applied in different positions of the gel and at different times. Smears, or lack of sharpness of the bands after long focusing time, denote improper sample handling and protein alteration.

7. After electrophoresis the pH gradient in the focused gel may be read using a contact electrode. The most general approach, however, is to cut a strip along the focusing path (0.5−2.0 cm wide, with an inverse relation to the gel thickness). Segments between 3 and 10 mm long are then cut from this and eluted for about 15 min in 0.3−0.5 ml of 10 mM KCl (or with the same urea concentration as is present in the gel). For alkaline pH gradients, the processing of the gel should be as quick as possible, the elution medium should be thoroughly degassed and air-tight vials flushed with nitrogen should be used. A narrow-bore combination electrode is suitable for the measurements. For identification purposes, a defined temperature for the coolant and for the pH measurement is sufficient to assess an operational pH. However, for a proper physicochemical characterization, the temperature differences should be corrected for as suggested in *Table 11* (which also gives corrections for the presence of urea).

Table 9. IEF electrode solutions

Solution	Application	Concentration
H_3PO_4	Anolyte for all pH ranges	1.0 M
H_2SO_4	Anolyte for very acidic pH ranges ($pH_a < 4$)[a]	0.1 M
CH_3COOH	Anolyte for alkaline pH ranges ($pH_a > 7$)[a]	0.5 M
NaOH	Catholyte for all pH ranges	1.0 M[b]
Histidine	Catholyte for acidic pH ranges ($pH_c < 5$)[a]	0.2 M
Tris	Catholyte for acidic and neutral pH ranges ($pH_c < 5$)[a]	0.5 M

[a] pH_a, pH_c represent the lower and higher extremes of the pH range, respectively.
[b] Store in air-tight plastic bottles.

Table 10. Electrophoresis conditions

Pre-electrophoresis

	Voltage (V)	Time (min)
Thick gels (>1.5 mm)	200	30
Thin gels (<1.5 mm)	400	30

Electrophoresis of samples

	Voltage (V)[a]	Time (h)
Thick gels (2 mm)	200	5
Thick gels (1 mm)	400	3
Thin gels (0.5 mm)	600	2
Thin gels (0.25 mm)	800	1.5

[a] Electrophoresis is carried out at constant power so as to give an initial voltage as indicated.

Table 11. Compensation factors

Effect of temperature[a]		Effect of urea[b]	
pH	Δ25 – 4°C	pH	Δ[c]
—	—	3.0	+0.09
3.5	+0.0	3.5	+0.08
4.0	+0.06	4.0	+0.08
4.5	+0.10	4.5	+0.07
5.0	+0.14	5.0	+0.06
5.5	+0.20	5.5	+0.06
6.0	+0.23	6.0	+0.05
6.5	+0.28	6.5	+0.05
7.0	+0.36	7.0	+0.05
7.5	+0.39	7.5	+0.06
8.0	+0.45	8.0	+0.06
8.5	+0.52	8.5	+0.06
9.0	+0.53	9.0	+0.06
9.5	+0.54	9.5	+0.05
10.0	+0.55	—	—

[a] From ref. 70
[b] From ref. 71
[c] For 1 M urea. The effect is proportional to urea concentration.

2.5 General protein staining

Table 12 gives a list, with pertinent references, of some of the most common protein stains used in IEF. Detailed recipes are given below.

2.5.1 Coomassie Blue G-250 (33)

Protocol 5. Staining procedure

1. Add 2.0 g of Coomassie Blue G-250 to 400 ml of 2 M H_2SO_4.
2. Dilute the suspension with 400 ml of water and stir for at least 3 h.

Protocol 5. *continued*

3. To the filtrate (through Whatman No. 1 paper) add 89 ml of 10 M KOH and 120 ml of 100% (w/v) TCA while stirring.

4. Keep the gel immersed in this mixture until the required stain intensity is obtained.

5. To remove all salts and to increase the colour contrast, rinse the gel extensively with water.

The advantages of this protocol are that only one step is required (i.e. no protein fixation, no destaining), peptides down to approximately 1500 molecular mass can be detected, there is little interference from carrier ampholytes and, finally, the staining mixture has a long shelf life. The small amount of dye that may precipitate with time can be removed by filtration or washed from the surface of the gels with liquid soap.

2.5.2 Coomassie Blue R-250/CuSO$_4$ (34)

Protocol 6. Staining procedure

1. Dissolve 1.09 g of CuSO$_4$ in 650 ml of water and then add 190 ml of glacial acetic acid.

2. Mix this solution with 250 ml of ethanol containing 0.545 g of Coomassie Blue R-250.

3. Stain the gel (without previous fixation) in this dye solution for between 30 min and a few hours, depending on its thickness.

4. Destain the gel in several changes of 600 ml of ethanol, 140 ml of glacial acetic acid and 1260 ml of water.

During immersion in the staining solution, unsupported gels shrink and their surface becomes sticky. Therefore avoid any contact with dry surfaces.

Table 12. Protein staining methods

Application	Sensitivity	References
General use	low	33
General use	medium	34
General use	high	35
General use	very high	36
In presence of detergents	medium	34
In presence of detergents	high	37

2.5.3 Coomassie Blue R-250/sulphosalicylic acid (35)

Heat the gels for 15 min at 60°C in a solution of 1.0 g of Coomassie Blue R-250 in 280 ml of methanol and 730 ml of water containing 110 g of TCA and 35 g of sulphosalicylic acid (SSA). Destain at 60°C in 500 ml of ethanol, 160 ml of glacial acetic acid and 1340 ml of water. Precipitation of the dye at the gel surface with this stain is a common problem (remove such precipitate with alkaline liquid soap).

2.5.4 Silver stain (36)

A typical method for silver staining is described in *Protocol 7*

Protocol 7. Silver staining procedure[a]

Reagent	Volume	Time
1. 12% TCA	500 ml	30 min
2. 50% methanol, 12% acetic acid	1 litre	30 min
3. 1% HJO$_4$	250 ml	30 min
4. 10% ethanol, 5% acetic acid	200 ml	10 min (repeat 3×)
5. 3.4 mM potassium dichromate, 3.2 mM nitric acid	200 ml	5 min
6. Water	200 ml	30 sec (repeat 4×)
7. 12 mM silver nitrate	200 ml	5 min in the light 25 min in the dark
8. 0.28 M sodium carbonate, 0.05% formaldehyde	300 ml 300 ml	30 sec (repeat 2×) several minutes
9. 10% acetic acid	100 ml	2 min
10. Photographic fixative (Kodak Rapid Fix)	200 ml	10 min
11. Water	several litres	extensive washes
12. Finally, remove any silver precipitate on the gel surface using a swab.		

[a] Each step in the protocol is given in the order in which it should be used.

2.5.5 Coomassie Blue G-250/urea/perchloric acid

Protocol 8. Staining procedure

1. Fix the protein bands for 30 min at 60°C in a solution of 147 g of TCA and 44 g of sulphosalicylic acid (SSA) in 910 ml of water.
2. Prepare the staining solution by dissolving 0.4 g of Coomassie Blue G-250 and 39 g of urea in about 800 ml of water, and adding, immediately before

Protocol 8. *continued*

use and with vigorous stirring, 29 ml of 70% $HClO_4$. Bring the volume to 1 litre with distilled water.

3. Stain the gels for 30 min at 60°C in this dye mixture.

4. Destaining requires 4–22 h, in 100 ml of acetic acid, 140 ml of ethanol, 200 ml of ethyl acetate and 1560 ml of water.

This staining protocol may be applied in the presence as well as in the absence of detergents.

2.6 Specific protein detection methods

Table 13 lists some of the most common detection techniques for specific types of proteins: it is given as an example and with appropriate references, but it is not meant to be exhaustive.

2.7 Quantitation of the focused bands

Some general rules about densitometry are worth repeating.

- The scanning photometer should have a spatial resolution of the same order as the fractionation technique.
- The stoichiometry of the protein–dye complex varies among different proteins.
- This same relationship is linear only over a limited range of protein concentration.
- At high absorbance, the limitations of the spectrophotometer interfere with the photometric measurement.

Bearing these points in mind, for an accurate, or at least meaningful, quantitation, a working range should be devised by electrophoresis and staining of scalar concentrations of the material under investigation. Under the actual experimental conditions,

Table 13. Specific protein detection techniques

Proteins detected	Technique	Reference
Glycoproteins	PAS (periodic acid – Schiff) stain	38
Lipoproteins	Sudan Black stain	39
Radioactive proteins	Autoradiography	40
	Fluorography	41
Enzymes	Zymograms[a]	42
Antigens	Immunoprecipitation *in situ*	43
	Print-immunofixation	44
	Blotting	45

[a] The concentration of buffer in the assay medium usually needs to be increased in comparison with zone electrophoresis to counteract the buffering action by carrier ampholytes.

Pier Giorgio Righetti et al.

all bands must give a signal not exceeding the linearity range. Set the monochromator at a wavelength different from λ_{max} whenever appropriate, and possible. If a band of special interest corresponds to a minor component, and it would be difficult to visualize it without overloading the major component(s), its amount should be referred either to the total protein load or to another minor component.

2.8 Troubleshooting

2.8.1 Waviness of bands near the anode

This may be caused by:

- Carbonation of the catholyte: in this case, prepare fresh NaOH with degassed distilled water and store properly in sealed plastic container.
- Excess catalysts: reduce the amount of ammonium persulphate.
- Too long sample slots: fill them with dilute carrier ampholytes.
- Too low a concentration of carrier ampholytes: check the gel formulation.

To alleviate the problem, it is usually beneficial to add low concentrations of sucrose, glycine or urea, and to apply the sample near the cathode. To salvage a gel during the run, as soon as the waves appear apply a new anodic strip soaked with a weaker acid (e.g. acetic acid instead of phosphoric acid) inside the original one, and move the electrodes closer to one another.

2.8.2 Burning along the cathodic strip

This may be caused by:

- The formation of a zone of pure water at pH 7: add to your acidic pH range a 10% solution of either the 3–10 or the 6–8 range ampholytes.
- Hydrolysis of the acrylamide matrix after prolonged exposure to alkaline pH: choose a weaker base, if adequate, and, unless a pre-run of the gel is strictly required, apply the electrode strips after loading the samples.

2.8.3 pH gradients different from expected

- For acidic and alkaline pH ranges, the problem is alleviated by the choice of anolytes and catholytes whose pH is close to the extremes of the pH gradient.
- Alkaline pH ranges should be protected from carbon dioxide by flushing the electrophoretic chamber with moisture-saturated nitrogen (or better with argon) and by surrounding the plate with pads soaked in NaOH. It is worth remembering that pH readings on unprotected alkaline solutions become meaningless within half an hour or so.
- A large amount of a weak acid or base, supplied as sample buffer, may shift the pH range (2-mercaptoethanol is one of such bases). The typical effect of the addition of urea is to increase the apparent pIs of the carrier ampholytes (see *Table 11*).

177

- It may correspond to cathodic drift; to counteract this:

  (a) Reduce the running time to the required minimum (as experimentally determined for the protein of interest, or for a coloured marker of similar molecular mass).

  (b) Increase the viscosity of the medium (with sucrose, glycerol, or urea).

  (c) Reduce the amount of ammonium persulphate.

  (d) Remove acrylic acid impurities by recrystallizing acrylamide and bisacrylamide, and by treating the monomer solution with mixed-bed ion-exchange resin.

  (e) For a final cure, incorporate into the gel matrix a reactive base, such as 2-dimethylamino-propyl-methacrylamide (Polyscience) (10); its optimal concentration (of the order of 1 μM) should be experimentally determined for the system being used.

2.8.4 Sample precipitation at the application point

If large amounts of material precipitate at the application point, even when the molecular mass of the sample proteins is well below the limits recommended in *Table 4*, the trouble is usually caused by protein aggregation.

- Try applying the sample in different positions on the gel, with and without pre-running: some proteins might be altered only by a given pH.

- If you have evidence that the sample contains high molecular mass components, reduce the value of %T of the polyacrylamide gel.

- If you suspect protein aggregation brought about by the high concentration of the sample (for example when the problem is reproduced by disc electrophoresis runs) do not pre-run and set a low voltage (100−200 V) for several hours to avoid the concentrating effect of an established pH gradient at the beginning of the run. Also consider decreasing the protein load and switching to a more sensitive detection technique. The addition of detergent and/or urea is usually beneficial.

- If the proteins are only sensitive to the ionic strength and/or the dielectric constant of the medium (in this case they perform well in disc electrophoresis and are precipitated if dialysed against distilled water), increasing the carrier ampholyte concentration, adding glycine or taurine, and sample application without pre-running may overcome the problem.

- The direct choice of denaturing conditions (8 M urea, detergents, 2-mercapto-ethanol) very often minimizes these solubility problems, dissociating proteins (and macromolecular aggregates) to polypeptide chains.

2.9 Preparative aspects

There are several preparative methodologies utilizing the CA-IEF technique. A classical one is in vertical columns filled with a sucrose density gradient, which is

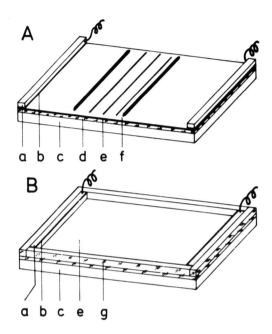

Figure 8. Preparative IEF. (**A**) Small-scale separations. (**B**) Trough for large-scale preparations. (a) electrode, (b) filter paper pad soaked with electrode solution, (c) cooling block, (d) glass plate, (e) gel layer, (f) focused proteins, (g) trough. (By permission from Radola, see ref. 46.)

then emptied and collected in fractions at the end of the IEF run. Although over the years LKB has sold more than 3000 columns (which should thus be available in any major biochemistry department) they are seldom utilized today. Thus we will restrict our description to a single method which has become quite popular, Radola's granulated gel technique, while referring the readers to a review elsewhere (4) for a complete survey of all other variants.

Radola (30) has described a method for large-scale preparative IEF in troughs coated with a suspension of granular gels, such as Sephadex G-25 Superfine (7.5 g per 100 ml) or Sephadex G-200 Superfine (4 g per 100 ml) or BioGel P-60 (4 g per 100 ml). The trough consists of a glass or quartz plate at the bottom of a Perspex (lucite) frame (*Figure 8*). Various sizes of trough may be used; typical dimensions are 40 × 20 cm or 20 × 20 cm, with a gel layer thickness of up to 10 mm. The total gel volume in the trough varies from 300 to 800 ml. The procedure is described in *Protocol 9*.

Protocol 9. Large-scale preparative IEF in troughs

   1. Pour a slurry of Sephadex (containing 1% Ampholine) into the trough and allow to dry in air to the correct consistency. Achieving the correct consistency of the slurry before focusing seems to be the key to successful results. Best results

Protocol 9. *continued*

seem to be obtained when 25% of the water in the gel has evaporated and the gel does not move when the plate is inclined at 45°. Alternative, faster protocols for obtaining a gel of adequate consistency have been proposed. One of them suggests removing excess liquid by blotting with paper wicks; another (mostly intended for analytical applications) disposes of the drying step and describes an arrangement for spreading the gel at its final concentration. Run the plate on a cooling block maintained at 2−10°C.

2. Apply the electric field via flat electrodes or platinum bands which make contact with the gel through absorbent paper pads soaked in 1 M sulphuric acid at the anode, and 2 M ethylenediamine at the cathode. In most preparative experiments initial voltages of 10−15 V/cm and terminal voltages of 20−40 V/cm have been used. As much as 0.05 W/cm^2 are well tolerated in 1 cm thick granulated gel layers.

3. Samples may be mixed with the gel suspension or added to the surface of preformed gels either as a streak or from the edge of a glass slide. Larger sample volumes may be mixed with the dry gel ($\sim$60 mg Sephadex G-200 per ml) and poured into a slot in the gel slab. Samples may be applied at any position between the electrodes. Ideally, the gel slab should be covered with a lid during electrophoresis to prevent it from drying out.

4. After focusing, locate the proteins by the paper print technique (32). Focused proteins in the surface gel layer are absorbed onto a strip of filter paper which is then dried at 110°C. The proteins may then be stained directly with dyes of low sensitivity such as Light Green SF or Coomassie Violet R-150 after removing ampholytes by washing in acid. Alternatively, proteins can be detected directly in gels cast on a quartz plate by densitometry at 270−280 nm. The pH gradient in the gel can be measured *in situ* with a combination microelectrode sliding on a calibrated ruler.

Radola's technique offers the advantages of combining high resolution, high sample load and easy recovery of focused components. As much as 5−10 mg protein per millilitre of gel suspension may be fractionated in wide pH range gels. Purification of 10 g of pronase E in 800 ml of gel suspension has been reported (47). At these high protein loads, even uncoloured samples can be easily detected, since they appear in the gel as translucent zones.

Proteins are easily recovered in a small volume at relatively high concentrations by elution. The absence of sucrose is a further advantage in the subsequent separation of proteins from ampholytes. As there is no sieving effect for macromolecules above the exclusion limits of the Sephadex, high molecular mass substances, such as virus particles, can be focused without steric hindrance. The system has high flexibility, since it allows analytical, small- and large-scale preparative runs in the same trough, merely by varying the gel thickness. When no suitable methods are available for detecting specific biological activities by the paper print technique, it is possible to fractionate the gel and test for activity in eluates. This also applies to the recovery of protein bands at the end of the IEF run. For this purpose, LKB has produced a fractionation grid (*Figure 9*) (together with a trough and a sample applicator) which allows recovery of 30 fractions, that is, one fraction each 8 mm along a separation axis of 25 cm. After scraping off each fraction with a spatula, the protein is eluted by placing the gel into a syringe equipped with glass-wool as the bottom filter, adding a volume of buffer, and ejecting the eluate with the syringe piston.

3. Immobilized pH gradients

3.1 General considerations

3.1.1 The problems of conventional IEF

Table 14 lists some of the major problems associated with conventional IEF using amphoteric buffers. Some of them are quite severe; for example (*Table 14,* point 1) low ionic strength often induces near-isoelectric precipitation and smearing of proteins, even in analytical runs at low protein loads. The problem of uneven conductivity is magnified in poor ampholyte mixtures, like the Poly-Sep 47 (a mixture of 47 amphoteric and non-amphoteric buffers, claimed to be superior to carrier ampholytes) (48). Due to their poor composition, huge conductivity gaps form along

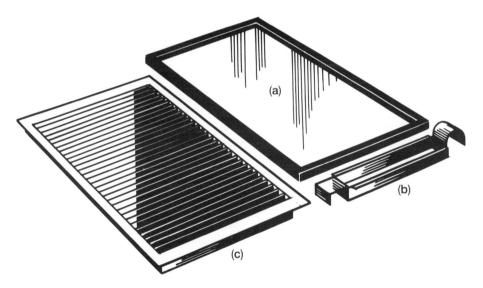

Figure 9. Trough (a), sample applicator (b) and fractionation grid (c) for preparative IEF in granulated gels using the LKB Multiphor apparatus. After IEF, the grid (c) is lowered onto the trough (a) thus fractionating the gel layer into 30 fractions. Each fraction is scooped up from each channel with a spatula, transferred to a small column and eluted in buffer. (By courtesy of LKB Produkter AB.)

Table 14. Problems with carrier ampholyte focusing

1. Medium of very low and unknown ionic strength.
2. Uneven buffering capacity.
3. Uneven conductivity.
4. Unknown chemical environment.
5. Not amenable to pH gradient engineering.
6. Cathodic drift (pH gradient instability).

the migration path, against which proteins of different pIs are stacked. The results are simply appalling (49). Cathodic drift (*Table 14,* point 6) is also a major unsolved problem of CA-IEF, resulting in extensive loss of proteins at the gel cathodic extremity upon prolonged runs. For all these reasons, as a result of an intensive collaboration with Dr Bjellqvist's group in Stockholm and Dr Görg's group in Munich (3) in 1982, we launched the technique of immobilized pH gradients (IPGs).

3.1.2 The Immobiline matrix

IPGs are based on the principle that the pH gradient, which exists prior to the IEF run itself, is copolymerized, and thus insolubilized, within the fibres of a polyacryl-amide matrix. This is achieved by using, as buffers, a set of seven non-amphoteric, weak acids and bases, having the following general chemical composition: $CH_2=CH-CO-NH-R$, where R denotes either three different weak carboxyl groups, with pKs 3.6, 4.4 and 4.6 or four tertiary amino groups, with pKs 6.2, 7.0, 8.5 and 9.3 (*Table 15*) available under the trade name Immobiline (from Pharmacia). During gel polymerization, these buffering species are efficiently incorporated into the gel (84−86% conversion efficiency at 50°C for 1 h). Immobiline-based pH gradients can be cast in the same way as conventional polyacrylamide gradient gels, using a density gradient to stabilize the Immobiline concentration gradient, with the aid of a standard, two-vessel gradient mixer (see Chapter 1, Section 9.3.3). As shown by their chemical formulae, these buffers are no longer amphoteric, as in conventional IEF, but are bifunctional. At one end of the molecule is located the buffering (or titrant) group, and at the other end is an acrylic double bond which disappears during immobilization of the buffer on the gel matrix. The three carboxyl Immobilines have rather small temperature coefficients (dpK/dT) in the 10−25°C range, due to their small standard heats of ionization ($\approx$ 1 kcal/mol) and thus exhibit negligible pK variations in this temperature interval. On the other hand, the four basic Immobilines exhibit rather large ΔpKs (as much as $\Delta pK = 0.44$ for the pK 8.5 species) due to their larger heats of ionization (6−12 kcal/mol). Therefore, for reproducible runs and pH gradient calculations, all the experimental parameters have been fixed at 10°C.

Temperature is not the only variable that affects Immobiline pKs (and therefore the actual pH gradient generated). Additives in the gel that change the water structure (e.g. urea) or lower its dielectric constant, and the ionic strength of the solution, alter their pK values. The largest changes, in fact, are due to the presence of urea:

Table 15. Apparent pK values of Immobiline chemicals[a]

pK	In water		In polyacrylamide gel[b] %T = 5, %C = 3		In polyacrylamide gel[b] %T = 5, %C = 3 glycerol (25% (w/v)		Physical state at room temperature
	10°C	25°C	10°C	25°C	10°C	25°C	
Acids with carboxyl as the buffering group							
3.6	3.57	3.58	—	—	3.68 ± 0.02	3.75 ± 0.02	solid
4.4	4.39	4.39	4.30 ± 0.02	4.36 ± 0.02	4.40 ± 0.03	4.47 ± 0.03	solid
4.6	4.60	4.61	4.51 ± 0.02	4.61 ± 0.02	4.61 ± 0.02	4.71 ± 0.02	solid
Bases with tertiary amines as the buffering group							
6.2	6.41	6.23	6.21 ± 0.05	6.15 ± 0.03	6.32 ± 0.08	6.24 ± 0.07	solid
7.0	7.12	6.97	7.06 ± 0.07	6.96 ± 0.05	7.08 ± 0.07	6.95 ± 0.06	solid
8.5	8.96	8.53	8.50 ± 0.06	8.38 ± 0.06	8.66 ± 0.09	8.45 ± 0.07	liquid
9.3	9.64	9.28	9.59 ± 0.08	9.31 ± 0.07	9.57 ± 0.06	9.30 ± 0.05	liquid

[a] pK values measured with glass surface electrode without any correction at an ionic strength of 10^{-2}.
[b] Mean value of ten determinations. Due to the slow response of the electrode the pK values for the amines are uncertain.

183

acidic Immobilines increase their pK in 8 M urea by as much as 0.9 pH units, while the basic Immobilines increase their pK by only 0.45 pH units (50). Detergents in the gel (2%) do not alter the Immobiline pK, suggesting that they are not incorporated into the surfactant micelle. For generating extended pH gradients, we use two additional chemicals which are strong titrants having pKs well outside the desired pH range. One is QAE (quaternary amino ethyl) acrylamide (p$K > 12$) and the other is AMPS (2-acrylamido-2-methyl propane sulphonic acid, p$K \approx 0.8$), both available from Polysciences Inc. (Warrington, PA).

Figure 10 depicts the process of focusing in an IPG matrix. The proteins are placed on a gel with a pre-formed, immobilized pH gradient (represented by carboxyl and tertiary amino groups grafted to the polyacrylamide chains; *Figure 10a*). When the

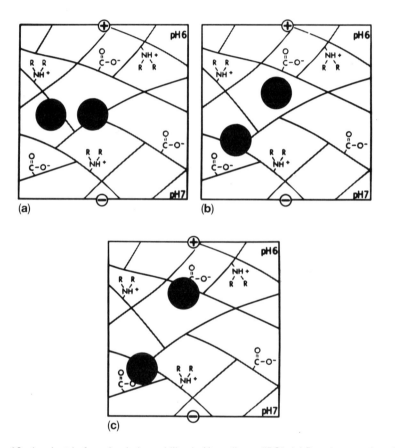

Figure 10. Isoelectric focusing in immobilized pH gradients (IPG). (a) Proteins are placed in a gel with a pre-formed IPG gradient. The gel contains the buffering and titrant groups covalently bonded to the polyacrylamide matrix. (b) and (c), when the field is applied, the sample proteins migrate to the point in the gel where they have zero net charge (the pI). Higher resolution may be obtained by preparing a narrow or ultranarrow pH gradient covering the specific region of interest. (By courtesy of LKB Produkter AB.)

184

field is applied (*Figure 10b*) only the sample molecules (and any ungrafted ion) migrate in the electric field. Upon termination of electrophoresis (*Figure 10c*) the proteins are separated into stationary, isoelectric zones. Due to the possibility of designing stable pH gradients at will, separations have been reported in only 0.1 pH unit-wide gradients over the entire separation axis leading to an extremely high resolving power (ΔpI = 0.001 pH unit, see Section 2.1.3).

3.1.3 Narrow and ultra-narrow pH gradients

We define gradients from 0.1 to 1 pH unit as narrow (towards the 1 pH unit limit) and ultra-narrow (close to the 0.1 pH unit limit) gradients. Within these limits we work on a tandem principle — that is we choose a buffering Immobiline, either a base or an acid, with its pK within the pH interval we want to generate, and a non-buffering Immobiline, then an acid or a base, respectively, with its pK at least 2 pH units removed from either the minimum or maximum of our pH range. The titrant will provide equivalents of acid or base to titrate the buffering group but will not itself buffer in the desired pH interval. For these calculations, we used to resort to modified Henderson — Hasselbalch equations and to rather complex monograms found in the LKB application note No. 321. In a later note (No. 324) 58 gradients are listed, each one pH unit wide, starting with the pH 3.8 − 4.8 interval and ending with the pH 9.5 − 10.5 range, separated by 0.1 pH unit increments. In *Table 16* are recipes giving the Immobiline volume which must be added to give 15 ml of mixture in the acidic (mixing) chamber to obtain pH$_{min}$ and the corresponding volume for the basic (reservoir) chamber of the gradient mixer needed to generate pH$_{max}$ of the desired pH interval. For 1 pH unit gradients between the limits pH 4.6 − 6.1 and pH 7.2 − 8.4 there are wide gaps in the pKs of neighbouring Immobilines and so three Immobilines need to be used to generate the desired pH$_{min}$ and pH$_{max}$ values (*Table 16*). As an example, take the pH 4.6 − 5.6 interval. There are no available Immobilines with pKs within this pH region, so the nearest species, pKs 4.6 and 6.2, will act as both partial buffers and partial titrants. A third Immobiline is needed in each vessel, a true titrant that will bring the pH to the desired value. As titrant for the acidic solution (pH$_{min}$) we use pK 3.6 Immobiline and for pH$_{max}$ we use pK 9.3 Immobiline (*Table 16*).

If a narrower pH gradient is needed, it can be derived from any of the 58 one pH intervals given in *Table 16* by a simple linear interpolation of intermediate Immobiline molarities. Suppose that from a pH 6.8 − 7.8 range, which is excellent for most haemoglobin (Hb) analyses, we want to obtain a pH gradient of pH 7.1 − 7.5, which will resolve neutral mutants that cofocus with HbA. *Figure 11* shows the graphical method. The limiting molarities of the two Immobilines in the 1 pH unit interval are joined by a straight line (because the gradient is linear), and then the new pH interval is defined according to experimental needs (in our case pH 7.1 − 7.5). Two new lines are drawn from the two new limits of the pH interval, parallel to the ordinates (broken vertical lines). Where they intersect the two sloping lines defining the two Immobiline molarities, four new lines (dashed) are drawn parallel to the abscissa and four new molarities of the Immobilines defining the new pH

Table 16. 1pH unit gradients: volumes of Immobiline for 15 ml of each starting solution[a,b]

Control pH at 20°C	Volume (μl) 0.2 M Immobiline pK — Acidic dense solution							pH range	mid point	Control pH at 20°C	Volume (μl) 0.2 M Immobiline pK — Basic light solution						
	3.6	4.4	4.6	6.2	7.0	8.5	9.3				3.6	4.4	4.6	6.2	7.0	8.5	9.3
3.84 ± 0.03	—	750	—	—	—	—	159	3.8–4.8	4.3	4.95 ± 0.06	—	750	—	—	—	—	591
3.94 ± 0.03	—	710	—	—	—	—	180	3.9–4.9	4.4	5.04 ± 0.07	—	810	—	—	—	—	667
4.03 ± 0.03	—	—	755	—	—	—	157	4.0–5.0	4.5	5.14 ± 0.06	—	—	745	—	—	—	584
4.13 ± 0.03	—	—	713	—	—	—	177	4.1–5.1	4.6	5.23 ± 0.07	—	—	803	—	—	—	659
4.22 ± 0.03	—	—	689	—	—	—	203	4.2–5.2	4.7	5.33 ± 0.08	—	—	884	—	—	—	753
4.32 ± 0.03	—	—	682	—	—	—	235	4.3–5.3	4.8	5.42 ± 0.10	—	—	992	—	—	—	871
4.42 ± 0.03	—	—	691	—	—	—	275	4.4–5.4	4.9	5.52 ± 0.12	—	—	1133	—	—	—	1021
4.51 ± 0.04	—	—	716	—	—	—	325	4.5–5.5	5.0	5.61 ± 0.14	—	—	1314	—	—	—	1208
4.64 ± 0.05	562	—	600	863	—	—	—	4.6–5.6	5.1	5.69 ± 0.04	—	—	863	863	—	—	105
4.75 ± 0.05	458	—	675	863	—	—	—	4.7–5.7	5.2	5.79 ± 0.04	—	—	863	863	—	—	150
4.86 ± 0.04	352	—	750	863	—	—	—	4.8–5.8	5.3	5.90 ± 0.04	—	—	863	863	—	—	202
4.96 ± 0.03	218	—	863	863	—	—	—	4.9–5.9	5.4	5.99 ± 0.03	—	—	863	863	—	—	248
5.07 ± 0.03	158	—	863	863	—	—	—	5.0–6.0	5.5	6.09 ± 0.04	—	—	863	803	—	—	338
5.17 ± 0.04	113	—	863	863	—	—	—	5.1–6.1	5.6	6.20 ± 0.04	—	—	863	713	—	—	443
5.24 ± 0.18	1251	—	—	1355	—	—	—	5.2–6.2	5.7	6.34 ± 0.04	337	—	—	724	—	—	—
5.33 ± 0.12	1055	—	—	1165	—	—	—	5.3–6.3	5.8	6.43 ± 0.03	284	—	—	694	—	—	—
5.43 ± 0.12	899	—	—	1017	—	—	—	5.4–6.4	5.9	6.53 ± 0.03	242	—	—	682	—	—	—
5.52 ± 0.09	775	—	—	903	—	—	—	5.5–6.5	6.0	6.63 ± 0.03	209	—	—	686	—	—	—
5.62 ± 0.07	676	—	—	817	—	—	—	5.6–6.6	6.1	6.73 ± 0.03	182	—	—	707	—	—	—
5.71 ± 0.06	598	—	—	755	—	—	—	5.7–6.7	6.2	6.82 ± 0.03	161	—	—	745	—	—	—
5.81 ± 0.06	536	—	—	713	—	—	—	5.8–6.8	6.3	6.92 ± 0.03	144	—	—	803	—	—	—
5.91 ± 0.05	486	—	—	689	—	—	—	5.9–6.9	6.4	7.02 ± 0.03	131	—	—	884	—	—	—
6.01 ± 0.05	447	—	—	682	—	—	—	6.0–7.0	6.5	7.12 ± 0.03	120	—	—	992	—	—	—
6.10 ± 0.04	416	—	—	691	—	—	—	6.1–7.1	6.6	7.22 ± 0.03	112	—	—	1133	—	—	—
6.11 ± 0.11	972	—	—	—	1086	—	—	6.2–7.2	6.7	7.21 ± 0.03	262	—	—	—	686	—	—
6.21 ± 0.09	833	—	—	—	956	—	—	6.3–7.3	6.8	7.31 ± 0.03	224	—	—	—	682	—	—
6.30 ± 0.08	722	—	—	—	857	—	—	6.4–7.4	6.9	7.41 ± 0.03	195	—	—	—	694	—	—
6.40 ± 0.07	635	—	—	—	783	—	—	6.5–7.5	7.0	7.50 ± 0.03	171	—	—	—	724	—	—
6.49 ± 0.06	565	—	—	—	732	—	—	6.6–7.6	7.1	7.60 ± 0.03	152	—	—	—	771	—	—
6.59 ± 0.05	509	—	—	—	699	—	—	6.7–7.7	7.2	7.70 ± 0.03	137	—	—	—	840	—	—

Table 16. continued

Control pH at 20°C	Acidic dense solution 3.6	4.4	4.6	6.2	7.0	8.5	9.3	pH range	mid point	Control pH at 20°C	Basic light solution 3.6	4.4	4.6	6.2	7.0	8.5	9.3
6.69 ± 0.05	465	–	–	–	683	–	–	6.8–7.8	7.3	7.80 ± 0.03	125	–	–	–	934	–	–
6.78 ± 0.04	430	–	–	–	684	–	–	6.9–7.9	7.4	7.90 ± 0.03	116	–	–	–	1058	–	–
6.88 ± 0.04	403	–	–	–	701	–	–	7.0–8.0	7.5	8.00 ± 0.03	108	–	–	–	1217	–	–
6.98 ± 0.04	381	–	–	–	736	–	–	7.1–8.1	7.6	8.09 ± 0.03	103	–	–	–	1422	–	–
7.21 ± 0.06	1028	–	–	–	750	750	–	7.2–8.2	7.7	8.36 ± 0.05	548	–	–	–	750	750	–
7.31 ± 0.06	983	–	–	–	750	750	–	7.3–8.3	7.8	8.46 ± 0.05	503	–	–	–	750	750	–
7.41 ± 0.05	938	–	–	–	750	750	–	7.4–8.4	7.9	8.56 ± 0.05	458	–	–	–	750	750	–
7.66 ± 0.15	1230	–	–	–	–	1334	–	7.5–8.5	8.0	8.76 ± 0.04	331	–	–	–	–	720	–
7.75 ± 0.12	1037	–	–	–	–	1149	–	7.6–8.6	8.1	8.85 ± 0.03	279	–	–	–	–	692	–
7.85 ± 0.10	885	–	–	–	–	1004	–	7.7–8.7	8.2	8.95 ± 0.03	238	–	–	–	–	682	–
7.94 ± 0.08	764	–	–	–	–	893	–	7.8–8.8	8.3	9.05 ± 0.06	206	–	–	–	–	687	–
8.04 ± 0.07	667	–	–	–	–	810	–	7.9–8.9	8.4	9.14 ± 0.06	180	–	–	–	–	710	–
8.13 ± 0.06	591	–	–	–	–	750	–	8.0–9.0	8.5	9.24 ± 0.06	159	–	–	–	–	750	–
8.23 ± 0.06	530	–	–	–	–	710	–	8.1–9.1	8.6	9.34 ± 0.06	143	–	–	–	–	810	–
8.33 ± 0.05	482	–	–	–	–	687	–	8.2–9.2	8.7	9.44 ± 0.06	130	–	–	–	–	893	–
8.43 ± 0.04	443	–	–	–	–	682	–	8.3–9.3	8.8	9.54 ± 0.06	119	–	–	–	–	1004	–
8.52 ± 0.04	413	–	–	–	–	692	–	8.4–9.4	8.9	9.64 ± 0.06	111	–	–	–	–	1149	–
8.62 ± 0.04	389	–	–	–	–	720	–	8.5–9.5	9.0	9.74 ± 0.06	105	–	–	–	–	1334	–
8.40 ± 0.14	1208	–	–	–	–	–	1314	8.6–9.6	9.1	9.50 ± 0.06	325	–	–	–	–	–	716
8.49 ± 0.12	1021	–	–	–	–	–	1133	8.7–9.7	9.2	9.59 ± 0.06	275	–	–	–	–	–	691
8.59 ± 0.10	871	–	–	–	–	–	992	8.8–9.8	9.3	9.69 ± 0.06	235	–	–	–	–	–	682
8.68 ± 0.08	753	–	–	–	–	–	884	8.9–9.9	9.4	9.79 ± 0.06	203	–	–	–	–	–	689
8.78 ± 0.07	659	–	–	–	–	–	803	9.0–10.0	9.5	9.88 ± 0.06	177	–	–	–	–	–	713
8.87 ± 0.06	584	–	–	–	–	–	745	9.1–10.1	9.6	9.98 ± 0.06	157	–	–	–	–	–	755
8.97 ± 0.05	525	–	–	–	–	–	707	9.2–10.2	9.7	10.08 ± 0.06	141	–	–	–	–	–	817
9.07 ± 0.04	478	–	–	–	–	–	686	9.3–10.3	9.8	10.18 ± 0.06	129	–	–	–	–	–	903
9.16 ± 0.07	440	–	–	–	–	–	682	9.4–10.4	9.9	10.28 ± 0.06	119	–	–	–	–	–	1017
9.26 ± 0.07	410	–	–	–	–	–	694	9.5–10.5	10.0	10.38 ± 0.06	111	–	–	–	–	–	1165

[a] From LKB Application Note 324 (1984). The pH range (given in the middle column) is the one existing in the gel during the run at 10°C. For controlling the pH of the starting solutions, the values (control pH) are given at 20°C.

[b] When using the standard gel cassette (*Figure 5*), the volumes given are sufficient to prepare two gels.

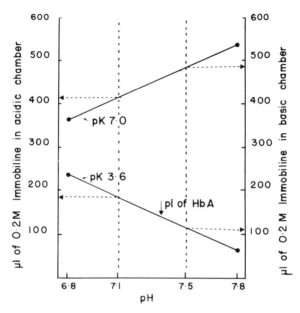

Figure 11. Graphic representation of the preparation of narrow (up to 1 pH unit) IPG gradients on the 'tandem' principle. The limiting molarities of pK 7.0 (buffering species) and pK 3.6 (titrant) Immobilines needed to generate a pH 6.8 – 7.8 interval, as obtained directly from *Table 16*, are plotted on a graph. These points are joined by straight lines, and the new molarities needed to generate any narrower pH gradients within the stated pH intervals are obtained by simple linear interpolation (broken vertical and horizontal lines). In this example a narrow pH 7.1 – 7.5 gradient is graphically derived. (By permission from Rochette *et al.*; see ref. 52.)

interval are read directly on the ordinates. This process can be repeated for any desired pH interval down to ranges as narrow as 0.1 pH units.

3.1.4 Extended pH gradients using strong titrants

Linear pH gradients are obtained by arranging for an even buffering power throughout. The latter could be ensured only by ideal buffers spaced apart by $\Delta pK = 1$. In practice, there are only six Immobiline species rather than seven since the pK 4.4 and 4.6 species are so close in pK. Therefore other approaches must be used to solve this problem. Two methods are possible. In one approach (constant buffer concentration), the concentration of each buffer is kept constant throughout the span of the pH gradient and 'holes' of buffering power are filled by increasing the amounts of the buffering species bordering the largest ΔpKs. In the other approach (varying buffer concentration) the variation in concentration of different buffers along the width of the desired pH gradient results in a shift in each buffer's apparent pK, together with the ΔpK values evening out. Recipes for pH gradients 2 and 3 pH units wide, whose intervals can be generated without the aid of strongly acidic and basic titrants, that is using the first approach, are given in *Table 17*. In *Table 18*

Table 17. Broad pH gradients: volumes of Immobiline for 15 ml of each starting solution[a,b]

Control pH at 20°C	Volume (μl) 0.2 M Immobiline pK Acidic dense solution							pH range	Mid point	Control pH at 20°C	Volume (μl) 0.2 M Immobiline pK Basic light solution					
	3.6	4.4	4.6	6.2	7.0	8.5	9.5				3.6	4.4	6.2	7.0	8.5	9.3
3.53 ± 0.06	299	–	223	157	–	–	–	3.5–5.0	4.25	5.06 ± 0.07	212	310	465	–	–	–
4.00 ± 0.06	569	–	99	439	–	–	–	4.0–6.0	5.0	6.09 ± 0.14	390	521	276	235	–	722
4.54 ± 0.06	415	–	240	499	–	–	–	4.5–6.5	5.5	6.53 ± 0.05	–	570	244	219	–	297
5.08 ± 0.03	69	–	428	414	–	–	–	5.0–7.0	6.0	7.01 ± 0.06	–	474	270	287	284	320
5.56 ± 0.03	–	–	450	354	113	–	–	5.5–7.5	6.5	7.51 ± 0.09	347	–	236	325	329	–
6.06 ± 0.08	435	–	–	323	208	44	–	6.0–8.0	7.0	8.11 ± 0.09	286	–	174	278	362	–
6.56 ± 0.13	771	–	–	276	185	538	–	6.5–8.5	7.5	8.66 ± 0.06	192	–	153	232	189	–
7.03 ± 0.24	1349	–	–	–	272	372	845	7.0–9.0	8.0	8.94 ± 0.07	484	–	–	925	139	546
7.50 ± 0.11	668	–	–	–	445	226	348	7.5–9.5	8.5	9.37 ± 0.06	207	–	–	329	366	346
8.10 ± 0.07	399	–	–	–	364	355	94	8.0–10.0	9.0	9.89 ± 0.05	91	–	–	–	–	289
4.01 ± 0.05	578	–	110	450	–	–	–	4.0–8.0	5.5	7.02 ± 0.14	302	738	151	269	346	876
5.03 ± 0.12	702	–	254	416	133	346	–	5.0–8.0	6.5	8.12 ± 0.07	175	123	131	345	–	–
6.04 ± 0.14	779	–	–	402	93	364	80	6.0–9.0	7.5	9.01 ± 0.06	241	–	161	449	237	225
6.98 ± 0.07	542	–	–	–	378	351	–	7.0–10.0	8.5	9.88 ± 0.05	90	–	–	324	350	280

[a] From LKB Application Note No. 324 (1984).
[b] When using the standard gel cassette (Figure 5), the volumes given are sufficient to prepare two gels.

189

Table 18. 4, 5 and 6 pH unit wide immobilized pH gradients[a]

pH 4 – 8 Gradients

Initial pH: 4.000
Final pH: 8.000
Notes: 4.068 – 8.153(7.813):
4.084 – 8.004(7.689)
Buffer concentrations

Initial pH: 4.015
Final pH: 8.008
Notes: 4.076 – 8.112: 4.091 – 7.986
Buffer concentrations

pK	0.80	Cham 1	11.085	Cham 2	0.000	pK	0.80	Cham 1	8.643	Cham 2	0.000
pK	3.57	Cham 1	0.864	Cham 2	0.864	pK	3.57	Cham 1	1.420	Cham 2	0.000
pK	4.51	Cham 1	5.259	Cham 2	5.259	pK	4.51	Cham 1	4.560	Cham 2	6.519
pK	6.21	Cham 1	4.167	Cham 2	4.167	pK	6.21	Cham 1	5.222	Cham 2	1.975
pK	7.06	Cham 1	2.379	Cham 2	2.379	pK	7.06	Cham 1	2.326	Cham 2	4.586
pK	8.50	Cham 1	6.432	Cham 2	6.432	pK	8.50	Cham 1	3.276	Cham 2	3.822
pK	12.00	Cham 1	0.000	Cham 2	0.924	pK	12.00	Cham 1	0.000	Cham 2	3.130

pH 4 – 9 Gradients

Initial pH: 4.000
Final pH: 9.000
Notes:
4.071 – 9.086(8.007):4.088 – 8.925(7.876)
Buffer concentrations

Initial pH: 4.019
Final pH: 8.996
Notes: 4.076 – 8.980: 4.081 – 8.825
Buffer concentrations

pK	0.80	Cham 1	11.649	Cham 2	0.000	pK	0.80	Cham 1	11.793	Cham 2	0.000
pK	3.57	Cham 1	0.657	Cham 2	0.657	pK	3.57	Cham 1	2.832	Cham 2	4.130
pK	4.51	Cham 1	5.394	Cham 2	5.394	pK	4.51	Cham 1	4.253	Cham 2	7.780
pK	6.21	Cham 1	3.888	Cham 2	3.888	pK	6.21	Cham 1	5.353	Cham 2	2.635
pK	7.06	Cham 1	2.943	Cham 2	2.943	pK	7.06	Cham 1	0.585	Cham 2	4.252
pK	8.50	Cham 1	4.734	Cham 2	4.734	pK	8.50	Cham 1	4.814	Cham 2	3.056
pK	9.59	Cham 1	1.857	Cham 2	1.857	pK	9.59	Cham 1	4.196	Cham 2	4.654
pK	12.00	Cham 1	0.000	Cham 2	3.399	pK	12.00	Cham 1	0.000	Cham 2	7.415

pH 5 – 9 Gradients

Initial pH: 5.000
Final pH: 9.000
Notes: 5.076 – 9.086(7.544):
5.085 – 8.926(7.457)
Buffer concentrations

Initial pH: 5.022
Final pH: 9.037
Notes: 5.100 – 9.080: 5.110 – 8.921
Buffer concentrations

pK	0.80	Cham 1	7.527	Cham 2	0.000	pK	0.80	Cham 1	11.404	Cham 2	0.000
pK	4.51	Cham 1	7.089	Cham 2	7.089	pK	4.51	Cham 1	7.989	Cham 2	3.498
pK	6.21	Cham 1	3.408	Cham 2	3.408	pK	6.21	Cham 1	3.102	Cham 2	3.493
pK	7.06	Cham 1	3.180	Cham 2	3.180	pK	7.06	Cham 1	1.908	Cham 2	2.933
pK	8.50	Cham 1	4.683	Cham 2	4.683	pK	8.50	Cham 1	11.020	Cham 2	4.091
pK	9.59	Cham 1	1.839	Cham 2	1.839	pK	9.59	Cham 1	1.694	Cham 2	2.692
pK	12.00	Cham 1	0.000	Cham 2	4.464	pK	12.00	Cham 1	0.000	Cham 2	0.438

pH 5 – 10 Gradients

Initial pH: 5.000
Final pH: 10.000
Notes: 5.076 – 9.892(8.171):
5.086 – 9.745(8.034)
Buffer concentrations

Initial pH: 4.993
Final pH: 10.034
Notes: 5.068 – 9.933: 5.075 – 9.779
Buffer concentrations

pK	0.80	Cham 1	9.102	Cham 2	0.000					
pK	4.51	Cham 1	7.179	Cham 2	7.179	pK	0.80	Cham 1	7.605	Cham 2	0.295
pK	6.21	Cham 1	3.210	Cham 2	3.210	pK	4.51	Cham 1	6.260	Cham 2	0.805
pK	7.06	Cham 1	3.531	Cham 2	3.531	pK	6.21	Cham 1	4.044	Cham 2	0.454
pK	8.50	Cham 1	3.729	Cham 2	3.729	pK	7.06	Cham 1	3.717	Cham 2	5.677
pK	9.59	Cham 1	4.275	Cham 2	4.275	pK	8.50	Cham 1	3.088	Cham 2	4.197
pK	12.00	Cham 1	0.000	Cham 2	5.922	pK	9.59	Cham 1	1.730	Cham 2	3.687

190

Table 18 (continued)

pH 6 – 10 Gradients

Initial pH: 6.000					Initial pH: 5.994					
Final pH: 10.000					Final pH: 9.996					
Notes: 6.039 – 9.886: 5.988 – 9.733					Notes: 6.035 – 9.876: 5.986 – 9.723					
Buffer concentrations					Buffer concentrations					
pK	0.80	Cham 1	12.060	Cham 2	0.000	pK	3.57	Cham 1	12.919	Cham 2	1.370
pK	0.80	Cham 1	1.323	Cham 2	1.323	pK	6.21	Cham 1	3.748	Cham 2	4.583
pK	6.21	Cham 1	3.351	Cham 2	3.351	pK	7.06	Cham 1	3.339	Cham 2	4.962
pK	7.06	Cham 1	3.603	Cham 2	3.603	pK	8.50	Cham 1	3.585	Cham 2	3.283
pK	8.50	Cham 1	3.711	Cham 2	3.711	pK	9.59	Cham 1	3.891	Cham 2	4.475
pK	9.59	Cham 1	4.299	Cham 2	4.299					
pK	12.00	Cham 1	0.000	Cham 2	0.000					

pH 4 – 10 Gradients

Initial pH: 4.000										
Final pH: 10.000					Initial pH: 4.026					
Notes: 4.079 – 9.891(8.658):					Final pH: 9.968					
4.097 – 9.742(8.502)					Notes: 4.105 – 9.826: 4.121 – 9.675					
Buffer concentrations					Buffer concentrations					
pK	0.80	Cham 1	13.656	Cham 2	0.000	pK	0.80	Cham 1	15.849	Cham 2	0.000
pK	4.51	Cham 1	5.712	Cham 2	5.712	pK	4.51	Cham 1	4.740	Cham 2	2.446
pK	6.21	Cham 1	3.609	Cham 2	3.609	pK	6.21	Cham 1	6.645	Cham 2	0.000
pK	7.06	Cham 1	3.342	Cham 2	3.342	pK	7.06	Cham 1	0.000	Cham 2	6.799
pK	8.50	Cham 1	3.765	Cham 2	3.765	pK	8.50	Cham 1	6.633	Cham 2	1.977
pK	9.59	Cham 1	4.305	Cham 2	4.305	pK	9.59	Cham 1	3.776	Cham 2	5.641
pK	12.00	Cham 1	0.000	Cham 2	4.431	pK	12.00	Cham 1	0.000	Cham 2	0.715

a From ref. 39. The 'same concentration' formulations are listed in the left column, the ones with 'different concentrations' in the right.

In each mixture record: Initial pH and Final pH refer to initial and final pHs in the gel phase at 10°C; Notes gives the pH of the limiting solutions (and of the buffer solutions prior to the addition of titrants) at 20°C (figure before the colon) and at 25°C (figures after the colon); buffer concentrations are expressed as mM/litre. To convert into μl/ml for 0.2 M, multiply by 5. Cham 1 and Cham 2 refer to the acidic and basic limiting solutions in chambers 1 and 2 respectively. The pKs listed correspond to the actual values in the gel phase (*Table 15*).

are given 4 and 5 pH unit gradients and the 6 pH unit gradient from pH 4 to 10, which spans the widest possible pH interval. The recipes in this table are calculated using the second approach. Two sets of formulae are given. On the left side of the table, the mixtures are for identical molarities of the buffering Immobilines used in the two chambers of the gradient mixer and then titrated with strong acid or bases to the two extremes. On the right side of the table, the mixtures are for the buffering species present in unequal concentrations in the two vessels [designated as Cham 1 (acidic) and Cham 2 (basic)]. These will generate a gradient of buffering species. These recipes ensure that preparation of any Immobiline gel will be trouble free, since all the complex computing routines have already been performed and no further calculations are required.

3.1.5 Extended pH gradients using only commercially available Immobilines

From the recipes given in *Tables 16 – 18* it may appear that the knowledge required for use of IPGs is complete. However, the two strong titrants (an acid, pK 0.8 and a base, pK >12, required for all gradients of 4 pH units or more) are not currently available from Pharmacia-LKB and there are no immediate plans for marketing them. Therefore, we have recalculated and listed in *Table 19* recipes for all those gradients spanning 4, 5, and 6 pH units but now using only the six available Immobiline buffers. How do the formulations in *Tables 18* and *19* compare? In terms of linearity of the pH profile, the quality of the gradients is comparable for the pH 5−9, 6−10, and 5−10 intervals. Worse results were observed, however, with the pH 4−8, 4−9 and 4−10 gradients (standard deviations 2−3 times larger in *Table 19* than in *Table 18*). In fact, the largest deviations are to be found in the acidic portion of the gradient, between pH 4 and 6. In addition, since all formulations are normalized to give the same average value of buffering power (β) = 3.0 mEq. l^{-1} pH^{-1}, the average ionic strength of the formulations in *Table 19* is in most cases lower (by about 20%) than the corresponding ones in *Table 18*. Despite these shortcomings, the recipes listed in *Table 19* are fully acceptable; given a standard deviation below 5% (in *Table 18* a standard deviation not exceeding 1% of the operative pH interval was achieved), it is still possible to measure pI values with good accuracy and, in any event, to obtain completely reproducible pH gradients from run to run.

In *Table 19* formulations are also given for the widest possible IPG span, a pH 3−10 interval, prepared with a two-chamber mixer either according to the 'same concentration' or to the 'different concentration' approaches. Both formulations perform well and with practically identical results in the analysis of complex protein mixtures. It should be noted, however, that in neither case could such a wide pH range be created without resorting to strong titrants.

3.1.6 Non-linear, extended pH gradients

IPG formulations in *Tables 16−19* have been given only in terms of rigorously linear pH gradients. Although this has been the only solution adopted so far, it might not be the optimal one in some cases. The pH slope might need to be altered in pH regions that are overcrowded with proteins. In conventional IEF, flattening of the pH gradient slope in some regions was obtained in three different ways:

- by adding an amphoteric spacer (separator IEF)
- by changing the gel thickness ('thickness-modified' slope)
- by changing the concentration of carrier ampholytes ('concentration modified' slope) (4).

Table 20 gives two examples of non-linear, pH 4−10 intervals, obtained without titrants (top part) or with only a strongly acidic titrant (bottom part). This has been calculated for a general case involving the separation of proteins in a complex mixture, such as cell lysates. We have computed the statistical distribution of the pIs of water-soluble proteins and plotted them in the histogram of *Figure 12*. From the histogram, given the relative abundance of different species, it is clear that an optimally resolving pH gradient should have a gentler slope in the acidic portion and a steeper profile in the alkaline region. Such a profile has been calculated by assigning to each 0.5 pH unit interval in the pH 3.5−10 region a slope inversely proportional to the relative

Table 19. 4, 5 and 6 pH unit wide gradients prepared using only six available Immobilines[a]

A. pH 4 – 8 Gradients. No titrants
Initial pH: 4.090
Final pH: 8.010
Notes: pH in solutions at 25°C: 4.12 – 7.97
Buffer concentrations

pK	3.57	Cham 1	7.850	Cham 2	0.000
pK	4.51	Cham 1	3.380	Cham 2	7.384
pK	6.21	Cham 1	3.136	Cham 2	4.792
pK	7.06	Cham 1	1.564	Cham 2	1.894
pK	8.50	Cham 1	2.261	Cham 2	4.454
pK	9.59	Cham 1	0.000	Cham 2	3.838

B. pH 5 – 9 Gradients. No titrants
Initial pH: 5.060
Final pH: 9.040
Notes: pH in solution at 25°C: 5.13 – 8.91
Buffer concentrations

pK	3.57	Cham 1	11.059	Cham 2	0.000
pK	4.51	Cham 1	7.759	Cham 2	3.317
pK	6.21	Cham 1	2.903	Cham 2	3.510
pK	7.06	Cham 1	1.835	Cham 2	2.833
pK	8.50	Cham 1	10.601	Cham 2	3.892
pK	9.59	Cham 1	1.630	Cham 2	3.071

C. pH 6 – 10 Gradients. No titrants
Initial pH: 5.980
Final pH: 10.000
Notes: pH in solution at 25°C: 5.97 – 9.72
Buffer concentrations

pK	3.57	Cham 1	12.546	Cham 2	1.329
pK	6.21	Cham 1	3.634	Cham 2	4.446
pK	7.06	Cham 1	3.237	Cham 2	4.813
pK	8.50	Cham 1	3.471	Cham 2	3.185
pK	9.59	Cham 1	3.761	Cham 2	4.341

D. pH 4 – 9 Gradients. No titrants
Initial pH: 4.140
Final pH: 8.920
Notes: pH in solution at 25°C: 4.18 – 8.65
Buffer concentrations

pK	3.57	Cham 1	11.048	Cham 2	1.964
pK	4.51	Cham 1	3.138	Cham 2	5.656
pK	6.21	Cham 1	3.092	Cham 2	4.801
pK	7.06	Cham 1	0.298	Cham 2	3.952
pK	8.50	Cham 1	3.336	Cham 2	0.942
pK	9.59	Cham 1	2.951	Cham 2	8.834

193

Table 19. continued

E. pH 5 – 10 Gradients. No titrants
Initial pH: 5.040
Final pH: 10.040
Notes: pH in solution at 25°C: 5.11 – 9.79
Buffer concentrations

pK	3.57	Cham 1	7.505	Cham 2	0.286
pK	4.51	Cham 1	6.718	Cham 2	0.788
pK	6.21	Cham 1	3.970	Cham 2	0.447
pK	7.06	Cham 1	3.646	Cham 2	5.593
pK	8.50	Cham 1	3.027	Cham 2	4.138
pK	9.59	Cham 1	1.689	Cham 2	3.637

F. pH 4 – 10 Gradients. No titrants
Initial pH: 4.150
Final pH: 9.950
Notes: pH in solution at 25°C: 4.16 – 9.66
Buffer concentrations

pK	3.57	Cham 1	14.699	Cham 2	0.000
pK	4.51	Cham 1	0.000	Cham 2	1.523
pK	6.21	Cham 1	6.073	Cham 2	0.667
pK	7.06	Cham 1	1.187	Cham 2	6.503
pK	8.50	Cham 1	4.452	Cham 2	2.089
pK	9.59	Cham 1	0.000	Cham 2	4.763

G. pH 3 – 10 Gradients. Same concentration
Initial pH: 3.000
Final pH: 10.000
Notes: pH in solution at 25°C: 3.02 – 9.75 (mixture: 7.76)
Buffer concentrations

pK	0.80	Cham 1	14.211	Cham 2	0.000
pK	3.57	Cham 1	4.137	Cham 2	4.137
pK	4.51	Cham 1	3.764	Cham 2	3.764
pK	6.21	Cham 1	4.017	Cham 2	4.017
pK	7.06	Cham 1	3.041	Cham 2	3.041
pK	8.50	Cham 1	3.780	Cham 2	3.780
pK	9.59	Cham 1	4.278	Cham 2	4.278
pK	12.00	Cham 1	0.000	Cham 2	6.647

H. pH 3 – 10 Gradients. Different concentrations
Initial pH: 2.970
Final pH: 9.960
Notes: pH in solution at 25°C: 3.00 – 9.66
Buffer concentrations

pK	0.80	Cham 1	17.083	Cham 2	0.000
pK	3.57	Cham 1	2.252	Cham 2	0.000
pK	4.51	Cham 1	5.013	Cham 2	2.596
pK	6.21	Cham 1	6.688	Cham 2	0.000
pK	7.06	Cham 1	0.000	Cham 2	7.115
pK	8.50	Cham 1	6.931	Cham 2	2.072
pK	9.59	Cham 1	3.946	Cham 2	5.899
pK	12.00	Cham 1	0.000	Cham 2	0.747

[a] Cham 1; the acidic dense solution in chamber 1 of the gradient maker. Cham 2; the basic light solution in chamber 2 of the gradient maker. See footnote to *Table 18* for other explanations.

Table 20. Non-linear 4 – 10 immobilized pH gradient[a]

'Ideal' pH 4 – 10
Initial pH: 4.190
Final pH: 9.980
Notes; pH in solution at 25°C: 4.24 – 9.70
Buffer concentrations

pK	3.57	Cham 1	9.321	Cham 2	0.577
pK	4.51	Cham 1	4.327	Cham 2	0.000
pK	6.21	Cham 1	8.943	Cham 2	2.596
pK	7.06	Cham 1	0.000	Cham 2	3.173
pK	8.50	Cham 1	0.000	Cham 2	1.154
pK	9.59	Cham 1	0.000	Cham 1	1.846

'Ideal' pH 4 – 10
Initial pH: 4.040
Final pH: 9.880
Notes: pH in solution at 25°C: 4.13 – 9.61 (with acidic titrant)
Buffer concentrations

pK	0.80	Cham 1	7.028	Cham 2	0.910
pK	4.51	Cham 1	7.659	Cham 2	0.000
pK	6.21	Cham 1	9.010	Cham 2	2.703
pK	7.06	Cham 1	0.000	Cham 2	3.604
pK	8.50	Cham 1	0.000	Cham 2	1.352
pK	9.59	Cham 1	0.000	Cham 2	2.523

[a] From ref. 53. See footnote to Table 18 for other explanations.

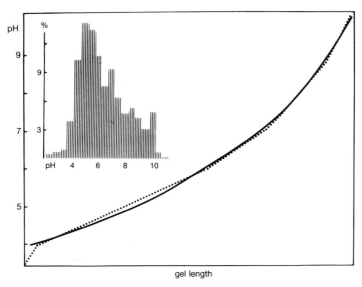

Figure 12. Non-linear 4 – 10 pH gradient: 'ideal' (....) and actual (_____, formulation D, including acidic titrant from Table 20) profiles. The shape of the 'ideal' profile was computed from data on the statistical distribution of protein pIs (8). The relevant histogram is redrawn in the figure inset (by permission from Gianazza et al.; see ref. 53).

abundance of proteins in that interval. This generated the ideal curve (dotted line) in *Figure 12*. Of the two formulations given in *Table 20*, the one including the strongly acidic titrant most closely followed the theoretically predicted course (solid line) (29). What is also important here is the establishment of a new principle in IPG technology, namely that the pH gradient and the density gradient stabilizing it need not be colinear, because the pH can be adjusted by localized flattening for increased resolution while leaving the density gradient unaltered. Though we have considered only the example of an extended pH gradient, narrower pH intervals can be treated in the same fashion.

3.2 IPG methodology

The overall procedure is outlined in *Protocol 10*. Note that the basic equipment required is the same as for conventional CA-IEF gels. Thus the reader should consult Sections 2.2.1 and 2.2.2. In addition, as we essentially use the same polyacrylamide matrix, the reader is referred to Section 2.3 for its general properties.

Protocol 10. IPG flow sheet

1. Assemble the gel mould; mark the polarity of the gradient on the back of the supporting plate.
2. Mix the required amounts of Immobilines (*Tables 16–19*). Fill to one-half of the final volume with distilled water.
3. Check the pH of the solution and adjust as required.
4. Add acrylamide (*Table 5*), glycerol (0.2–0.3 ml/ml to the 'dense' solution only), TEMED (*Table 5*) and bring to the final volume with distilled water.
5. For ranges removed from neutrality, titrate to ~pH 7.5, using Tris for the acidic and acetic acid for the alkaline solution.
6. Transfer the denser solution to the mixing chamber and the lighter solution to the reservoir of a gradient mixer. Centre the mixer on a magnetic stirrer and check for the absence of air bubbles in the connecting duct.
7. Add ammonium persulphate to the solutions.
8. Allow the gradient to pour into the mould from the gradient mixer.
9. Allow the gel to polymerize for 1 h at 50°C.
10. Disassemble the mould and weigh the gel.
11. Wash the gel for 1 h in 1 litre of distilled water on a shaking platform.
12. Dry the gel back to its original weight using a non-heated fan.
13. Transfer the gel to the electrophoretic chamber (at 10°C) and apply the electrodic strips (see *Protocol 4*).
14. Load the samples and start the run.
15. After the electrophoresis, stain the gel to detect the separated proteins.

3.2.1 Casting an Immobiline gel

When preparing for an IPG experiment, two pieces of information are required: the total liquid volume needed to fill the gel cassette, and the required pH interval. Once the first is known, this volume is divided into two halves: one half is titrated to one extreme of the pH interval, the other to the opposite extreme. As the analytical cassette usually has a thickness of 0.5 mm and, for the standard 12 × 25 cm size, (see *Figure 5*) contains 15 ml of liquid to be gelled, in principle two solutions, each of 7.5 ml, should be prepared. However, because the volume of some Immobilines to be added to 7.5 ml might sometimes be rather small (i.e. <50 μl), *Tables 16* and *17* give the required volume (μl) of stock (0.2 M) Immobiline solutions to be added to 15 ml of each starting solution. Clearly this volume will be enough for preparing two gel slabs. The Immobiline solutions (mostly the basic ones) tend to leave droplets on the plastic disposable tips of micropipettes. For accurate dispensing, therefore, we suggest rinsing the tips once or twice with distilled water after each measurement. The polymerization cassette is filled with the aid of a two-vessel gradient mixer and thus the liquid elements which fill the vertically-standing cassette have to be stabilized against remixing by a density gradient. In *Tables 16* and *17* the two solutions are called 'acidic dense' and 'basic light' solutions. This choice is, however, a purely conventional one, and can be reversed, provided one marks the bottom of the mould as the cathodic side. In order to understand the sequence of steps needed, let us refer to *Protocol 11* (as a general example of an IPG protocol) and to *Figure 13* for the final gel assembly.

i. Preparation of the gel mould

Protocol 11. Assembling the mould

1. Wash the glass plate bearing the U-frame with detergent and rinse with distilled water.
2. Dry with paper tissue.
3. To mould sample application slots in the gel, apply suitably-sized pieces of Dymo tape to the glass plate with the U-frame; a 5 × 3 mm slot can be used for sample volumes between 5 and 20 μl (this step is only necessary when preparing a new mould or re-arranging an old one; see *Figure 5a*). To prevent the gel from sticking to the glass plates with U-frame and slot former, coat them with Repel-Silane according to *Protocol 1*. Make sure that no dust or fragments of gel from previous experiments remain on the surface of the gasket, since this can cause the mould to leak.
4. Use a drop of water on the Gel Bond PAG film to determine the hydrophilic side. Apply a few drops of water to the plain glass plate and carefully lay the sheet of Gel Bond PAG film on top with the hydrophobic side down (see *Figure 5b*). Avoid touching the surface of the film with your fingers. Allow the film to protrude 1 mm over one of the long sides of the plate, as a support

Protocol 11. *continued*

for the tubing from the gradient mixer when filling the cassette with gel solution (but only if you have a cover plate without any V-indentations). Roll the film flat to remove air bubbles and to ensure good contact with the glass plate.

5. Clamp the glass plates together with the Gel Bond PAG film and slot former on the inside, by means of clamps placed all along the U-frame, opposite to the protruding film. To avoid leakage, the clamps must be positioned so that the maximum possible pressure is applied (see *Figure 5c*).

Figure 13 gives the final assembly for cassette and gradient mixer. Note that inserting the capillary tubing conveying the solution from the mixer into the cassette is greatly facilitated when using a cover plate bearing three V-shaped indentations. As for the gradient mixer, it should be noted that one chamber contains a magnetic stirrer, while in the reservoir is inserted a plastic cylinder having the same volume, held by a trapezoidal rod. The latter, in reality, is a 'compensating cone' needed to raise the liquid level to such an extent that the two solutions (in the mixing chamber and in the reservoir) will be hydrostatically equilibrated. In addition, this plastic rod can also be utilized for manually stirring the reservoir after addition of TEMED and persulphate.

ii. Polymerization of a linear pH gradient

It is preferable to use 'soft' gels, i.e. with a low %T. Originally, all recipes were given for 5%T matrices, but today we prefer 4%T or even 3%T gels. These 'soft' gels can be easily dried without cracking and allow better entry of larger proteins. In addition, the local ionic strength along the polymer coil is increased, and this permits sharper protein bands due to increased solubility at the pI. A linear pH gradient is generated by mixing equal volumes of the two starting solutions in a gradient mixer. It is essential, for any gel formulation removed from neutrality (pH $6.5-7.5$), to titrate the two solutions to neutral pH, so as to ensure reproducible polymerization conditions and avoid hydrolysis of the four alkaline Immobilines. If the pH interval used is acidic, add Tris, if it is basic add acetic acid. We recommend that a minimum of 15 ml of each solution (enough for two gels) is prepared and that the volumes of Immobiline needed are measured with a well-calibrated micro-syringe to ensure high accuracy. Prepare the acidic, dense solution and the basic, light solution for the pH gradient as described in *Protocol 10* (stock acrylamide solutions are given in *Table 2*; for the catalysts, refer to *Table 21*). If the same gradient is to be prepared repeatedly, the buffering and non-buffering Immobiline and water mixtures can be prepared as stock solutions and stored according to the recommendations for Immobiline. Prepared gel solutions must not be stored. However, gels with a pH less than pH 8 can be stored in a humidity chamber for up to one week after polymerization.

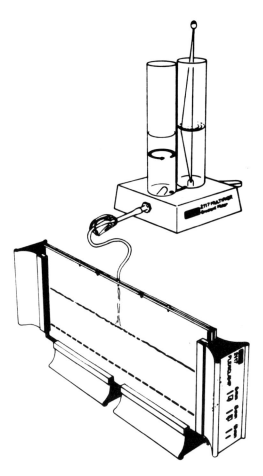

Figure 13. Set-up for casting an IPG gel. A linear pH gradient is generated by mixing equal volumes of a dense and light solution, titrated to the extremes of the desired pH interval. Note the 'compensating' rod in the reservoir, used as a stirrer after addition of catalysts and for hydrostatically equilbrating the two solutions. Insertion of the capillary conveying the solution from the mixer to the cassette is greatly facilitated by using modern cover plates, bearing three V-shaped indentations. (By courtesy of LKB Produkter AB.)

Protocol 12. Polymerization of linear pH gradient gel

  1. Check that the valve in the gradient mixer and the clamp on the outlet tubing are both closed.

  2. Transfer 7.5 ml of the basic, light solution, to the reservoir chamber.

  3. Slowly open the valve just enough to fill the connecting channel with the solution, and quickly close it again. Then transfer 7.5 ml of the acidic, dense solution to the mixing chamber.

Protocol 12. *continued*

4. Place the prepared mould upright on a levelled surface. The optimum flow rate is obtained when the outlet of the gradient mixer is 5 cm above the top of the mould. Open the clamp of the outlet tubing, fill the tubing halfway with the dense solution, and close the clamp again.

5. Switch on the stirrer, and set to a speed of about 500 r.p.m.

6. Add the catalysts to each chamber as specified in *Table 21*.

7. Insert the free end of the tubing between the glass plates of the mould at the central V-indentation (*Figure 13*).

8. Open the clamp on the outlet tubing, then immediately open the valve between the dense and light solutions so that the gradient solution starts to flow down into the mould under gravity. Make sure that the levels of liquid in the two chambers fall at the same rate. The mould should be filled within 5 min. To assist the mould to fill uniformly across its width, the tubing from the mixer may be substituted with a 2- or 3-way outlet assembled from small glass or plastic connectors (e.g. spare parts of chromatographic equipment) and butterfly needles.

9. When the gradient mixer is empty, carefully remove the tubing from the mould. After leaving the cassette to rest for 5 min, place it on a levelled surface in an oven at 50°C. Polymerization is allowed to continue for 1 h. Meanwhile, wash and dry the mixer and tubing.

10. When polymerization is complete, remove the clamps and carefully take the mould apart. Start by removing the glass plate from the supporting foil. Then hold the remaining part so that the glass surface is on top and the supporting foil underneath. Gently peel the gel away from the slot former, taking special care not to tear the gel around the slots.

11. Weigh the gel and then place it in 500 ml of distilled water for 1 h to wash out any remaining ammonium persulphate, TEMED and amounts of unreacted monomers and Immobilines. Change the water once or twice during washing.

12. After washing the gel, carefully remove any excess water from the surface with a moist paper tissue. To remove the water absorbed by the gel during the washing step, leave it at room temperature until the weight has returned to within 5% of the original weight. To shorten the drying time, use a non-heating fan placed at about 50 cm from the gel to increase the rate of evaporation. Check the weight of the gel after 5 min and from this, estimate the total drying time. The drying step is essential, as a gel containing too much water will 'sweat' during the electrofocusing run and droplets of water will form on the surface. However, if the gel dries too much, the value of %T will be increased, resulting in longer focusing times and a greater sieving effect.

Table 21. Working concentration for the catalysts

Concentration rating	TEMED		Ammonium persulphate (40% w/v) (μl/ml)
	Acidic pH	Basic pH	
Lower limit	0.5	0.3	0.6
Standard, $\%T = 5^a$	0.5	0.3	0.8
Standard, $\%T = 3$	0.7	0.5	1.0
For 5 – 10% alcohol	0.7	0.5	1.0
Higher limit[b]	0.9	0.6	1.4

[a] From ref. 3
[b] From LKB Application Note no. 321.

3.2.2 Reswelling dry Immobiline gels

Pre-cast, dried Immobiline gels, encompassing a few acidic ranges, are now available from Pharmacia-LKB. They all contain $4\%T$ and they span the following pH ranges: pH 4 – 7, pH 4.2 – 4.8, pH 4.5 – 5.4, pH 5.0 – 6.0, and pH 5.6 – 6.6. Pre-cast, dried IPG gels in the alkaline region have not been introduced as yet, possibly because at high pHs the hydrolysis of both the gel matrix and the Immobiline chemicals bound to it is much more pronounced.

It has been found that the diffusion of water through Immobiline gels does not follow a simple Fick's law of passive transport from high (the water phase) to zero (the dried gel phase) concentration regions, but it is an active phenomenon: even under isoionic conditions, acidic ranges cause swelling 4 – 5 times faster than alkaline ones (46). Given these findings, it is preferable to reswell dried Immobiline gels in a cassette similar to the one for casting the IPG gel. *Figure 14* shows the reswelling system produced by Pharmacia-LKB. The dried gel is inserted in the cassette, which is clamped and allowed to stand on the short side. The reswelling solution is gently injected into the chamber via a small hole in the lower right side using a cannula, until the cassette is completely filled. As the system is volume-controlled, it can be left to reswell overnight, if needed. Gel drying and reswelling is the preferred procedure when an IPG gel containing additives is needed. In this case it is always best to cast an 'empty' gel, wash it, dry it and then reconstitute it in the presence of the desired additive (e.g. urea, alkyl ureas, detergents, carrier ampholytes and mixtures thereof).

3.2.3 Electrophoresis

A list of the electrode solutions in common use can be found in *Table 22*. A common electrophoresis protocol consists of an initial voltage setting of 500 V, for 1 – 2 h, followed by an overnight run at 2000 – 2500 V. Ultranarrow gradients are further subjected to a couple of hours at 5000 V, or higher at about 1000 V/cm across the region containing the bands of interest.

201

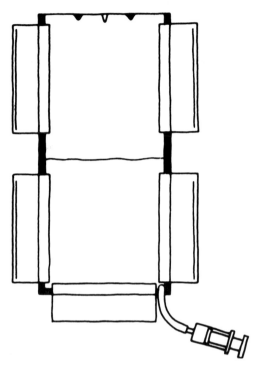

Figure 14. Reswelling cassette for dry IPG gels. The dried IPG gel (on its plastic backing) is inserted in the cassette, which is then gently filled with any desired reswelling solution via a bottom hole. (By courtesy of LKB Produkter AB.)

Table 22. IPG electrode solutions

Substance	Application	Concentration
Glutamic acid	anolyte	10 mM
Lysine	catholyte	10 mM
Carrier ampholytes[a,b]	both electrolytes	0.3 – 1%
Distilled water[b]	both electrolytes	–

[a] Of the same or of a narrower range than the IPG.
[b] For mixed-bed gels or for samples with high salt concentration.

3.2.4 Staining and pH measurements

IPGs tend to bind strongly to dyes, so the gels are better stained for a relatively short time (30 – 60 min) with a stain of medium intensity, e.g. the second method listed in *Table 12*. For silver staining according to *Protocol 7*, the gels should be cast on silanized glass rather than on Gel Bond PAG.

Accurate pH measurements are virtually impossible by equilibration between a gel slice and excess water, and not very reliable with a contact electrode. One can preferably either refer to the banding pattern of a set of marker proteins, or elute carrier ampholytes from a mixed bed gel (see Section 3.2.6; the correction factors to be applied for temperature of the run different from 10°C can be calculated from *Table 15*; for the effect of urea, see ref. 50).

3.2.5 Storage of the Immobiline chemicals

Up to now, the seven different Immobilines have been supplied as dry powders from the manufacturer (except for pKs 8.5 and 9.3 which are in a liquid state). Stock solutions are prepared by adding 25 g of distilled water (i.e. add reconstituting liquid by weight rather than volume, the former being more accurate), thus obtaining 0.2 M solutions of each species. They are then dispensed in 5 ml aliquots and stored frozen at −20°C. Each 5 ml aliquot has a shelf life at +4°C of at least 4 weeks whilst frozen batches are stable for at least 6 months.

There are two major problems with the Immobiline chemicals, especially with the alkaline ones; hydrolysis and spontaneous auto-polymerization. Hydrolysis is quite a nuisance because then only acrylic acid is incorporated into the IPG matrix, with a strong acidification of the calculated pH gradient. Hydrolysis is an auto-catalysed process for the basic Immobilines, since it is pH-dependent (54). For the pK 8.5 and 9.3 species, such a cleavage reaction on the amido bond can occur even in the frozen state, at a rate of about 20% per year (55). Auto-polymerization (56,57), is also quite deleterious for the IPG technique. Again, this reaction occurs particularly with alkaline Immobilines, and is purely auto-catalytic, as it is greatly accelerated by deprotonated amino groups. Oligomers and *n*-mers are formed which stay in solution and can even be incorporated into the IPG gel since, in principle, they still contain a double bond at one extremity (unless they anneal to form a ring). These auto-polymerization products range in size from simple dimers and trimers to molecules having the same elution volume of a 64-kd protein. Analysis of pK 9.3 Immobiline stored frozen revealed, after more than 6 months of storage, the presence of about 20% polymer. These products of auto-polymerization, when added to proteins in solution, are able to bridge them *via* two unlike binding surfaces. A lattice is formed and the proteins (especially larger ones, like ferritin, α_2-macroglobulin, thyroglobulin) are precipitated out of solution. This precipitation effect is quite strong and begins even at the level of short oligomers (>decamer) (57). There is an easy test to check for the presence of such polymers called the 'ferritin precipitation test' (56). As a short-term remedy, we have described an easy method for oligomer removal, based on adsorption onto hydrophobic polymer phases (e.g. the XAD-2 polymer or a C_{18}-bonded phase (58).

These problems with the four basic Immobilines could potentially remove one of the major advantages of the IPG technique, namely its high reproducibility from run to run. As a remedy to these drawbacks, in collaboration with Pharmacia-LKB, we have explored new ways to stabilize the four alkaline Immobilines permanently. It appears that, when dissolved in anhydrous propanol (containing a maximum of 60 p.p.m. water), these species are stabilized against both hydrolysis and auto-polymerization for a virtually unlimited period of time (less than 1% degradation per year even when stored at +4°C). Thus, whilst at the time of writing all Immobilines are still supplied ready to be reconstituted in water, it is possible that, in the future, the four basic Immobilines will be available directly as 0.2 M solutions in the correct solvent. The acidic Immobilines, being much more stable, will continue to be available as aqueous solutions.

3.2.6 Mixed-bed CA-IPG gels

In CA-IPG gels the primary, immobilized pH gradient is admixed with a secondary, soluble carrier ampholyte-driven pH gradient. It sounds strange that, given the problems connected with the carrier ampholyte buffers (discontinuities along the electrophoretic path, pH gradient decay, etc.), which the IPG technique was supposed to solve, one should resurrect this past methodology. In fact, when working with membrane proteins (59) and with microvillar hydrolases, partly embedded in biological membranes (11), we found that the addition of carrier ampholytes to the sample and IPG gel would increase protein solubility, possibly by forming mixed micelles with the detergent used for membrane solubilization (59) or by directly complexing with the protein itself (11). It is a fact that, in the absence of carrier ampholytes, these same proteins essentially fail to enter the gel and mostly precipitate or give elongated smears around the application site (in general cathodic sample loading). It has also been found that, on a relative hydrophobicity scale, the four basic Immobilines (pKs 6.2, 7.0, 8.5 and 9.3) are decidedly more hydrophobic than their acidic counterparts (pKs 3.6, 4.4 and 4.6). Upon incorporation in the gel matrix, the phenomenon becomes cooperative and could lead to the formation of hydrophobic patches on the surface of such a hydrophilic gel as polyacrylamide. As the strength of a hydrophobic interaction is directly proportional to the product of the cavity area times its surface tension, it is clear that experimental conditions which lead to a decrement of molecular contact area weaken such interactions. Thus, our original idea of carrier ampholytes as solubilizing ions in IPG matrices has been extended to the hypothesis of carrier ampholytes as shielding molecules, coating, on one side, the polyacrylamide matrix studded with Immobilines (especially the basic ones) and, on the other side, the protein itself. This strongly quenches the direct hydrophobic protein−IPG matrix interaction, effectively detaches the protein from the surrounding polymer coils and allows good focusing into sharp bands. This phenomenon can be appreciated by considering *Figure 15*. When horse-spleen ferritin is focused alone in an IPG gel, it gives smears in the proximity of the application site. However, upon addition of 3−4% carrier ampholytes (1.5−2% actual gel concentration, as the IPG matrix covers a 1 pH unit interval while the carrier ampholytes span 2 pH units) a sharp array of ferritin bands is developed in the gel, with essentially no protein remaining in the sample slot. For this to happen, the carrier ampholyte shielding species should already be impregnated in the Immobiline gel and present in the sample solution as well. If added afterwards, for example by electrophoretic migration from the electrode strips, they will be ineffective since, once the hydrophobic protein matrix interaction has occurred, the surface which the carrier ampholytes were supposed to mask will not be available any longer for such shielding action. In other words, carrier ampholytes can only prevent the phenomenon and cannot cure it *a posteriori*. The users of the mixed CA-IPG technique should be aware of another fundamental fact: the shielding mechanism is most effective in the unfocused state. If the protein is applied to a pre-focused carrier ampholyte bed, severe precipitation and poor protein migration and banding patterns ensue.

A note of caution should be mentioned concerning the indiscriminate use of the CA-IPG technique: at high carrier ampholyte levels ($>1\%$) and high voltages (>100 V/cm) these gels start exuding water with dissolved carrier ampholytes, leading to severe risks of short-circuits, sparks and burning on the gel surface. The phenomenon is minimized by chaotropic substances (e.g. 8 M urea), by polyols (e.g. 30% sucrose) and by lowering the carrier ampholyte molarity in the gel (60). As an answer to the basic question of when and how much carrier ampholytes to add, we suggest the following guidelines:

- If your sample focuses well as such, ignore the mixed-bed technique (which presumably will be mostly needed with hydrophobic proteins and in alkaline pH ranges).

- Add only the minimum amount of carrier ampholytes (in general around 1%) needed to avoid sample precipitation in the sample slot and for producing sharply focused bands.

Finally, while the hypothesis of hydrophobic protein−IPG matrix interaction seems the most likely, it cannot be excluded that, for some samples, added carrier

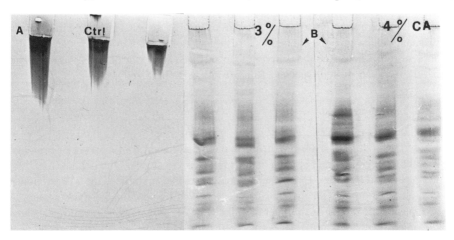

Figure 15. IEF of ferritin in IPGs. (A) 4%T polyacrylamide gel containing an IPG from pH 4 to 6 (formed using the pKs 3.6, 4.6, 6.2, and 9.3 Immobilines). (B) 4%T gel containing an IPG from pH 4 to 5, formed with pK 4.6 and 8.5 Immobilines. In (B) the gel was cut in half; the halves were impregnated with 3% and 4% carrier ampholytes, respectively, in the pH 4 – 6 range. The sample load (from left to right in each gel) was 200, 150, and 100 μg horse spleen ferritin. Focusing was performed overnight at 2000 V (final) and 10°C. Staining was with Coomassie Blue R-250. (By permission from Rabilloud *et al.*; see ref 57.)

ampholytes might simply act as buffering ions in the bulk water solution. In this case they may prevent protein denaturation due to abrupt pH changes in the sample layer as salt components are split by the current into a strongly alkaline (towards the cathode) and a strongly acidic (towards the anode) boundary.

3.3 Troubleshooting

One could cover pages with a description of all the troubles and possible remedies in any methodology. However, as Pharmacia-LKB has already listed all the major troubles encountered with the IPG technique in their literature, we refer readers to *Table 23* for all the possible causes and remedies suggested. We highlight the following points.

- When the gel is gluey and there is poor incorporation of Immobilines, the biggest offenders are generally the catalysts (e.g. ammonium persulphate too old, crystals wet due to adsorbed humidity, wrong amounts of catalysts added to the gel mix). Check in addition the polymerization temperature and the pH of the gelling solutions.

- Bear in mind the last point in *Table 23*: if you have done everything right, and still you do not see any focused protein, you might have simply positioned the platinum wires on the gel with the wrong polarity. Unlike conventional IEF gel, in IPGs the anode has to be positioned at the acidic (or less alkaline) gel extremity, while the cathode has to be placed at the alkaline (or less acidic) gel end.

3.4 Some analytical results with IPGs

We will limit this section to some examples of separations in narrow and ultra-narrow pH intervals, where the tremendous resolving power (ΔpI) of IPGs can be fully appreciated. The ΔpI is the difference, in surface charge, in pI units, between two barely resolved protein species. Rilbe (61) has defined ΔpI as:

$$\Delta(\text{pI}) = 3\sqrt{\frac{D[d(\text{pH})/dx]}{E[-du/d(\text{pH})]}}$$

where D and $du/d(\text{pH})$ are the diffusion coefficient and titration curve of proteins, E is the voltage gradient applied and $d(\text{pH})/dx$ is the slope of the pH gradient over the separation distance. Experimental conditions that minimize ΔpI will maximize the resolving power. Ideally, this can be achieved by simultaneously increasing E and decreasing $d(\text{pH})/dx$, an operation for which IPGs seem well suited. As stated previously (see Section 2.1.3), with conventional IEF it is very difficult to engineer pH gradients that are narrower than 1 pH unit. One can push the ΔpI, in IPGs, to the limit of 0.001 pH unit; the corresponding limit in CA-IEF is only 0.01 pH unit. We began to investigate the possibility of resolving neutral mutants, which carry a point mutation involving amino acids with non-ionizable side chains and are, in fact, described as 'electrophoretically silent' because they cannot be distinguished by conventional electrophoretic techniques. The results were quite exciting. The first

Table 23. Troubleshooting guide

Symptom	Cause	Remedy
Drifting of pH during measurement of basic starting solution	Inaccuracy of glass pH electrodes (alkaline error)	Consult information supplied by electrode manufacturer
Leaking mould	Dust or gel fragments on on the gasket	Carefully clean the gel plate and gasket
The gel consistency is not firm, gel does not hold its shape after removal from the mould	Inefficient polymerization	Prepare fresh ammonium persulphate and check that the recommended polymerization conditions are used
Plateau visible in the anodic and/or cathodic section of the gel during electrofocusing, no focusing proteins seen in that part of the gel	High concentration of salt in the system	Check that the correct amounts of ammonium persulphate and TEMED are used
Overheating of gel near sample application when beginning electrofocusing	High salt content in the sample	Reduce salt concentration by dialysis or gel filtration
Non-linear pH gradient	Back-flow in the gradient mixer	Find and mark the optimal position for the gradient mixer on the stirrer
Refractive line at pH 6.2 in the gel after focusing	Unincorporated polymers	Wash the gel in 2 litres of distilled water; change the water once and wash overnight
Curved protein zones in that portion of the gel which was at the top of the mould during polymerization	Too rapid polymerization	Decrease the rate of polymerization by putting the mould into the freezer for 15 min before filling it with the gel solution, or place the solutions in a refrigerator for 15 min before casting the gel
Uneven protein distribution across a zone	Slot or sample application not perpendicular to running direction	Place the slot or sample application pieces perfectly perpendicular to the running direction
Diffuse zones with unstained spots, or drops of water on the gel surface during electrofocusing	Incomplete drying of the gel after the washing step	Dry the gel until it is within 5% of its original weight
No zones detected	Gel is focused with the wrong polarity	Mark the polarity on the gel when removing it from the mould

mutant we had on hand was a haemoglobin A (HbA) variant. It was well separated by IPGs at both an analytical and preparative level and was found to be a known variant (Hb San Diego, carrying a Val→Met substitution in β-109; *Figure 16a*) (52). This variant could not possibly be separated by any other electrophoretic technique, including IEF in carrier ampholytes. Another variant in the sequence of HbA turned out to be Hb Beirut, a Val→Ala mutant in β-126; like Hb San Diego, it was slightly more acidic than HbA (*Figure 16b*). Note that these mutants were found not by electrophoretic nor chromatographic techniques, but accidentally, by physiological activity tests (altered oxygen binding curves in the blood of the carriers).

Next, we studied known mutants of fetal haemoglobin (HbF), which previously had been analysed by HPLC of denatured globin chains. Obviously, separation of intact, native tetramers would be desirable, both because additional sample manipulation steps would be eliminated and because intact molecules could be recovered for biological activity studies. As shown in *Figure 17a*, HbF Sardinia (which carries an Ile→Thr substitution in γ-75) is well resolved from the wild-type HbF in a shallow IPG range spanning only 0.25 pH units (62). Unlike the previous two samples, the neutral mutation results in Hb tetramers having a higher net positive charge and thus a higher pI. There is, however, a more subtle mutation that could not be resolved in the present case. As shown in *Figure 17b*, the lower band is actually an envelope of two components, called Aγ and Gγ, carrying a Gly→Ala mutation in γ-136. These two tetramers, normal components during fetal life, are found in approximately an 80:20 ratio. If the pH gradient is further decreased to 0.1 pH unit (over a standard 10-cm migration length), even these two tetramers can be separated (*Figure 17b*) with a resolution close to the practical limit of ΔpI equal to 0.001 (63).

3.5 Preparative aspects of IPGs

We have described in a number of papers several different procedures for performing preparative IPG runs (for a review, see refs 64 and 65). The zone of interest is located at the end of the run (e.g. by staining two lateral strips of the gel) and the protein eluted from it by electrophoresis (into a hydroxyapatite slurry, or in an ion-exchange resin, or simply in a dialysis bag). It appears that, just as analytically IPGs afford a resolution about one order of magnitude higher than conventional IEF, so, preparatively, IPGs seem to bear a comparably higher loading capacity. One possible reason for this is that a protein in an IPG matrix is not isoionic, in the sense that it forms a salt with the surrounding Immobiline ions bound to the polyacrylamide coils. Thus the protein solubility at the pI is substantially increased due to the fact that the Immobiline matrix provides counter-ions (different from protons) to the isoelectric protein and allows for salt formation.

A second cause for incremented protein loads can be seen in *Figure 18* which shows that the protein load is strongly dependent on the amount of polyacrylamide (%T) utilized for preparing the supporting gel. At a relatively high matrix content (6%T) the maximum tolerated protein load in an isoelectric zone is only 30 mg/ml of gel volume whereas, in highly dilute gels (2.5%T), this upper limit is significantly increased, up to 90 mg/ml of gel volume. This has been interpreted as a competition

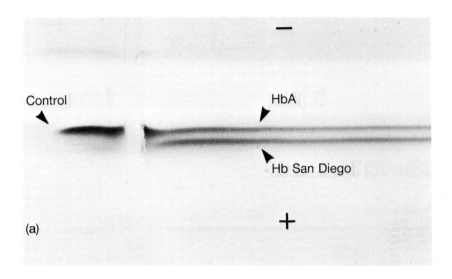

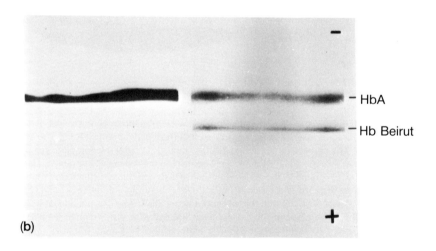

Figure 16. (a) IPG separation of HbA from Hb San Diego. The 125 × 110 mm, 1-mm-thick gel contained 5%*T* and Immobilines of p*K* 7.0 and 3.6 in a ratio that generated a pH span from 6.9 to 7.7. Approximately 8 mg of protein was loaded in the right trench; 1.5 mg of normal adult Hb was applied in the left pocket. The Δp*I* between HbA and Hb San Diego was estimated to be 0.01 pH unit (by permission from Rochette *et al.*; see ref. 52). (b) Separation of Hb Beirut from HbA. The gel contained 4%*T*, 4%*C* monomers and an IPG from pH 7.2 to pH 7.35. The gel was run for 6 h at 4000 V and 10°C. The total sample load was 200 μg per track. (By permission from Righetti *et al.*; see ref. 62.)

209

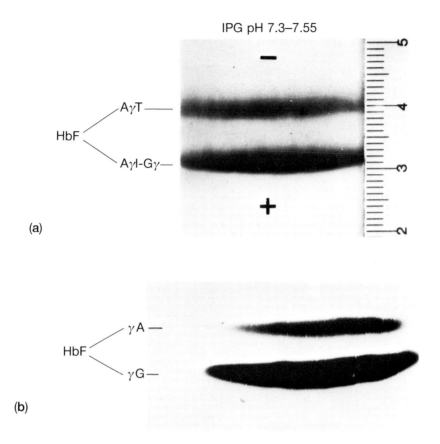

Figure 17. (a) IEF of a cord-blood lysate from a newborn heterozygous for HbF Sardinia. The gel contained 4%*T*, 4%*C* monomers and an IPG from pH 7.3 to pH 7.55. The sample load was 200 μg. The gel was run for 5 h at 4000 V and 10°C. Note the complete separation of the AγT tetramer (HbF Sardinia) from the zone containing the Aγl and Gγ tetramers. Because of the very shallow pH gradient, the normal components of cord blood collected at the anode (by permission from Righetti *et al.*; see ref. 62). (b) IPG separation of native Aγ and Gγ fetal haemoglobin tetramers. The 0.5 mm thin gel contained a 3%*T*, 4%*C* matrix and an IPG from pH 7.35 to pH 7.55 over 20 cm; the gel was equilibrated with 6% Pharmalytes pH 5 – 8. Focusing was performed overnight at 4000 V and 10°C. The Gγ/Aγ tetramer ratio was approximately 80:20, as expected from genetic analysis of newborns (by permission from Cossu and Righetti; see ref. 63).

for the available water between the two polymers, the polyacrylamide coils and the protein invading them. They both sequester and coordinate water in their hydration shell. However, in high %*T* gels the matrix coils, being the most abundant species, sequester most of the available water, leaving little liquid volume for the protein to be dissolved in. Such highly diluted gels have two additional advantages, namely:

- by diluting the matrix, while keeping constant the amount of Immobiline (the conventional ~ 10 mM buffering ion), the charge density on the polymer coil is in fact increased, and this results in sharper protein zones and increased protein loading capacity;

- below 3%*T*, the viscoelastic forces of the gel are weakened, allowing the osmotic forces in the protein zone to predominate and so draw more water from the surrounding gel regions. This results in a further increment in load capacity within a given protein zone due to local gel swelling and concomitant increase in cross-sectional area.

On the basis of the above observations, we have carried the preparative IPG technique to what could be the ultimate evolutionary step. *Figure 18* suggests that, if the gel matrix were totally abolished, possibly much greater protein loads could be tolerated. In addition, as discussed in Section 3.2.6, hydrophobic proteins are not compatible at all with high levels of Immobilines, as they are bound to the matrix by hydrophobic interaction and are poorly extracted after the fractionation process. Both observations lead to only one possible conclusion: why not try to perform a preparative IPG run in the absence of both gel matrix and Immobilines. This concept has materialized in the novel technique, called 'segmented immobilized pH gradients', as exemplified in *Figure 19*. The IPG gel is composed of two (or more) segments

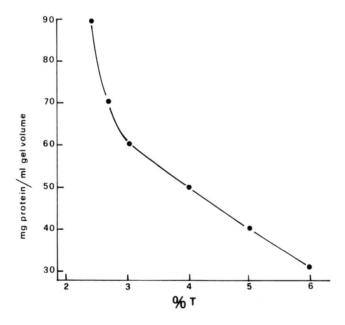

Figure 18. Relationship between loading capacity (in terms of mg protein per ml gel volume) and %*T* value of the gel matrix. Notice that, whereas in the range 3 – 6%*T* the protein load decreases linearly, in softer gels (<3%*T*) it increases exponentially (by permission from Righetti and Gelfi; see ref. 15).

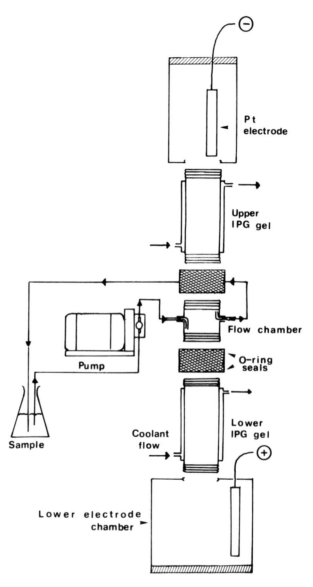

Figure 19. Diagram of the recycling, segmented IPG apparatus. The central flow chamber is coupled, via two symmetric O-ring seals, to upper and lower IPG glass cylinders (each 2 cm diam., 8 cm in height) which serve as containers of cathodic and anodic segments of a polyacrylamide gel containing immobilized pH gradients of appropriate pH intervals. The upper eletrode chamber is connected to the upper IPG gel segment via a water tight O-ring seal, whereas the bottom segment (here depicted with a screw-on connection) can in reality be plunged directly into the anolyte solution. The sample is recycled from a refrigerated reservoir by a peristaltic pump, in general operated at maximum speed (5 ml/min). (By permission from Faupel *et al.*; see ref. 66.)

Pier Giorgio Righetti et al.

separated by liquid interlayers, which function as recycling chambers. The floor
and the ceiling of such flow chambers are IPG gel extremities, arranged so as to
have pIs just lower (in the anodic side) and just higher (in the cathodic side) than
the protein of interest to be purified. The electric field and the liquid flow are or-
thogonally coupled, and the protein feed kept in a large, thermostatted, and stirred
reservoir. As the protein solution is continuously recycled, only the protein of interest is maintained isoelectric, by a continuous titration process, by the two IPG
walls, while all the non-isoelectric impurities are swept away and collect either in
the anodic or cathodic IPG segments. This newly patented process has a number
of advantages:

- it can handle very large liquid volumes;
- it can tolerate very large protein amounts, as only a fraction of it enters the IPG
  gel;
- it allows very high protein recoveries, as the component of interest never enters
  the IPG gel and thus does not have to be extracted from it.

The principle of this novel approach is further explained in *Figure 20*. The
extremities of the IPG gel segments delimiting the recycling chamber can be envisaged
as isoelectric membranes, endowed with a strong buffering capacity, having pI values

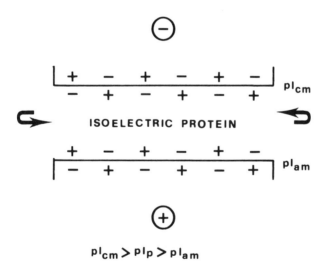

$$pI_{cm} > pI_p > pI_{am}$$

Figure 20. The principle of the recycling IPG apparatus. It is hypothesized that the two IPG
gel extremities facing the recycling chamber act like two isoelectric membranes titrating the
protein of interest to its isoelectric point (pI), thus keeping its mobility constant and equal
to zero throughout the purification process. For this to occur, it is necessary that $pI_{cm} > pI_p$
$> pI_{am}$, where the subfixes to pI indicate cathodic membrane, protein and anodic membrane,
respectively. In addition, the two Immobiline 'membranes' satisfy the condition of having high
buffering capacity at their pI values. The curved arrows indicate protein recycling in the central
chamber. (By permission from Righetti *et al.*; see ref. 67.)

213

on either side of and near to the pI of the component to be purified. The latter, by a continuous titration process, stays isoelectric for the duration of the experiment and it is therefore unable to leave the sample reservoir. Initially, we performed these preparative runs in vertical chambers delimited by segments of Immobiline gradients (66). Subsequently we adopted horizontal configuration and simply utilized isoelectric Immobiline membranes to face the flow chamber, instead of true pH gradients (67). The latter approach has the advantage that the impurities can cross such isoelectric membranes and collect in the electrolyte compartments without clogging the IPG gel.

As the technique is quite novel, we will not give practical guidelines to its use here, but refer the readers to the original articles (66−68).

4. Conclusions

Conventional IEF in amphoteric buffers is now a well established technique and further major improvements are unlikely. In contrast, the IPG methodology, which was first described in 1982, is probably certain to be improved over the years. Nevertheless, after a sluggish start, probably due to some inherent problems in the technique (see Sections 3.2.5 and 3.2.6) IPGs appear today to be almost trouble-free and seem to work properly in most applications. The field is entirely open, and a good harvest should follow.

Acknowledgements

Our research reported here is supported by Progetto Finalizzato 'Biotecnologie e Biostrumentazione' from Consiglio Nazionale delle Ricerche (CNR, Roma). We thank colleagues who have collaborated with us over the years (e.g. Drs A.Görg, B.Bjellqvist, P.K.Sinha, and T.Rabilloud, to name just a few) whose work has greatly helped in establishing the modern techniques.

References

1. Giddings, J. C. (1987). *J. Chromatogr.,* **395**, 19.
2. Svensson, H. (1961). *Acta Chem. Scand.,* **15**, 325; (1962). *ibid.,* **16**, 456.
3. Bjellqvist, B., Ek, K., Righetti, P. G., Gianazza, E., Görg, A., Postel, W., and Westermeier, R. (1982). *J. Biochem. Biophys. Methods,* **6**, 317.
4. Righetti, P. G. (1983). *Isoelectric Focusing: Theory, Methodology and Applications.* Elsevier, Amsterdam.
5. Gianazza, E., Chillemi, F., Gelfi, C., and Righetti, P. G. (1979). *J. Biochem. Biophys. Methods,* **1**, 237.
6. Gianazza, E., Chillemi, F., and Righetti, P. G. (1980). *J. Biochem. Biophys. Methods,* **3**, 135.
7. Gianazza, E., Chillemi, F., Duranti, M., and Righetti, P. G. (1984). *J. Biochem. Biophys. Methods,* **8**, 339.
8. Gianazza, E. and Righetti, P. G. (1980). *J. Chromatogr.,* **193**, 1.

9. Krishnamoorthy, R., Bianchi-Bosisio, A., Labie, D., and Righetti, P. G. (1978). *FEBS Lett.*, **94**, 319.
10. Righetti, P. G. and Macelloni, C. (1982). *J. Biochem. Biophys. Methods*, **6**, 1.
11. Sinha, P. K. and Righetti, P. G. (1986). *J. Biochem. Biophys. Methods*, **12**, 289.
12. Vesterberg, O. (1969). *Acta Chem. Scand.*, **23**, 2653.
13. Bianchi-Bosisio, A., Snyder, R. S., and Righetti, P. G. (1981). *J. Chromatogr.*, **209**, 265.
14. Galante, E., Caravaggio, T., and Righetti, P. G. (1975). In *Progress in Isoelectric Focusing and Isotachophoresis*. (ed. P. G. Righetti), p. 3. Elsevier, Amsterdam.
15. Righetti, P. G. (1980). *J. Chromatogr.*, **190**, 275.
16. Bianchi-Bosisio, A., Loehrlein, C., Snyder, R. S., and Righetti, P. G. (1980). *J. Chromatogr.*, **189**, 317.
17. Görg, A., Postel, W., and Westemeier, R. (1978). *Anal. Biochem.*, **89**, 60.
18. Loening, U. E. (1967). *Biochem. J.*, **102**, 251.
19. Ui, N. (1971). *Biochim. Biophys. Acta*, **299**, 567.
20. Hjelmeland, L. M. and Chrambach, A. (1981). *Electrophoresis*, **2**, 1.
21. Righetti, P. G., Brost, B. C., and Snyder, R. S. (1981). *J. Biochem. Biophys. Methods*, **4**, 347.
22. Gelfi, C. and Righetti, P. G. (1981). *Electrophoresis*, **2**, 213.
23. Gianazza, E., Pagani, M., Luzzana, M., and Righetti, P. G. (1975). *J. Chromatogr.*, **109**, 357.
24. Arnaud, P., Galbraith, R. M., Page, F. W., and Black, C. (1979). *Clin. Genet.*, **15**, 406.
25. Låås, T. and Olsson, I. (1981). *Anal. Biochem.*, **114**, 167.
26. Altland, K. and Kaempfer, M. (1980). *Electrophoresis*, **1**, 57.
27. Caspers, M. L., Posey, Y., and Brown, R. K. (1977). *Anal. Biochem.*, **79**, 166.
28. Righetti, P. G., Tudor, G., and Gianazza, E. (1982). *J. Biochem. Biophys. Methods*, **6**, 219.
29. Yao, J. G. and Bishop, R. (1982). *J. Chromatogr.*, **234**, 459.
30. Radola, B. J.(1980). *Electrophoresis*, **1**, 43.
31. Pflug, W. (1985). *Electrophoresis*, **6**, 19.
32. Eckersall, P. D. and Conner, J. G. (1984). *Anal. Biochem.*, **138**, 52.
33. Blakesley, R. W. and Boezi, J. A. (1977). *Anal. Biochem.*, **82**, 580.
34. Righetti, P. G. and Drysdale, J. W. (1974). *J. Chromatogr.*, **98**, 271.
35. Neuhoff, V., Stamm, R., and Eibl, H. (1985). *Electrophoresis*, **6**, 427.
36. Merril, C. R., Goldman, D., Sedman, S. A., and Ebert, M. H. (1981). *Science*, **211**, 1438.
37. Vesterberg, O. (1972). *Biochim. Biophys. Acta*, **257**, 11.
38. Hebert, J. P. and Strobbel, B. (1974). *LKB Application Note # 151.*
39. Godolphin, W. J. and Stinson, R. A. (1974). *Clin. Chim. Acta*, **56**, 97.
40. Laskey, R. A. and Mills, A. D. (1975). *Eur. J. Biochem.*, **56**, 335.
41. Laskey, R. A. (1980). In *Methods in Enzymology*. (ed. L. Grossman and K. Moldave), Vol. 65, p. 363. Academic Press, New York.
42. Harris, H. and Hopkinson, D. A. (1976). *Handbook of Enzyme Electrophoresis in Human Genetics*. Elsevier, Amsterdam.
43. Richtie, R. F. and Smith, R. (1976). *Clin. Chem.*, **22**, 497.
44. Arnaud, P., Wilson, G. B., Koistinen, J., and Fudenberg, H. H. (1977). *J. Immunol. Methods*, **16**, 221.
45. Towbin, H., Staehelin, T., and Gordon, J. (1979). *Proc. Natl. Acad. Sci. USA*, **76**, 4350.
46. Gelfi, C. and Righetti, P. G. (1984). *Electrophoresis*, **5**, 257.

47. Radola, B. J. (1973). *Biochim. Biophys. Acta,* **295**, 412; (1974). *ibid,* **386**, 181.
48. Cuono, C. B. and Chapo, G. A. (1982). *Electrophoresis,* **3**, 65.
49. Dunn, M. J. (1987). *J. Chromatogr.,* **418**, 145.
50. Gianazza, E., Artoni, G., and Righetti, P. G. (1983). *Electrophoresis,* **4**, 321.
51. Gianazza, E., Celentano, F., Dossi, G., Bjellqvist, B., and Righetti, P. G. (1984). *Electrophoresis,* **5**, 88.
52. Rochette, J., Righetti, P. G., Bianchi-Bosisio, A., Vertongen, F., Schneck, G., Boissel, J. P., Labie, D., and Wajcman, H. (1984). *J. Chromatogr.,* **285**, 143.
53. Gianazza, E., Giacon, P., Sahlin, B., and Righetti, P. G. (1985). *Electrophoresis,* **6**, 53.
54. Pietta, P. G., Pocaterra, E., Fiorino, A., Gianazza, E., and Righetti, P. G. (1985). *Electrophoresis,* **6**, 162.
55. Astrua-Testori, S., Pernelle, J. J., Wahrmann, J. P., and Righetti, P. G. (1986). *Electrophoresis,* **7**, 527.
56. Righetti, P. G., Gelfi, C., Bossi, M. L., and Boschetti, E. (1987). *Electrophoresis,* **8**, 62.
57. Rabilloud, T., Gelfi, C., Bossi, M. L., and Righetti, P. G. (1987). *Electrophoresis,* **8**, 305.
58. Rabilloud, T., Pernelle, J. J., Wahrmann, J. P., Gelfi, C., and Righetti, P. G. (1987). *J. Chromatogr.,* **402**, 105.
59. Rimpilainen, M. and Righetti, P. G. (1985). *Electrophoresis,* **6**, 419.
60. Astrua-Testori, S. and Righetti, P. G. (1987). *J. Chromatogr.,* **387**, 121.
61. Rilbe, H. (1973). *Ann. NY Acad. Sci.,* **209**, 11.
62. Righetti, P. G., Gianazza, E., Bianchi-Bosisio, A., and Cossu, G. (1986). In *The Hemoglobinopathies* (ed. T. H. J. Huisman), p. 47. Churchill Livingstone, Edinburgh.
63. Cossu, G. and Righetti, P. G. (1987). *J. Chromatogr.,* **398**, 211.
64. Righetti, P. G. (1984). *J. Chromatogr.,* **300**, 165.
65. Righetti, P. G. and Gianazza, E. (1987). *Methods Biochem. Anal.,* **32**, 215.
66. Faupel, M., Barzaghi, B., Gelfi, C., and Righetti, P. G. (1987). *J. Biochem. Biophys. Methods,* **15**, 147.
67. Righetti, P. G., Barzaghi, B., Luzzana, M., Manfredi, G., and Faupel, M. (1987). *J. Biochem. Biophys. Methods,* **15**, 189.
68. Righetti, P. G., Wenisch, E., and Faupel, M. (1989). *J. Chromatogr.,* **475**, 293.
69. Righetti, P. G., Wenisch, E., Jungbauer, A., Katinger, H., and Faupel, M. (1990). *J. Chromatogr.,* **500** (in press).
70. Fredriksson, S. (1977). In *Isoelectric Focusing and Isotachophoresis* (B. J. Radola, ed. and D. Graesslin), pp. 71−83. Walter de Gruyter, Berlin.
71. Gelsema, W. J., De Ligny, C. L., and Van Der Veen, N. G. (1979). *J. Chromatogr.,* **171**, 171.

3

Two-dimensional gel electrophoresis

DAVID RICKWOOD,
J. ALEC A. CHAMBERS, and S. PETER SPRAGG

1. Introduction

Since the first edition of this book there has been an increasing awareness of the limitations of one-dimensional electrophoretic separations for the analysis of complex protein mixtures. Irrespective of the basis of separation, whether it be molecular weight or isoelectric point, there are likely to be problems in the interpretation of gel patterns as a result of the comigration of polypeptides. Although suspected, the scale of such problems was not revealed until the first two-dimensional separation methods were developed in the middle 1970s. Since the recognition of the importance of two-dimensional gel electrophoretic methods a great deal of work has been carried out to refine both the methodology of the separation technique itself and, equally important, to improve the analysis of two-dimensional gels. There has been increasing emphasis on the separation and characterization of polypeptides on the basis of their isoelectric point and molecular mass. Gels have tended to become smaller and staining methods more sensitive thus allowing the analysis of much smaller samples. These trends are reflected in the revisions made to this chapter.

All proteins migrate in an electric field at a speed that is dependent on their size, conformation and electrical charge. These latter two properties of proteins can be modified by the presence of denaturing agents and detergents that interact with the proteins. The choice of the optimal conditions for a separation is also very dependent on the nature of the proteins being analysed. For example, some types of proteins, such as proteins of the nucleus and membranes, are extremely prone to aggregate.

All two-dimensional methods should be designed so that the polypeptides are separated on the basis of a different molecular property in each dimension. Systems that do not achieve this produce an essentially diagonal pattern of polypeptides and the degree of resolution of the individual components obtained is often worse than that by a one-dimensional separation of proteins. The commonest two-dimensional electrophoresis method for analysing mixtures of polypeptides is to separate the proteins in the first dimension on the basis of charge by isoelectric focusing and then to separate the polypeptides in the second dimension in the presence of SDS

217

(SDS-PAGE) which, in most cases, gives a separation primarily on the basis of molecular mass of the polypeptides. Because of its wide range of applications, a significant part of this chapter is devoted to two-dimensional techniques that utilize isoelectric focusing and SDS-PAGE and Appendix 5 lists the range of applications for which this system has been used. However, other systems that have been developed for particular types of protein are also described.

One of the major problems of working with two-dimensional gels is the analysis and comparison of what can be very complex patterns that vary in only a relatively small area of the gel. Analysis of two-dimensional gels in its simplest form can be carried out by superimposing one photographic image over another. However, better results can usually be obtained by computer analysis of the gels as described in Section 6 of this chapter.

This is perhaps an appropriate point in the chapter to warn the novice that although two-dimensional gel electrophoresis is a very powerful technique it is by no means infallible. Even after two-dimensional gel electrophoresis, some proteins may co-migrate because either they are very similar proteins or because they are bound together very tightly. In addition, it is possible for protein mixtures to appear more heterogeneous than they really are as a result of artefactual changes during the preparation of the sample.

2. Apparatus for two-dimensional electrophoresis

2.1 Introduction

Very little specialized apparatus is necessary when running gels on a relatively small scale, say three or four gels a week. Often it is fairly simple to adapt one-dimensional apparatus for two-dimensional electrophoresis on this scale. On the other hand, where large numbers of gels are being run on a routine basis, then it is worthwhile considering investing in one of the specialized types of apparatus that have been devised for running several gels at a time. The other important benefit that can be gained from using apparatus for multiple gels is that it enables one to obtain more reproducible gel patterns.

2.2 Apparatus for first-dimensional gels

Most of the original publications describing two-dimensional electrophoresis used electrophoresis in rod gels for the first dimension. Usually authors have adapted the readily-available simple type of apparatus that can take 8–12 gel rods (Chapter 1, Section 5.1.1), the exact dimensions of which vary depending on the method. The diameter of gels can vary from less than a millimetre to 5 mm. It is also often feasible to separate the proteins in the first dimension using slab gels (see Chapter 1). The advantage of this is that all the samples are separated in the same gel and thus under identical conditions giving more reproducible separations than can be obtained using rod gels. However, rod gels must be used for those methods that recommend mixing of the sample with the gel mixture prior to electrophoresis. Irrespective of the format of gel chosen for the first dimension, it is essential that the gel is sufficiently strong to withstand the manipulations required in preparing it for the second dimension.

2.3 Apparatus for second-dimensional gels

The second-dimensional separations are always carried out in slab gels. There is quite a lot of variation in the gel thickness used by researchers. In some early methods, rather thick gels, up to 3.5 mm thick, were used, but now there is general recognition that gels 0.5 − 1.5 mm thick are not only easier to dry down after electrophoresis but they are also easier to keep cool during electrophoresis. The other very important factor that determines the efficiency of cooling is the material used for the slab gel plates. Perspex (Lucite) plates are much more durable than glass plates but they also have a much lower thermal conductivity leading to much less efficient cooling of the gel. Thus, whenever possible, glass plates should be used for forming the gel mould. It is extremely important to ensure efficient cooling of the gel in order to obtain distortion-free and reproducible gel patterns.

The design of the slab gel plates also varies depending on both the design of the slab gel apparatus and the method used to apply the first-dimensional gel to the slab gel. *Figure 1* shows some of the methods that have been used to allow the first-dimensional gel to be loaded into position for electrophoresis in the second dimension. Thicker first-dimensional gels are usually best accommodated by using one or two

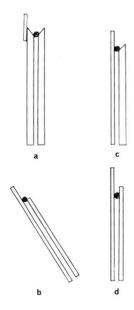

Figure 1. Methods of applying the first-dimensional gel to the second-dimensional slab gel. (a) Use of bevelled plates (0.6 cm thick), (b) use of normal plates (0.3 cm thick), (c) a combination of bevelled and normal plates, (d) first-dimensional gel squashed between 0.3 cm thick plates.

bevelled plates (*Figure 1a* and *b*). However, the use of these thicker plates does seriously reduce the efficiency with which the plates can be cooled. Using smaller diameter first-dimensional gels allows one to use gel plates of the normal thickness (*Figure 1c*). However, in all these methods the first-dimensional rod gel is situated very close to the top of the second-dimensional slab gel. This can present problems if the rod gel is to be sealed in position using the polyacrylamide stacking gel mixture since oxygen in the air inhibits the polymerization of the gel mixture. In such cases the first-dimensional gel may not be completely enclosed by the polymerized gel. One way to overcome this problem is to seal the first-dimensional gel in position with agarose. However, care is needed since the hot agarose gel mixture may adversely affect the sample proteins in the gel. An alternative method is to sandwich the gel between the two plates when the second-dimensional gel is being assembled (*Figure 1d*). This procedure can be used with even quite thick gels since they can be easily squashed in the correct position. Nevertheless, one disadvantage of this method is that it is not possible to make up the second-dimensional gel until the first-dimensional gel is available.

Typical apparatus for running slab gels is described in Chapter 1, Section 5.1.2 However, this type of apparatus can only accommodate one or two slab gels. In the authors' laboratories, four sets of apparatus (i.e. eight gels) are run in parallel, an arrangement which will suit many users. However, in order to maximize the comparability of gels, it is best to run the gels at the same time under the same conditions. This is particularly important if one wishes to analyse the gels using automated computer methods (see Section 6). In such cases it is advisable to use one of the several designs of apparatus for running two-dimensional gels such as that shown in *Figure 2*. Sources of other designs for multiple gel electrophoresis are given in *Table 1*. Whilst the running of gels in parallel greatly improves the reproducibility of gel patterns, it is important to realise that such methods create problems of processing the gels through staining, destaining, photography and drying procedures as well as, in some cases, autoradiography. The reader should take this into account before embarking on the running of large numbers of gels.

3. General aspects of two-dimensional gel electrophoresis

3.1 Introduction

The art of running two-dimensional gels can only be learned from experience gained in the laboratory. The actual details of each procedure will depend on the exact type of two-dimensional technique used. This section deals specifically with the experimental procedures and general techniques common to most two-dimensional separations. In addition, the reader should consult Chapter 1 which deals with general techniques as applied to one-dimensional gels.

One of the most important decisions is choice of the types of separation to be used in the first and second dimensions. Isoelectric focusing followed by SDS-PAGE has

David Rickwood et al.

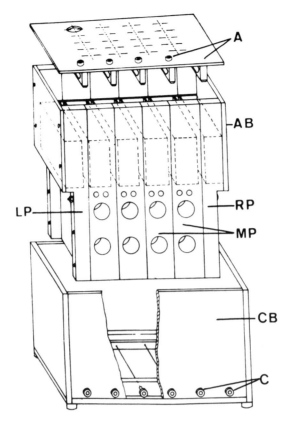

Figure 2. Apparatus for running multiple slab gels. The perspective view of the apparatus is shown. A, Perspex plate with anode terminals; MP, middle part; RP and LP, right and left side parts, respectively; AB, anode buffer chamber; CB, cathode buffer chamber; C, cathode terminals. (Reproduced from ref. 1 with permission.)

Table 1. Examples of apparatus for running multiple gel slabs

Authors	Number of gels	Gel dimensions (cm)	References
E. Kaltschmidt and H. G. Wittmann	5	19 × 20	1
B. Dean	8	13 × 16	2
N. G. Anderson and N. L. Anderson	10	17 × 17	3
J. I. Garrels	4	20 × 17	4
M. I. Jones, W. E. Massingham, and S. P. Spragg	10	17 × 17	5

become almost the standard method. However, as described in the later sections of this chapter, other methods have been devised and undoubtedly still more new types of separation will be developed in the future.

221

3.2 Solutions and apparatus

Acrylamide solutions are very toxic and so they must always be handled with the greatest care. Always wear a mask when weighing out the solid and ensure that the solutions do not come into contact with the skin. The solutions required for two-dimensional gel electrophoresis are similar to those used in one-dimensional separations. However, because of the overall greater complexity of two-dimensional separations, it is advantageous to prepare as many of the solutions as possible in advance. Most solutions can be stored frozen at −20°C or at 5°C in the presence of a bacterial inhibitor such as chloroform. However, some solutions do need to be freshly prepared. For example, ammonium persulphate solutions become less efficient at polymerizing gels even if they are stored in the cold. Therefore, it is best to prepare this solution on the same day as the gels are to be prepared. Urea in solution can also lead to problems since it reacts in solution to form cyanate ions which can both degrade proteins and modify their charges. The rate of cyanate formation increases with temperature and so protein samples should not be heated in urea solutions. Cyanate ions can be neutralized by compounds with free amino groups. It is good practice to prepare a stock solution of 8 M urea, filter it through Whatman No. 1 paper and pass it over a column of mixed-bed ion-exchange resin immediately before use. Where possible, urea solutions should be kept cool and buffered with Tris to minimize the accumulation of cyanate ions.

Although originally 2-mercaptoethanol was used as a reducing agent for polyacrylamide gel electrophoresis, at the concentrations at which it is effective (∼ 10 mM) it inhibits the polymerization of acrylamide solutions. In addition, any mercaptoethanol remaining in the gel after running seriously interferes with subsequent silver staining of the gel. An alternative reducing agent that has been widely used is dithiothreitol (DTT) which is not only less unpleasant to use but is effective at concentrations less than 1 mM and does not affect the polymerization of gels. Oxygen is also an inhibitor of polymerization of acrylamide solutions. Degassing solutions not only enhances the subsequent rate of polymerization but also greatly reduces the risk of air bubbles forming in the gel mixture as a result of the heat generated during polymerization.

As noted in Section 2, the apparatus used for two-dimensional electrophoresis is usually fairly conventional. The most important thing in preparing the apparatus for both the first- and second-dimensional gels is to ensure that all surfaces in contact with the gel are scrupulously clean. Gel plates and tubes need to be washed in chromic acid or strong detergent, rinsed in distilled water and then wiped clean with ethanol. Some methods recommend that gel tubes and plates are siliconized before use. However, this does create problems since gels, particularly low percentage polyacrylamide tube gels, tend to slip out of the tube. Furthermore, the poor adhesion of the gel to siliconized glass can lead to leakage of the electrolyte from the upper chamber leading to a drastic decrease in the quality of the separation obtained. When using siliconized gel tubes, the first of these problems can be overcome by covering the bottom of the gel tube with nylon gauze or perforated Parafilm or equivalent material before running the gel. This will ensure that the gel is retained in its tube during electrophoresis.

3.3 Preparation of the sample

The amount of sample, the method of its preparation and the solution that the sample is loaded in vary greatly depending on the type of two-dimensional separation. Hence this section deals only with the general guidelines to be followed when preparing samples. The reader should also consult Sections 6.6 and 6.7 of Chapter 1 which are also relevant to this topic. Where appropriate, more detailed descriptions of the preparation of samples are described in the sections in which the detailed protocols are given.

One of the most important rules for the preparation of samples is to ensure that there is no particulate matter present in the sample solution prior to gel electrophoresis. Such material can arise from protein aggregation and can lead to serious streaking of protein spots in the gel as a result of slow dissolution of the aggregates. Particulate material can be removed usually by centrifuging samples at 13 000 *g* for 5 min in a microcentrifuge although other workers prefer to centrifuge for up to 2 h at 100 000 *g*. However, it is important to ensure that this procedure does not lead to the preferential loss of one species of protein from the sample; this may be a significant problem if you use the latter method.

Another problem can be the presence of nucleic acids in the sample since these often interact with polypeptides and thus adversely affect the resolution of the protein components during subsequent electrophoresis. This problem can be particularly serious in separations that employ electrofocusing in one of the dimensions. As a general rule it is best to ensure that the sample is contaminated with as little nucleic acid material as possible. In some cases nucleic acids can be removed, by selective extraction or precipitation procedures. For example, histones can be acid extracted from chromatin (6) and the rRNA of ribosomes can be precipitated with LiCl leaving the proteins in solution (7). High molecular weight nucleic acids can be removed by ultracentrifugation but usually isopycnic centrifugation in gradients of CsCl or Cs_2SO_4 containing a denaturant such as 5 M urea or 6 M guanidine hydrochloride is more useful (8,9). Typically, solid CsCl is added to the protein sample of 6 M guanidine hydrochloride to give a final concentration of 50% (w/w) CsCl. The solution is then centrifuged at 100 000 *g* for 24 h at 5°C. Centrifugation pellets the RNA. The DNA bands in the bottom half of the gradient and the protein is concentrated at the top of the gradient. Nucleic acid contamination of the protein sample after this procedure is minimal. The protein sample is then dialysed into the medium of choice for loading onto the first-dimensional gel. When using these gradient techniques, it is best to arrange conditions so that the proteins band close to the top of the gradient rather than form a pellicle at the surface since such aggregates are often very difficult to dissociate even in urea. If pellicle formation is a problem then try loading less protein onto the gradients. An alternative method is to use a matrix such as hydroxyapatite that binds nucleic acids preferentially (10). However, this does introduce the possibility of selective loss of acidic proteins in addition to the normal losses of protein that accompany column chromatography. Nucleases, preferably insolubilized by attachment to Sepharose or a similar type of matrix, can be used to digest nucleic acids in the sample. However, nucleic acids bound to proteins are not always very susceptible to nucleases and so the usefulness of this procedure is limited. In addition, if proteases are present in the sample then these can degrade the proteins present in the sample during the incubation with the nucleases.

It is often a good idea to add a marker to the sample to mark the progress of electrophoresis. In zone electrophoresis one usually adds Bromophenol Blue to mark the ion front and so avoid loss of proteins off the bottom of the gel. In the case of isoelectric focusing, one can add internal markers to indicate the pH gradient that is formed. These are either dyes or specific commercially-available proteins. To determine the isoelectric point of a particular sample protein, the sample should be loaded at either end of the gel in separate runs or the gels run for different lengths of time to ensure that the proteins have reached their equilibrium positions.

3.4 Preparation and electrophoresis of gels

3.4.1 Preparation of first-dimensional gels

This is the simplest part of two-dimensional gel separations. The gels are prepared, usually in tubes, as described in Section 6 of Chapter 1 and the proteins are fractionated using one of the techniques described in the later sections of this chapter. It must be emphasized that the final quality of the two-dimensional separation is very dependent on the degree of resolution obtained in the first dimension. Hence it is often worth trying variations of the techniques described in this chapter in order to optimize the resolution. After electrophoresis the gels (still in their tubes) can be frozen at $-20°C$ as long as they are marked to indicate their identity and orientation. In fact freezing the gels usually facilitates their subsequent removal from the gel tubes. At high concentrations of urea, crystals of urea form upon freezing so that the gel should be allowed to warm up before running the gel in the second dimension.

3.4.2 Preparation of the first-dimensional gel for the second dimension

Remove the gel from the tube by cracking the gel tube in a vice. Alternatively extrude the gel from the tube using gentle pressure after the gel has been detached from the glass by rimming round with a needle as described in Section 7.1 of Chapter 1. If the gel has not been frozen, freezing and thawing will usually aid removal of the gel. It is very important that proteins from the experimenter's skin are not transferred to the outside of the gel during this procedure since these will then migrate in the second dimension and may confuse the interpretation of the final pattern of proteins. Therefore, disposable plastic gloves should be worn throughout all these manipulations.

Once the gel has been removed it is of paramount importance to mark the orientation of the gel immediately. Do not rely on the original appearance of the gel, which may change during subsequent equilibration, nor your memory which all too often leads to mistakes and confusion. One of the best methods for marking the orientation of the gel is to insert a small piece of fine gauge wire (~ 0.1 mm diam.) into the gel. If you do this, do not leave the ends protruding from the gel or these may catch and the wire can then be dislodged from the gel. An alternative method is to inject a small volume of indian ink into one of the gel. If the first-dimensional gel has an ion front, then it is convenient to put the marker at the position of the ion front, usually shown by the position of the marker dye. Then, after equilibration but before the subsequent second-dimensional separation, the gel can be cut at the position of the ion front allowing you to normalize the position of identical spots even if the first-dimensional gels have run different distances. In equilibrium systems, for example isoelectric focusing, where there is no ion front, one should adopt a fixed convention for marking the gel; for example, by always marking the bottom end of the gel. During these manipulations either keep the gel in the palm of a gloved hand or, if necessary, transfer it to a trough of aluminium foil or Parafilm or equivalent surface. Never allow the gel to come into contact with an absorbent surface such as tissue since the gel will stick to it quite tenaciously.

At this point the gel can be stored frozen in a trough of aluminium foil or Parafilm at −20°C or immediately equilibrated ready for the second dimension. The choice of equilibration conditions is still a subject of some controversy. Recommendations vary from no equilibration at all to equilibration times of up to 2 h. Independent workers have not been able to support the suggestions that equilibration is unnecessary (11) but it is known that the longer the length of equilibration time the greater is the loss of polypeptides from the gel, particularly of low molecular mass species. Losses from the gel can exceed 25% and so can be particularly serious if only small quantities of proteins have been loaded onto the gel. Greater diffusion of the polypeptides in the first-dimensional gel will also decrease the resolution of the polypeptide spots in the second dimension. The diffusion of bands can be reduced by fixing the gel in, for example, methanol−acetic acid solution (5:1:5 methanol/acetic acid/water by vol.) but this does appear to result in a higher retention of proteins in the first-dimensional gel during electrophoresis in the second dimension. Strict adherence to the equilibration conditions used in published methods will usually ensure that you obtain reasonable results. However, it may well be advantageous to try variations of the recommended equilibration protocol since, unless it is explicitly stated, it is unlikely that the procedure used has been optimized for that particular two-dimensional separation.

3.4.3 Preparation of the second-dimensional gel

It is usually best to set up the apparatus for the second dimension and start preparing the gel in good time, usually before the first-dimensional separation has finished. In most cases one uses a polyacrylamide slab gel prepared in the usual way (see Section 6.5 of Chapter 1). In the authors' experience discontinuous gel systems give better resolution. The resolving gel can be of uniform concentration and these do give very consistent results, but, for more complex samples, better results can be obtained by using a gradient gel. The gradient can be either linear or exponential and can be prepared using the appropriate type of gradient maker as described in Section 9.3 of Chapter 1. Discontinuous gradients consisting of two or more gel concentrations are not recommended because artefactual bands appear at the interfaces between the different concentrations. The major problem with gradient gels is in ensuring that the gradients are exactly reproducible not only between different gels made on the same day but also between gels made over a period of weeks or months. Such reproducibility between gels is an important part of achieving success in the art of two-dimensional electrophoresis. As a precautionary check of the migration of polypeptides in the second dimension, it is advisable to run an appropriate set of markers on each gel. If a uniform resolving gel does give sufficient resolution then use this simpler system to ensure better reproducibility of gel patterns.

3.4.4 Electrophoresis in the second dimension

After appropriate equilibration, the first-dimensional gel is loaded onto the second-dimensional gel. The gel can be sealed in position with low percentage acrylamide stacking gel. It is important that the percentage of acrylamide in the stacking gel is less than that of the first-dimensional gel, otherwise a significant fraction of the sample remains trapped in the first-dimensional gel. As mentioned earlier, one problem with sealing the first-dimensional gel in position at the top of the gel plates with acrylamide solution is that the gel solution at the top is inhibited from polymerizing by the oxygen in the air and so this may leave the top of the gel exposed and adversely affect the separation obtained. One simple way around this problem is to cast the stacking gel on top of the resolving gel, place the first-dimensional gel on top of the stacking gel and then seal it in position with a solution of hot 1% agarose in the stacking gel buffer. To obtain the best results using this procedure it is important to remove all excess liquid from both the top of the stacking gel and the first-dimensional gel before applying the agarose solution. If necessary one can use low melting point agarose to ensure that the gel is not exposed to excessive temperatures. Whichever method is used it is important to ensure that the first-dimensional gel is as parallel as possible to the top of the resolving gel and that there are no air bubbles around the gel. Before the sealing gel sets, form a sample well at one end of the slab gel and at the same height as the first-dimensional gel into which marker proteins can be loaded to calibrate the gel. Alternatively, place an agarose disc containing the molecular mass markers next to the first-dimensional gel before sealing it in place. When the sealing gel is set, place the second-dimensional gel into a vertical electrophoresis apparatus and electrophorese until the marker dye in the marker protein mixture reaches the bottom of the gel.

3.4.5 Analysis of the distribution of polypeptides

Chapter 1 (Section 7) of this book describes in detail the various procedures for detecting polypeptides in polyacrylamide gels. A list of some methods used for two-dimensional gels is given in *Table 2*. The reader will notice that the methods used are very similar to those given in Chapter 1. The main differences occur when a particular procedure is required as a result of a particular type of separation (e.g.

Table 2. Staining techniques for two-dimensional polyacrylamide gels

Staining procedure	Destaining procedure	Additional information	References
A. 3–4 h in 0.1% Coomassie Blue in methanol:water:acetic acid (5:5:1)	Overnight by diffusion against methanol:water:acetic acid (5:5:1)	—	9
B. 20 min in 0.1% Coomassie Blue in 50% TCA	Several changes of 7% acetic acid	Removal of ampholytes	11
C. 15 min in 0.55% Amido Black in 50% acetic acid	40 h in 1% acetic acid	—	1
D. 3 h in 0.25% Coomassie Blue in methanol:water:acetic acid (5:5:1)	Several changes of 5% methanol, 10% acetic acid	—	12
E. Overnight in 25% isopropyl alcohol, 10% acetic acid, 0.025–0.05% Coomassie Blue, followed by 6–9 h in 10% isopropyl alcohol, 10% acetic acid, 0.0025–0.005% Coomassie Blue	Several changes of 10% acetic acid	An additional optional staining step overnight in 10% acetic acid containing 0.0025% Coomassie Blue helps intensify the gel pattern	13
F. 1–4 h in 0.1% Coomassie Brilliant Blue R-250 in 7.5% acetic acid, 50% methanol in water	Overnight in 7.5% acetic acid, 50% methanol in water	—	14
G. 3 h at 80°C, or overnight at room temperature in 0.1% Amido Black in 0.7% acetic acid, 30% ethanol in water	Several changes of 7% acetic acid, 20% ethanol in water	—	15
H. 60 min in 50% methanol, stain with fresh alkaline 0.8% AgNO$_3$ solution. Wash with water for 5 min. To develop soak in fresh 0.02% HCHO, 0.005% citric acid for 10 min	Wash gel in water, then transfer to 50% methanol	More sensitive than Coomassie Blue methods.	16

the removal of carrier ampholytes in systems that use isoelectric focusing as one of the dimensions). Where such method-specific procedures are required they will be noted in the following sections.

4. Two-dimensional separations of proteins on the basis of their isoelectric points and mobility in SDS – polyacrylamide gels

4.1 Introduction

Of the several different types of two-dimensional gel separations that have been devised for analysing proteins, this method has become pre-eminent both in terms of its range of applications and the amount of work that has been done in developing the method. This is amply demonstrated by Appendix 5 that lists the applications of this type of two-dimensional separation. One reason for the popularity of this method is that, because the two dimensions separate proteins on the basis of independent properties (their isoelectric points in one dimension and mobility in an SDS-polyacrylamide gel in the other) one usually obtains an excellent resolution of even very complex protein mixtures. The isoelectric point of a protein also provides valuable information on its likely amino acid composition and even makes it possible to study some types of post-translational modifications of proteins. In addition, as described in Section 4.7 of Chapter 1, the mobility of proteins in SDS – polyacrylamide gels is primarily a function of their molecular mass. Hence it is possible to obtain an indication of the molecular mass of the separated protein. One version of this two-dimensional method is sometimes referred to in the literature as the ISO-DALT method because it separates proteins on the basis of their isoelectric points and molecular mass in daltons. Since the first edition of this book there have been many papers on suggested improvements to this technique, many of which have been of marginal importance to the average user. Some of these variations on the original method are listed in *Table 3* but for the most part they are not widely used, with perhaps just two exceptions: the use of Immobilines for the isoelectric focusing dimension (see Section 3.1.2 of Chapter 2) and the trend towards capillary gels for the first dimension. In this chapter the authors will not attempt to cover all the variations of these techniques in great detail since this has been done in other reviews (24, 25). Instead we will provide a practical guide to the reader for the methods which should have the widest application in terms of the quality of results with a minimum of special equipment or skills.

Almost all methods use isoelectric focusing as the first dimension. Although methods that use SDS-PAGE in the first dimension have been described (23), they have not been widely used by other groups. The reason for this is that SDS binds extremely tightly to many proteins and because of its net negative charge it changes the effective isoelectric point of the proteins to which it is bound. This type of method will not be discussed further in this chapter.

Table 3. Variations of methods for the separation of proteins on the basis of isoelectric point and SDS-PAGE

Procedure	References
Loading samples into the isoelectric focusing gel	10,17
Microgels for small samples	18,19
Use of Immobilines for the isoelectric focusing gel	20,21
Use of buffers instead of carrier ampholytes	22
SDS-PAGE followed by isoelectric focusing	23

4.2 First-dimensional separation by isoelectric focusing

4.2.1 Introduction

The use of isoelectric focusing for separating proteins on the basis of their isoelectric points has been described in comprehensive detail in Chapter 2. A pH gradient can be generated by applying an electric field to a gel containing a mixture of amphoteric components. The original compounds, called carrier ampholytes, were originally marketed by LKB Producta (now Pharmacia-LKB) under the name of Ampholines and consisted of a mixture of low molecular mass components with both basic amino and acidic carboxyl groups. Since that time a number of other types of carrier ampholytes have become commercially available. A brief summary of the commonest of these is given in Section 2.3.3 of Chapter 2. In addition, partly because of the very high prices of the commercial carrier ampholytes, a great deal of effort has been put into the development of buffer mixtures that can be used instead of the synthetic ampholytes. Overall, however, mixtures of buffers have proved to be much less satisfactory than the commercial carrier ampholytes (26) and they will not be considered further in this chapter.

The pH gradient forms in less than 1 h when a potential difference is applied across the ends of an isoelectric focusing gel containing carrier ampholytes. Polypeptides move more slowly than carrier ampholytes and usually find themselves in an environment where they have a net negative or positive charge and they then migrate through the gel until they reach a position where they have no net charge. Although size may affect the rate at which polypeptides migrate through the gel, their final position is determined only by their isoelectric point. However, note that when separating polypeptides under denaturing conditions the isoelectric points may be different from those of the native proteins. It is also a wise precaution always to use the same brand of carrier ampholytes to obtain reproducible results. In fact everything possible should be done to avoid the introduction of unnecessary variables, such as the source of chemicals, which may have undesirable effects on the reproducibility of two-dimensional gel patterns. Finally, the researcher should aim to adjust the pH gradient of the gel so that the proteins of interest are not focused close to the ends of the gel where the gradient may be less stable.

The method described in the following sections is based on the procedure devised by O'Farrell (11). Most of the variations published subsequently were aimed at improving both the resolution and reproducibility of the isoelectric focusing dimension. The most important of these are described in Section 4.4.

4.2.2 Preparation of the sample

Sample preparation for two-dimensional gel electrophoresis requires balancing the requirements for efficient solubilization of the sample proteins with compatibility with the first-dimensional system; for example, isoelectric focusing requires a minimum of salt in the sample. The presence of non-proteinaceous material such as nucleic acids and particulate material in the sample can all interfere with the migration of proteins.

i. Solubilization of samples

The factors to be considered when choosing a solubilization procedure for isoelectric focusing include:

- compatibility with isoelectric focusing
- maximal solubilization of proteins
- minimal solubilization of likely contaminants
- minimal proteolysis and chemical modification.

The solubilization systems used are typically based upon physical disruption of the material using an appropriate homogenizing technique, usually in the presence of a large excess of solutions containing very high concentrations of urea (>8 M) and a non-ionic detergent, often Triton X-100 or the closely-related Nonidet P-40 (NP-40). Alternatively, solubilization with SDS can be used followed by displacement of the SDS by the combined effects of high concentrations of urea and a non-ionic detergent. The latter system was originally devised by O'Farrell (11) and was further developed into a procedure for membrane proteins by Ames and Nikaido (13). The presence of a thiol reagent such as 10 mM DTT or 1% 2-mercaptoethanol is also useful for solubilizing proteins. References for several sample solubilization procedures are listed in *Table 4*. The procedure originally described by O'Farrell is probably a good point from which to start. Some more unusual solubilization procedures are cited by Dunn and Burghes (24).

The choice of method will depend upon the system being analysed. When you have chosen a system, it is extremely important that it is optimized so that there is always an excess of the protein solubilizing agents to ensure maximum solubilization of the sample. Typically this means a ratio of urea to protein of 2.5 and, in the case of SDS, a ratio of detergent to protein of 4, both values weight/weight. SDS is useful for solubilizing samples which contain very insoluble proteins or proteins that are prone to aggregation. In some cases it is the only way to solubilize samples completely. However, the use of SDS can lead to the formation of artefactual bands in the first dimension or the loss of protein. Therefore, if at all possible, it is best to avoid using SDS for solubilizing samples. If it is necessary to use SDS, then care

Table 4. Methods used to solubilize samples for two-dimensional gel electrophoresis

Sample	Solubilizing buffer	Treatment	References
Bacterial proteins	9.5 M urea, 2% NP-40, 5% 2-mercaptoethanol	Freeze-thaw, sonicate	11
Bacterial membranes	50 mM Tris—HCl (pH 6.8), 2% SDS and 0.5 mM $MgCl_2$[a]	Incubate at 70°C for 30 min	13
Nerve cells	14 μl of 1% SDS, 10% 2-mercaptoethanol	Homogenize in glass homogenizer	27
Serum proteins	9 M urea, 0.1 M DTT	None (soluble proteins)	3
Animal cells	3% SDS, 10% 2-mercaptoethanol[c]	Shear lyse cells and freeze, then lyophilize	4
Membranes, seed proteins	9.5 M urea, 5 mM K_2CO_3 (pH 10.3)	Sonicate at low frequency	28
Fungal mycelium, membranes	4 M guanidine thiocyanate, 1% 2-mercaptoethanol[d]	Blend using a suitable blender	3,25

[a] Then add 2 volumes of 9.5 M urea, 8% NP-40, 5% 2-mercaptoethanol, 2% Ampholines.
[b] Then add 5 μl of 10% NP-40 and 4% Ampholines and 4 mg of solid urea.
[c] Then incubate with DNase and RNase.
[d] Then equilibrate the sample with 8 M urea, 0.1% NP-40, 1 mM DTT.

must be taken to avoid artefacts. This is usually done by adding a large excess of non-ionic detergent to the sample prior to electrophoresis, usually an 8-fold excess of NP-40 which, in combination with high concentrations (6−9.5 M) of urea, effectively suppresses the formation of SDS-induced artefactual gel patterns (13).

It is difficult to bring about effective solubilization of protein samples without releasing other macromolecular components that may be present. The major problem is nucleic acids and the predominant one is RNA. Some approaches to the removal of nucleic acids have been described in Section 3.3.

ii. Avoidance of proteolysis
Proteolysis can be a significant problem in some tissues. Incomplete denaturation of tissue proteins tends to exacerbate the problem. Many proteases are fairly robust and will be active under partially denaturing conditions because the high concentration of exposed peptide bonds is equivalent to a very high substrate concentration. Protease inhibitors should be used only with the greatest caution because of their potential for introducing artefacts. For example, covalent modifiers of proteases such as PCMB and phenylmethylsulphonyl fluoride (PMSF) are not completely specific and so can modify other proteins, leading to charge heterogeneity and multiple spots on two-dimensional gels. The best way to avoid proteolysis is by using as strong denaturing conditions as possible; it is often worth trying several different procedures to see which is the most effective for your sample.

iii. Avoidance of modification of proteins
All possible steps should be taken to minimize sample protein modification. Most solubilization media are based on high concentrations of urea. Heating samples in the presence of urea ($>60°C$) can lead to rapid carbamylation of proteins with the loss of charge on lysine residues. This is the most obvious cause of protein modification but the inclusion of any reactive compound in the extraction medium should be considered carefully in this context.

iv. Removal of insoluble and particulate material
As discussed in Section 3.3, it is very important to remove any particulate material from the sample. This is usually done by centrifugation. Conditions used vary from 10 000 g for 30 min to 100 000 g for 2 h. The conditions chosen will depend on the sample. The longer run conditions will also remove high molecular weight DNA if it is present. However, one should be aware that very large proteins and protein complexes can also be lost under such conditions.

v. Removal of soluble interfering substances
The major interfering contaminant is nucleic acid, especially RNA. When nucleic acids bind to proteins they cause streaking of spots in two-dimensional gel electrophoresis and when silver staining is used they cause a high background. Unfortunately there is no simple effective way to remove them. Section 3.3 gives a number of possible methods that can be used to remove DNA and RNA. In the authors' experience the most effective method is to separate the protein from nucleic

acids on gradients of Cs_2SO_4 containing 6 M guanidine hydrochloride (9). This method is similar to that described in more detail in Section 3.3.

Excess lipid in the sample can interfere with solubilization of proteins. One of the functions of the non-ionic detergent in the solubilization medium is to stabilize exposed hydrophobic regions of denatured proteins. If there is an excess of membrane or stored lipid in the sample the detergent will preferentially form mixed micelles with these lipids and so not be available for protein binding. This can be compensated for by raising the concentration of detergent in the solubilization medium. As a rough guide, one can assume that each gram of membrane lipid will bind one gram of detergent. Nevertheless, extremely lipid-rich materials may have to be defatted using organic solvents before solubilization.

Irrespective of the method used to prepare the samples, it is important that the samples do not contain significant concentrations of salt since this interferes with isoelectric focusing. It is therefore good practice as a precautionary measure to dialyse the samples before use. Even if a large number of samples have to be dialysed, multiple sample dialysers are now available that can handle more than 20 samples in parallel. Dialysis should be carried out at 4°C against at least one change of 8 M urea, 5 mM DTT, 0.1% NP-40. After this the dialysate can be brought to 9.5 M urea by the addition of solid urea and the NP-40 concentration adjusted to 1% by the addition of a 10% stock solution.

In some cases it will be important to know how much material is loaded onto the gel. Unfortunately many components of solubilization media interfere with common protein assays such as Lowry or Biuret. However, when necessary, the assay of Schaffner and Weissman (29) or a modified Bradford assay (30) can be used.

4.2.3 Preparation of the first-dimensional gel

The first dimension has been the subject of many variations. Although it is beyond the scope of this chapter to cover all such variations in detail, several of the more widely-used ones will be mentioned for consideration.

Usually the first-dimensional gel is run in tubes. The filling of the narrow tubes required for this first step can be troublesome. The procedure described here is probably the commonest and does not require special equipment. Alternative procedures involving upward displacement of the gel mixture from a reservoir or trough into the gel tubes either by light suction or overlaying the gel solution with water have been described (e.g. ref. 25) but these procedures require the removal of excess acrylamide gel from the outside of the gel tubes after polymerization.

The solutions for preparation of the first-dimensional isoelectric focusing gels are listed in *Table 5*. The gels are prepared as described in *Protocol 1*.

Protocol 1. Preparation of the first-dimensional electrofocusing gel

1. Clean the glass tubes (2.5 mm inner diam., 13 cm long) and seal the bottom of each of these with two layers of Parafilm.
2. Mount the sealed tubes upright in a stand.

Protocol 1. *continued*

3. Prepare the gel solution (*Table 5*) and degas it immediately before adding the ammonium persulphate.

4. Transfer the solution to the tubes using a Pasteur pipette. Put the pipette as far down the tube as you can and withdraw it slowly, delivering the solution with a smooth continuous action. Fill the tubes to within 5 mm of the top of the tube. Avoid introducing air bubbles. If there are any in the tube they can usually be brought to the top by gently tapping the side of the tube (polymerized gels that contain air bubbles should be discarded).

5. Overlayer the gel with water and allow it to polymerize for 1 h.

6. At the end of the hour, remove the water by aspiration and overlayer the gel for a further hour with 25 μl of sample buffer (*Table 5*).

7. While the gel is polymerizing, prepare the electrolyte solutions:
   Anolyte: 10 mM H_3PO_4
   Catholyte: 20 mM NaOH

8. Fill the lower reservoir (the anode chamber) with the H_3PO_4 solution and lower the rod gels (without their Parafilm seals) into it. If small air bubbles are trapped underneath the gel tubes, remove them using a Pasteur pipette with a bent tip.

9. Remove the sample buffer overlay from the gels and fill the upper reservoir with a volume of NaOH solution equal to that of the H_3PO_4 in the lower reservoir.

10. Connect the gel chamber to the power supply. Before applying the sample, pre-electrophorese the gel at 200 V for 15 min, then 300 V for 30 min and finally 400 V for 30 min. This is to pre-form the pH gradient and also helps to remove some of the ionic residues arising from the ammonium persulphate.

11. After pre-electrophoresis, turn the voltage down to zero and disconnect the apparatus from the power pack.

Note that there have been many variations of pre-electrophoresis conditions. If a power supply that can supply constant power is available, it may be used as follows. Set the initial voltage to 1000 V, the initial current to 2 mA and power to 2 W. Let the pre-electrophoresis continue until the voltage becomes limiting. If the power supply does not indicate which variable is limiting, one can tell when either the current becomes too low to measure or the power consumption drops below 2 W. This should take about 30 min. At this point the gradient is formed and the sample can be applied.

Table 5. Recipes for the first-dimensional electrofocusing gel (O'Farrell method)

First-dimensional gel mixture
Each 10 ml of the gel mixture contains:

5.5 g of urea (ultrapure)
1.33 ml of 28.38% acrylamide, 1.62% bisacrylamide
2 ml of 10% Nonidet P-40 (NP-40)
0.4 ml of 40% Ampholines (pH 5 – 7)
0.1 ml of 40% Ampholines (pH 3.5 – 10)
1.95 ml of water
5 μl of TEMED
10 μl of 10% ammonium persulphate

Sample buffer
9.5 M urea
5% 2-mercaptoethanol
2% Nonidet P-40
1.6% Ampholines (pH 5 – 7)
0.4% Ampholines (pH 3.5 – 10)

4.2.4 Loading the sample and electrophoresis of gels

Protocol 2. Loading and electrophoresis of electrofocusing gels

1. Remove the NaOH solution from the upper reservoir by aspiration and apply the sample to the gel. Usually a sample volume of 25 μl containing about 20 μg of protein is applied. If the gel is slightly shorter than normal, a larger volume can be loaded. The volume and concentration of sample to be loaded depends on the method of detection to be used. A complex sample to be stained with Coomassie Blue will require 250 μg or more of protein while a labelled sample of low complexity and high specific activity may require only 100 ng.

2. Overlay the sample with 10 μl of overlay buffer containing 9 M urea and 1% carrier ampholytes (use the same mixture as for the sample buffer).

3. Carefully cover the gels with fresh 20 mM NaOH and then refill the reservoir.

4. Reconnect the apparatus to the power pack. Electrophorese the proteins at either 300 V for 16 h or 400 V for 12 h. In the interests of reproducibility it is best to use one set of conditions for all subsequent sets of first-dimensional gels. Many different run conditions (defined as product of voltage and run time in hours) have been used (see *Table 6*).

5. At the end of this time, increase the voltage to 800 V for 1 h. This helps to sharpen the focused protein bands.

6. At the end of the run, turn off the power supply, remove the gels and either store them on ice ready for running in the second dimension immediately or store at −20°C or −80°C until needed. Although gels can be extruded from

Protocol 2. *continued*

the tubes and stored wrapped in foil there is less chance of damaging the gels if they are stored in the gel tube. Gels can be stored for several weeks without loss of resolution. Make sure the gels are clearly labelled at all times in terms of both their identity and orientation.

Table 6. Conditions used for isoelectric focusing separations[a]

Run time (V. h.)	Run conditions	References
1000	1 h each of 50 V, 100 V, 150 V, 200 V, 250 V	30
3600	150 V for 24 h	11
6000	300 V for 18 h then 400 V for 1.5 h	13
16 000	400 V.h pre-run, 800 V focusing	31
19 000	19 h for 1000 V	4
25 000	500 V for 5 h then 1250 V for 18 h	17
34 500	Constant power, 2000 V_{max}, 1000 $V_{initial}$	32

4.3 Second-dimensional separation by SDS-polyacrylamide gel electrophoresis

4.3.1 Introduction

In contrast to the many variations of procedure reported for the isoelectric focusing dimension, there have been few modifications proposed for the second-dimensional SDS-PAGE step. The major choice is whether to use a gel system with a uniform separating gel or to have a gradient separating gel. Since the latter usually gives better results and has a wider range of applications, this is the system that will be described here. The most important aspect is to ensure that the polyacrylamide gradients in all the gels are absolutely identical. This is best achieved by casting all the gels at the same time using the same solution. An apparatus that can be used for this is shown in *Figure 3*.

4.3.2 Preparation of the second-dimensional gel

It is often most convenient to prepare the second-dimensional separating gel several hours in advance or even the day before it is to be used. However, the stacking gel must be poured shortly before loading the first dimensional gel onto it.

The original procedure calls for $10-16\% T$ exponential gradient gel. The use of a standard gradient maker to make an exponential gradient is as described in Chapter 1 (Section 9.3.3). For a separating gel 16.4 cm wide by 14.6 cm high and 0.8 mm thick, the volumes of the gel solutions required are listed in *Table 7*. Adjust the volumes proportionately if you require a gel of different dimensions.

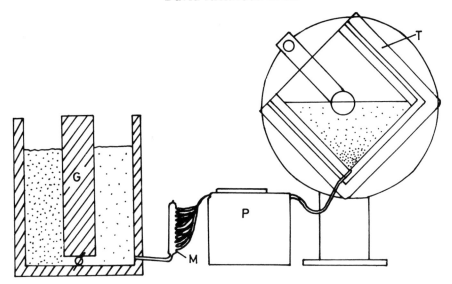

Figure 3. Apparatus for pouring multiple slab gradient gels. The figure shows a rotatable gel-casting tank (T) containing the assembled gel moulds, connected to a 10-channel peristaltic pump (P), which feeds from a 10-channel outlet manifold (M) from the gradient maker (G), which can be either of the simple type described in Section 9.3.3 of Chapter 1 or one of the commercially-available programmable gradient makers. Adapted from ref. 3.

Table 7. Recipes for the second-dimensional gel (O'Farrell method)

Resolving gel
Light solution:
4 ml of 0.4% SDS, 1.5 M Tris – HCl (pH 8.8)
5.3 ml of 29.2% acrylamide, 0.8% bisacrylamide
6.7 ml of water
8 µl of TEMED
25 µl of 10% ammonium persulphate

Dense solution:
2 ml of 0.4% SDS, 1.5 M Tris – HCl (pH 8.8)
4.3 ml of 29.2% acrylamide, 0.8% bisacrylamide
1.7 ml of 75% glycerol
4 µl of TEMED
10 µl of 10% ammonium persulphate

Stacking gel:
1.25 ml of 0.4% SDS, 0.5 M of Tris – HCl (pH 6.8)
0.75 ml of 29.2% acrylamide, 0.8% bisacrylamide
3.0 ml of water
5 µl of TEMED
15 µl of 10% ammonium persulphate

Sealing gel:
0.1% agarose[a] in 0.125 M Tris – HCl (pH 6.8). Melt the agarose in the Tris buffer and allow to cool to 45°C (LGT agarose), or 55°C for normal agarose, before adding a tenth volume of 20% SDS.

---

[a] A low gelling temperature agarose (LGT agarose) such as SeaPlaque, Marine Colloids, is preferred.

The procedure for preparation of the second-dimensional gels is described in *Protocol 3*.

Protocol 3. Preparation of the second-dimensional SDS-polyacrylamide gel

1. Degas the solutions immediately before adding ammonium persulphate.
2. With the tap between the two chambers of the exponential gradient maker closed, pour 16 ml of 'light' solution into the reservoir chamber and 5 ml of 'heavy' solution into the mixing chamber.
3. Seal the mixing chamber with a rubber bung and then release the excess pressure using a syringe needle. Begin stirring the mixing chamber so that mixing is efficient but try to avoid introducing air bubbles.
4. Open the tap between the chambers and start pumping the gel mixture between the gel plates. A pumping rate of about 5 ml/min is recommended. Stop pouring when the gel is about 2.5 cm from the top of the plates.
5. Overlayer the gel with either water or water-saturated butan-2-ol.
6. Drain any residual acrylamide mixture from the gradient maker and clean the gradient maker by pumping several volumes of water through it.
7. When the gel has polymerized, pour off the overlay, wash the surface of the gel two or three times with distilled water and put a thin layer (5 mm) of gel buffer containing 0.1% SDS on the gel. If the gel is to be stored, wrap the mould containing the gel in plastic film and store at 4°C.
8. While the isoelectric focusing gels are equilibrating (see Section 4.3.3) prepare the stacking gel.
9. Pour off the overlay buffer, wash the surface of the separating gel with water and drain. The surface of this gel must be drained well to ensure a good bond between the stacking and separating gel.
10. Prepare the stacking gel according to *Table 7*, pour the stacking gel to within about 1 mm of the top of the plates and overlayer it as in step 5.
11. When the stacking gel has polymerized, pour off the overlay and remove any residual liquid by blotting or aspiration. The gel is now ready to receive the first-dimensional gel.

4.3.3 Recovery and equilibration of the isoelectric focusing gel for the second dimension

Although described sequentially here, recovery and equilibration of the isoelectric focusing gel is begun before the stacking gel of the second-dimension gel is prepared. The procedure for equilibration of the isoelectric focusing gel is described in *Protocol 4*.

Protocol 4. Preparation of the electrofocusing gel for the second dimension

1. Before beginning equilibration of the isoelectric focusing gel, prepare the sealing gel used to attach this gel to the second-dimensional gel (*Table 7*).

2. Remove the first-dimensional gel from its glass tube. If it has been frozen or kept on ice, the gel is mechanically quite firm and can be easily removed from the tube by rimming with water and, if necessary, by pushing it out with gentle air pressure (use a 20 ml syringe attached to the gel tube with silicone rubber tubing). Make sure that the acid end of the gel is marked (see Section 3.4.2). Extrude the gel into a small disposable plastic container of distilled water. A plastic Petri dish or weighing dish is suitable. Leave the gel in the water for about 30 sec or until any precipitated urea has redissolved.

3. Drain off the water by pouring the contents of the dish through a small (~ 5 cm diam.) domestic sieve (a fine plastic mesh is preferred).

4. Transfer the gel (by inverting the sieve) to a second Petri dish and cover it with equilibration buffer [2.5% SDS, 5 mM DTT, 125 mM Tris−HCl (pH 6.8), 10% glycerol 0.05% Bromophenol Blue]. About 20−25 ml will be needed for each gel. Do not put more than one gel in each Petri dish.

5. Leave the gel to incubate at room temperature for about 30 min with occasional gentle shaking. You may find it desirable to use a longer incubation time but this will lead to the loss of more protein from the gel. The most important function of the equilibration is to replace as much of the protein-associated urea with SDS as possible. This is slower than replacing the ampholytes in the gel from the first-dimensional separation with pH 6.8 buffer needed for the second-dimensional separation. Some proteins may only allow a very slow exchange so that it is best to start equilibrating the gel before pouring the stacking gel.

4.3.4 Loading the isoelectric focusing gel onto the second-dimensional gel

After the first-dimensional gel has been equilibrated, it should be loaded onto the second-dimensional gel as described in *Protocol 5*.

Protocol 5. Loading of the electrofocusing gel onto the second-dimensional gel

1. Drain the first-dimensional gel of buffer using a plastic sieve.

2. Tip the gel onto a piece of Parafilm or other hydrophobic surface and straighten it out with a spatula or glass rod. If you are using a Studier-type of gel apparatus (Chapter 1, Section 5.1.2) for the second-dimensional separation, tip the gel mould so that it is facing you at a fairly acute angle with the shorter plate

Protocol 5. *continued*

uppermost, resting it on a test-tube rack for example. Tip the Parafilm trough holding the first-dimensional gel and the gel should roll into the gap above the second-dimensional gel without too much prodding. Some sample buffer will drain from the rod gel at this stage. Remove this from the stacking gel by aspiration with a syringe. If you are using a multiple-gel apparatus, the gel is simply laid on the bevelled edges. Sometimes it may be simpler to lay it starting at one end and working towards the other.

3. Make a small well or lay a small agarose slice containing molecular mass markers beside the first-dimensional rod gel.

4. As soon as the isoelectric focusing gel is in position, take the molten agarose sealing solution (*Table 7*) and take up about 5 ml in a syringe with a wide-bore needle (19 gauge). Make sure there are no air bubbles in the needle by squirting the agarose through it and then insert the needle below the rod gel. Inject enough of the agarose gel to fill the space between the stacking gel and the rod gel. Take care at all times to avoid introducing air bubbles into the sealing gel. If this does occur, in most cases you can carefully remove them using the syringe needle. The rod gel need not be completely covered with the agarose gel sealing solution but it should have enough of a covering so that it is anchored firmly in place when the agarose sets.

5. Allow the agarose solution to gel. This take only a few minutes for conventional agarose but somewhat longer for low gelling temperature (LGT) agarose.

4.3.5 Electrophoresis and analysis of the gel

The electrophoresis buffer used is 0.192 M glycine, 25 mM Tris base, 0.1% SDS. This should be prepared as a 10× stock solution. Mounting of the slab gel on the apparatus and electrophoresis of the gel are as described for any other gradient gel (Section 6.7 of Chapter 1). The recommended running conditions are at a constant current of 20 mA. In this case the gel should run in about 4 h or until the Bromophenol Blue has just run off the gel. If a constant current power supply is not available, run at a constant voltage that gives an initial current somewhat higher than 20 mA (usually around 30−40 mA). The current will drop as the gel runs. The gel will take longer to run at constant voltage.

After running, the gel is processed for detecting polypeptides by dye, silver staining, autoradiography or fluorography as required (see Section 7 of Chapter 1). For staining procedures it is necessary to use those procedures that will fix the proteins and remove the carrier ampholytes (*Table 2*). A typical two-dimensional separation using this procedure is shown in *Figure 4*.

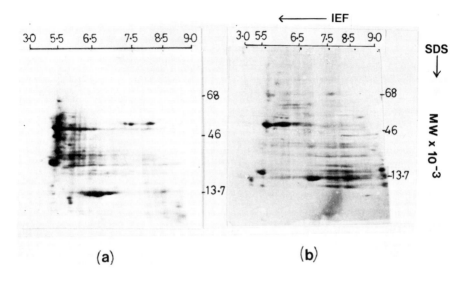

Figure 4. Two-dimensional separation of proteins using the O'Farrell method. Total cellular proteins were labelled with [³⁵S]methionine, separated using the O'Farrell technique described in the text, and the polypeptides were located by autoradiography.

4.4 Modifications to the basic method

4.4.1 Variations of the isoelectric focusing conditions

The system described in Section 4.3 gives very good resolution of most of the proteins in most tissues and organisms. However, the pH range is rather limited and rarely extends much above pH 8.0. This problem is a reflection of the limitations of carrier ampholytes, which tend to have gaps in conductivity at extremely alkaline pHs, and the longer time required for proteins with basic isoelectric points to focus. Both of these factors result in the need to run the gel longer, which in turn results in cathodic drift of the gradient, leading to a loss of the basic end of the gradient. In some cases it may be neccessary to adjust the pH range of the gradient to examine a particular region of the two-dimensional map. For example, the pH range pH 5.0–6.5. This can be achieved by changing the carrier ampholyte mixture. The greater the proportion of a given carrier ampholyte in the mixture the greater is its contribution to the pH gradient. Thus to obtain any desired pH range one can try mixing together various proportions of carrier ampholytes in the pH range of interest. It is often worth trying mixtures of ampholytes from different manufacturers since they all have somewhat different properties.

Stabilizing the basic end of the pH gradient
Much effort has been put into understanding the mechanism(s) by which the basic end of the pH gradient migrates off the end of the gel during isoelectric focusing

(cathodic drift). Although it is still difficult to generate a pH gradient extending to a pH more basic than pH 8.5, several approaches are now available to stabilize a pH gradient above pH 7. In order of increasing complexity they are as follows.

(a) Some improvement can be achieved if the NaOH electrolyte solution also contains 1 mM $Ca(OH)_2$. This prevents acidification of the NaOH solution by absorption of CO_2 since calcium carbonate forms preferentially and precipitates. Unfortunately the effect appears to be marginal.

(b) Omitting carrier ampholytes from sample solutions. The basic version of the two-dimensional system described in previous sections involves samples being applied to the first-dimensional gel in a buffer containing carrier ampholytes. In the authors' experience these appear to exacerbate the phenomenon of cathodic drift. If they are left out of both the sample buffer and the overlay for the first dimension, then the basic end of the gel is extended by about 1 pH unit (17).

(c) The use of weak electrolytes. Chrambach and co-workers have observed that replacing completely ionized electrolytes such as NaOH and H_3PO_4 with weak electrolytes such as Good buffers greatly reduces cathodic drift. This has been applied to first-dimensional gels; the authors of this chapter found that a catholyte of 100 mM Bis-Tris, 50% (w/v) sucrose and an anolyte of 100 mM Mops, 50% (w/v) sucrose is a generally useful electrolyte pair capable of generating gradients from pH 4.0−8.5 that are stable for 25 000 V. h. Sucrose in the electrolyte solutions steepens the gradient but has the disadvantage of making it extremely difficult to make the sample dense enough to underlayer on the gel. Therefore, mixing the sample in the rod gel mixture before polymerization has to be used (see Section 4.4.3). The basic concept is sound. By trying several pairs of electrolytes one may find a pair that improve the pH gradient in the region of interest. However, some pairs of electrolytes do appear to be incompatible (17, 33−35).

(d) The use of Immobiline carrier ampholytes. These are covalently bound to the gel matrix and therefore cannot drift towards the cathode. More complete directions for their use and a discussion of the problems associated with them are given in Chapter 2.

Clearly, because of cathodic drift, large changes to the standard system are required to get a stable pH gradient in the moderately alkaline pH regions. Therefore alternative approaches are required for extremely basic proteins. In the following section a method based on non-equilibrium isoelectric focusing (NEPHGE) is described while Section 5 includes descriptions of some of the methods used for specific types of basic proteins.

4.4.2 Analysis of basic proteins by NEPHGE

Non-equilibrium pH gradient electrophoresis (NEPHGE) was developed by O'Farrell for analysing extremely basic proteins (36). The major difference between this and typical two-dimensional systems is in the first dimension. Instead of applying the sample to the basic end of the gel, it is applied at the acid end. Under these conditions all the proteins are positively charged and migrate towards the basic end of the gel. If the gel were allowed to run to equilibrium, most of the basic proteins would still migrate off the end of the gel. To avoid this the gel is not run to equilibrium. Rather, fairly short run times are used. Thus for a typical first-dimensional isoelectric focusing gel the run time is 5–10 000 V. h, whereas for a NEPHGE run one would use as little as 1000 V. h depending upon the proteins being analysed.

Instructions for preparing a NEPHGE first-dimensional gel are described in *Table 8*. Preparation of the isoelectric focusing gel is essentially the same as for a normal gel but slightly more TEMED and ammonium persulphate are used because polymerization is less efficient at alkaline pHs. The recipe calls for either broad range (pH 3.5–10) or narrow range alkaline (pH 7–10) carrier ampholytes. If the intention is to analyse only very basic proteins then use the pH 7–10 range. Use pH 3.5–10 carrier ampholytes if you intend to look at a broader range of peptides. Load the gel and run it without pre-electrophoresis to form the pH gradient. The gel is run from the acid end to the basic, that is the gel is run in the opposite direction compared to a typical isoelectric focusing gel. The phosphoric acid is placed in the upper chamber and the NaOH solution in the lower. Obviously one must reverse the connections to the power supply also.

The second dimension is the same as that described in Section 4.3.2. The following points are important for NEPHGE. Firstly, this is not an equilibrium system. To obtain reproducible gel patterns one must use reproducible run conditions. These are defined by the V. h product for the run and the temperature. It will be necessary to try several values to optimize the separation. Secondly, because this is not an equilibrium system, the position of a protein band at a specific pH after a given run time does not reflect its isoelectric point.

4.4.3 Variations in sample application

End-loading of samples is certainly the commonest way to load samples. However, there are times when this is not the best way to do it. As outlined in Chapter 2, there are times when the sample may precipitate at the loading point or the volume

---

Table 8. Recipe for the first-dimensional gel for NEPHGE

Each 10 ml of gel mixture contains:

5.5 g of urea (ultrapure)
1.33 ml of 28.38% acrylamide, 1.62% bisacrylamide (30%*T*, 5.7%*C*)
2.0 ml of 10% Nonidet P-40 (NP-40)
0.5 ml of 40% carrier ampholytes[a]
1.93 ml of water
14 μl of TEMED
20 μl of 10% ammonium persulphate

[a] If basic proteins are to be analysed, use LKB Ampholine 7–10. If a wider range of pIs is to be studied, use pH 3.5–10.

that can be applied limits the amount of radioactivity that can be end-loaded onto the gel to values that are too small to be detectable in a reasonable time. This can be avoided by mixing the sample into the gel before polymerization (10, 17). Details are given in *Table 9*. Although the polymerization of acrylamide is a free radical reaction, there is comparatively little covalent binding of protein to the gel matrix and no significant changes are observed in the protein pattern as compared to the pattern generated if the sample is end-loaded. In such gels the basic end of the pH gradient is stabilized and, although it is necessary to run these gels somewhat longer to ensure good focusing (10 000 V. h), the pH gradient is reasonably broad. Short run times will give a very broad gradient but not all proteins will focus well under these conditions.

It should noted that some samples are not compatible with being mixed with the gel mixture. For example, reticulocyte lysates appear to inhibit polymerization. This may be due to quenching of the polymerization reaction by haemin.

4.4.4 Changing the scale of a gel: very large and very small gels

Gels with similar dimensions to those described are suitable for most separations. Generally speaking, a second-dimensional gel that can fit onto a 20 × 25 cm sheet of X-ray film after drying down should supply enough resolution and be quick enough to run for most purposes. However, under certain circumstances one may require either enhanced resolution of a very complex sample or rapid processing of many samples that are well defined or of reasonably low complexity. It is possible to make much larger gels to solve the first problem and use very small gels to assist with the second case.

Large first-dimensional gels about 30 cm long and second-dimensional gels with a 40 cm long separating gel have been in use since the large polyacrylamide gels for DNA sequencing became available. The procedures are essentially the same as those for the gels described in preceding sections of this chapter but some adjustments

---

Table 9. First-dimensional gels when the sample is mixed with the gel

Gel preparation
0.5 ml of sample in 8 M urea, 0.1% NP-40, 1 mM DTT
0.133 ml of 28.38% acrylamide, 1.62% bisacrylamide
0.3 g of urea (ultrapure)
0.2 ml of 10% Nonidet P-40
0.05 ml of 40% carrier ampholyte
2 μl of TEMED
5 μl of 10% ammonium persulphate

Possible electrolytes and running conditions

	Anolyte	Catholyte	Run conditions
(a)	10 mM H_3PO_4	20 mM NaOH	see *Protocol 2*
(b)	5% H_3PO_4	5% 1,2 diaminoethane	150 V, 24 h
(c)	100 mM Mops,	100 mM Bis-Tris,	500 V for 2.5 h then
	5% sucrose	50% sucrose	1250 V for 10 – 25 V. h

have to be made. The first-dimensional gel resembles a large capillary gel and so appropriate care has to be taken in dealing with it. There are also differences in the running conditions for the electrofocusing. In addition, adjustments have also to be made to the second-dimensional electrophoresis to avoid overheating. Detailed descriptions of this method have been published elsewhere (32, 37). Using such a system, resolution is improved to the point where it is possible to detect 15−50% of all possible translation products of a mammalian tissue.

At the other extreme, when using micro-scale two-dimensional gels one suffers from the loss of resolution that follows from scaling down. however, one does gain considerably in speed and economy. Such gels can be recommended for two cases:

- When large numbers of samples have to be screened and the samples are of a reasonably low complexity, for example, screening column fractions for a protein that cannot be otherwise assayed.

- When material is limiting. For example, when analysing the proteins of small numbers of cells; the advantage is that the individual spots are more concentrated compared to a standard gel. Analyses can be successfully conducted on a few tens of nanograms of proteins. A more complete description of procedures for micro-scale two-dimensional gels is given by Poehling and Neuhoff (18).

4.4.5 Variations in the second dimension

Depending upon the circumstances, any form of polyacrylamide slab gel can be used in the second dimension. Essentially a wide range of variations of gradient shape and range can prove useful depending on the nature of the sample. In some cases urea may be the preferred denaturant for the second dimension. In this case the equilibration buffer should be modified by substituting 8 M urea for the SDS. Similar changes should be made if any other system is to be used.

4.5 Factors affecting separation: problems and trouble-shooting

Two-dimensional gel electrophoresis is very demanding of samples, chemicals, and apparatus. Although the system can tolerate some deviation from optimal conditions, once the limits are reached problems can multiply. The following is a list of precautions to be taken and observations that may be diagnostic of certain problems. The section on trouble-shooting of one-dimensional gels in Chapter 1 and of iso-electric focusing gels in Chapter 2 are also highly pertinent.

4.5.1 Samples

Ideally the sample must be only protein. High concentrations of nucleic acids, salts, lipids and carbohydrates can all interfere with the system. The effects of such contaminants can affect the separation in a number of ways, as discussed below.

i. Particulate material

Particulate material forming in samples after the initial removal of cell debris may indicate a very serious problem with sample solubilization. If there is particulate material in the sample and streaking in both dimensions then you may be using too low a ratio of solubilizing agent to sample. One possible remedy is to try using larger ratios of extraction buffer to sample.

ii. Salts

The effects of small mobile ions on the isoelectric focusing dimension are twofold. The first is that they act as counter-ions to charged groups on the protein and can therefore alter its net charge and pI. The second is that they, rather than the proteins and carrier ampholytes, will carry the current and so slow down the migration of proteins to their isoelectric points. In extreme cases this can lead to physical damage to the gel (carbonization or gel breakage). If this is a systematic problem with sample preparation it may be apparent as a higher than expected initial current. If there is a problem with only one sample out of several it may not be noticed until the gel is ready for analysis. If sample preparation does involve exposure to high salt concentrations, care must be taken to desalt the sample by dialysis.

iii. Nucleic acids

Nucleic acids are large polyanions. Their isoelectric points lie well outside the range of isoelectric focusing gels but they can interact ionically with ampholytes and proteins. This affects the shape of the pH gradient and the behaviour of proteins. Many proteins will bind nucleic acids in a non-specific way. This means that a protein species, instead of being a homogeneous population, can become a heterogeneous one, individual molecules having widely varying masses and charges of nucleic acids bound to them. This reduces the sharpness of spots considerably and can be seen as streaking, especially in the isoelectric focusing dimension. An associated problem is that the mixture of solubilized protein, nucleic acids and carrier ampholytes can aggregate and form a precipitate at the alkaline end of the gel during sample loading. This is the cause of the smear seen at the alkaline end of the gel after electrophoresis in the second dimension. Nucleic acids may also stain with silver stains used for protein detection leading to a high background.

iv. Lipids and carbohydrates

These are infrequent contaminants but if present can cause problems. Carbohydrates can have similar effects, although less severe, to nucleic acids. They form aggregates that can bind proteins that then do not enter the gel. High levels of lipids can have severe effects also. Mixed micelles of lipids, free fatty acids, and the non-ionic detergent used can reduce the stabilization of denatured proteins by the detergent. In severe cases this can lead to removal of free detergent from the gel in a manner similar to the problems attributed to the use of SDS. If you have free lipid floating on your sample after centrifugation, you may have saturated any detergent present in the medium. Raise either the ratio of solubilization buffer to sample or the concentration of detergent in the buffer to avoid this. Some lipid materials, particularly bacterial lipopolysaccharides, can also cause severe problems by forming very large mixed micelles that do not focus.

4.5.2 Chemicals

Even trace quantities of contaminants can severely degrade the performance of the system. Problems resulting from the use of low quality materials can include poor polymerization, gels sticking to glassware, and poorly reproducible gel patterns. Similarly great care must be taken in preparing stock solutions and gel mixtures accurately. Always use the best reagents you can find. Reagents advertised as suitable for electrophoresis are supplied by several companies and such grades are recommended for acrylamide, bisacrylamide, ammonium persulphate, SDS, and TEMED. The quality of acrylamide used is particularly important for the first dimension. Although slightly cheaper grades may be acceptable for the second dimension, it is a false economy to use a cheaper grade for isoelectric focusing. The authors recommend routinely deionizing all acrylamide solutions by stirring them with $1-2$ g of mixed-bed ion-exchange resin [e.g. Amberlite MB3, Bio-Rad AG 501X8(D)] for 1 h then filtering through Whatman No. 1 filter paper. If silver staining is to be used, filter the solution through a 0.2 μm nitrocellulose filter as well. If the acrylamide solution is at all coloured after this treatment it should not be used. Do not store the acrylamide solution with ion-exchanger beads in it as the acrylamide will slowly polymerize on the beads. Good quality grades of Tris and glycine from major suppliers are usually adequate; Trizma from Sigma is frequently recommended as the best quality Tris for electrophoresis. Ammonium persulphate solutions are best prepared fresh on the day they are to be used, although other workers have recommended storing the solution in aliquots in tightly sealed glass bottles at $-20°C$.

The quality of both detergents and carrier ampholytes are very important for two-dimensional gels. The highest quality detergents should be used at all times. Care should be taken in the choice of ampholytes to ensure that they generate a pH gradient of the appropriate range and shape. The most important point to consider with both detergents and carrier ampholytes is the batch-to-batch variation. Different batches of non-ionic detergents can have different levels of ionic contaminants and the result of using such a batch is essentially the same as having too much salt in the sample; that is, poor focusing or streaking in the first dimension. If this happens when a new batch of detergent is used, then it will be necessary to deionize the detergent by stirring it with a mixed-bed ion-exchange resin (see above). There are also large differences in the various grades of SDS available. It is therefore important to use only those grades recommended for gel electrophoresis and again it is worth using only a single supplier for this detergent.

Batch-to-batch reproducibility is extremely important when considering the choice of carrier ampholyte. However, it cannot override the importance of obtaining a gradient of the right range and shape. Therefore it is important to compare the resolution offered by a new batch of ampholyte against that previously used. A poor batch of ampholyte will distort the pH gradient with the proteins either being concentrated in a fairly narrow band or a narrow region of the normal pattern will be stretched out over the whole gel. In extreme cases a conductivity gap can form (due to a lack of carrier ampholyte in a given pH range). The very high local voltage potentials generated across a small distance can physically damage the gel (burning or breaking). If there is a serious loss of resolution then it is better to reject the batch. On the other hand, if a batch gives equally good or better resolution than that in use it is a good idea to buy as much of the production batch as you can reasonably expect to use and store it carefully under nitrogen or argon in the cold.

High-quality water is essential for successful two-dimensional gels. Water should be distilled (doubly distilled is preferred) or passed through a high quality purification system such as Millipore Milli-Q system. This is particularly important when gels are to be silver-stained since trace quantities of Cl^- in the water can efficiently precipitate Ag^+ with an extreme loss of staining sensitivity. Generally speaking, a conductivity of less than 10 μS is recommended (1 μS is required for silver staining).

Oxygen quenches the acrylamide polymerization process but degassing of gel solutions will prevent this. If polymerization is poor or slow one of the first things to do is to degas your solutions more extensively. Always degas the gel solutions before adding ammonium persulphate. Otherwise the polymerization reaction may go to completion whilst degassing is in progress or proceed so rapidly that there can be heating of the gel and swirl formation during polymerization. This is a particular problem when using uniform concentration slab gels. After solutions have been degassed, try to avoid re-introducing oxygen (by vigorous shaking or pouring from great heights). Degassing urea-containing gel solutions has to be done carefully or the urea may precipitate. The authors have found that it is not really necessary to do so: there is very little free water in the gel to dissolve oxygen in.

4.5.3 Apparatus

i. Glass and plasticware

Failure to keep glass and plasticware clean can result in premature or irregular polymerization because of the build-up of acrylamide residues that can act as nuclei for rapid polymerization. Furthermore, some microorganisms can grow on traces of acrylamide and the release of cellular protein when they are exposed to SDS can cause background and artefactual spots. Glass plates and tubes for isoelectric focusing must be kept as clean as possible. The accumulation of acrylamide residues on any of these materials can act as a catalyst for premature polymerization and gels may also stick irreversibly to dirty plates. Conversely, gels do not grip greasy gel tubes or plates well and so the gels, particularly low percentage isoelectric focusing gels, can slip out during electrophoresis. Cleaning is best done by soaking gel plates overnight in chromic (chromosulphuric) acid, then rinsing extensively with tap water and distilled water, and finally washing with ethanol prior to air-drying. If chromic acid is not available, a solution of a strong decontaminating detergent such as Contrad 70, Decon 90, or RBS 35 should be used. For very heavy soiling, such as heavy vacuum grease sometimes used to make a watertight seal, a 1:1 mixture of RBS 35 concentrate and methanol is extremely effective.

Gel plates should be stored either in a dust-free environment or under detergent or chromic acid and washed and dried immediately before use. If detergents are used, check the pH frequently. If it drops below pH 9 the detergent should be replaced and the plates soaked overnight in fresh detergent.

Plexiglass and acrylic components of gel apparatus, gradient makers and gradient maker lines should all be cleaned thoroughly on a routine basis. Washing the gradient maker and lines through with large volumes of distilled water will suffice. They should however be regularly cleaned with a strong detergent. A multiple gel apparatus should be detergent cleaned after very use.

ii. Clamping gels to gel stands

When using a Studier type of gel apparatus (Chapter 1, Section 5.1.2), there can be a surprising loss of resolution if the gel plates are clamped too tightly to the gel stand (25). Although no explanation has been offered, it is possible that the combination of twisting and bending of the plates can cause microscopic separation between the three layers of the gel (sealing, stacking, and separating) with irregular transfer and lateral diffusion at the interfaces. A more obvious explanation is the occasional separation of a sealing gel from the stacking gel. Large air bubbles can be formed and there is no transfer in the affected regions. Transfer of proteins to the second dimension is complete by the time the Bromophenol Blue has entered the separating gel.

iii. Power supplies

Power supplies capable of delivering the voltages and currents required for two-dimensional gel systems are readily available from many suppliers. A supply capable of giving constant current and constant voltage is required, fitted with a voltmeter and ammeter to measure the outputs. The power supply must also be used properly to prevent damage. Electrical systems have their own equivalent of inertia (Lenz's Law). The practical result of this is that if a power supply is switched off at the end of a run without first turning the output controls to zero and it is then turned on again at the same setting, then the rapid change in the magnetic field in the transformer windings causes a voltage spike. If this is repeated on a daily basis then the cumulative effect will damage even the most robust power supply. The effects of this can be seen as an unstable output or high currents that can cause heating of gels. This is of course incompatible with reproducible gels. The problem can be prevented by using power supplies that have an interlock or smoothing circuits that prevent this or by developing the good habit of turning the power supply down before turning the power pack off.

5. Separation of special classes of proteins

5.1 Introduction

Most proteins can be fractionated using the combination of electrofocusing and SDS gel electrophoresis. However, some types of proteins are routinely separated using other two-dimensional separation systems either because the samples are unsuited for the more usual method or because the other methods have become established for historical reasons. As an example of the latter case, ribosomal proteins are always separated using a standard system which was one of the first two-dimensional techniques devised and now, universally accepted, allows the comparison of the results of all workers using many different types of ribosomes.

While the methods described in this section have been applied to particular types of proteins this does not mean that such techniques will not be suitable for other types of protein. Thus, should the frequently used combinations of electrofocusing and SDS gel electrophoresis prove unsuitable, one of these other methods may give the degree of resolution required.

5.2 Ribosomal proteins

The method of Kaltschmidt and Wittmann (1) with modifications by Howard and Traut (14) is routinely used to separate eukaryotic and prokaryotic ribosomal proteins. It consists of a discontinuous gel in the first dimension followed by a continuous one in the second, both dimensions containing high concentrations of urea. The first-dimensional rod gel consists of the sample set in an agarose large pore gel in the middle of a smaller pore acrylamide resolving gel. Originally Kaltschmidt and Wittmann used a large pore acrylamide sample gel but loss of sample due to the immobilization of protein in the acrylamide sample gel gave rise to the use of the modification devised by Howard and Traut of using an agarose sample gel. Under these conditions proteins in the sample gel having overall positive or negative charges will stack and migrate in opposite directions.

The composition of gels and buffers for the two-dimensional separation of ribosomal proteins is given in *Table 10.* Their preparation and use is described in *Protocol 6.*

Table 10. Gels for the separation of ribosomal proteins (1, 14)

First-dimensional gel (pH 8.6)	Second-dimensional gel (pH 4.5)
54.0 g of urea	36 g of urea
6.0 g of acrylamide	18 g of acrylamide
0.2 g of bisacrylamide	0.5 g of bisacrylamide
1.2 g of EDTA $-$ Na$_2$	0.96 ml of 5 M KOH
4.8 g of boric acid	0.58 ml of TEMED
7.3 g of Tris base	5.3 ml of glacial acetic acid
0.45 ml of TEMED	water to make 96.7 ml
water to make 148.5 ml	

Protocol 6. Preparation and electrophoresis of first-dimensional gels

1. Prepare the gel mixture (*Table 10*) and initiate polymerization by the addition of 1.5 ml of fresh 7% ammonium persulphate. The solution is sufficient for 20 rod gels.

2. Seal glass tubes (18 cm long and 0.5 cm i.d.) at one end with either Parafilm or similar material, or with a rubber bung (see Chapter 1). Fill half of the tube with the resolving gel, overlayer it with water, and polymerize for 40 min.

3. After polymerization, remove the overlay.

4. Mix the sample containing 50−100 μg of protein in 0.1−0.15 ml, at 65°C with an equal volume of 1% agarose in running buffer (pH 8.6). The running buffer is prepared from:
   - 360 g of urea
   - 2.4 g of EDTA-Na$_2$
   - 9.6 g of boric acid
   - 14.53 g of Tris base
   - water to make 1 litre.

5. Layer the sample in agarose onto the resolving gel, overlayer it with water and leave to set.

6. After removal of the overlay, fill the tube to within 0.3 cm of the top with more resolving gel and again overlayer with water and leave to polymerize.

7. Carry out electrophoresis in the first dimension, using the running buffer (pH 8.6) with the cathode in the top reservoir, the buffer of which also contains 0.5% pyronine G as a tracking dye. Initially, carry out electrophoresis at 3 mA per gel tube for 30 min to allow the proteins to stack, then increase the current to 6 mA per gel and continue electrophoresis for 5−6 h until the tracking dye reaches the end of the gel.

8. After electrophoresis, carefully extrude the rod gels and equilibrate each in 150 ml of equilibration buffer (pH 5.2) containing 8 M urea, 0.074% glacial acetic acid, and 12 mM KOH, for a total of 60 min, with at least two changes of buffer.

9. Transfer the rod gel to the slab gel apparatus and seal it in place with the second-dimensional resolving gel (pH 4.5), the recipe for which is given in *Table 10*. Degas the solution prior to use and initiate polymerization by the addition of 33 μl of fresh 10% ammonium persulphate for 10 ml of gel solution.

10. Layer the tracking dye (0.1% pyronine G in 20% glycerol) over the top of the rod gel under the running buffer, which contains 14 g of glycine and 1.15 ml of glacial acetic acid per litre (pH 4.0).

11. Electrophorese the slab gel with the anode at the top for 30−60 min at 40 V to allow the proteins to stack, then increase the voltage to 80−150 V for

Protocol 6. *continued*

6 – 12 h. The actual voltage used makes little difference to the separation of the proteins unless it results in a very high current, hence the main criterion used for picking a particular voltage is usually convenience.

12. After running, remove the gel from the gel plates and stain it (see *Table 2*, procedure C or F). A typical separation is shown in *Figure 5*.

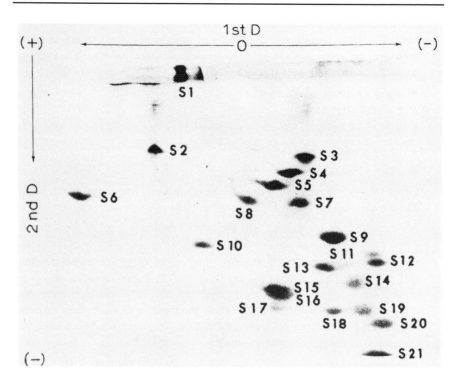

Figure 5. Two-dimensional separation of ribosomal proteins. *E.coli* MRE 600 30S ribosomal subunit proteins were extracted with 3 M LiCl, 4 M urea and separated by two-dimensional gel electrophoresis as described in the text. (Reproduced from ref. 14 with permission.)

An alternative to placing the sample in an agarose sample gel is to load equal amounts of protein onto two shorter resolving gels (each 9 cm long). Electrophoresis of one of the gels is then undertaken from the anode to the cathode, with 0.5% pyronine G as tracking dye, and the other from cathode to anode with 0.1% Bromophenol Blue as tracking dye, under the same conditions as previously described. After equilibrating the gels as before, they are placed in the slab gel apparatus with the two sample origins adjacent to each other in the centre of the slab gel and electrophoresis in the second dimension is carried out as previously described.

Because this separation utilizes a continuous buffer system in the second dimension it is essential to reduce the ionic content of the rod gel using the sequential equilibration in buffer prior to the second-dimensional separation. Otherwise, zone sharpening, which is seen when proteins migrate from a region of low to high conductivity, will not occur.

Certain adaptations of this, including variations of pH and the composition of the resolving gel in the first dimension and the inclusion of SDS in the second dimension, have been used successfully to separate ribosomal proteins. Details will not be given here but brief descriptions of, and references to, them can be found in *Table 11*.

5.3 Histones

The difficulties of using electrofocusing as a separation method for histones has led to the adaptation of alternative techniques for two-dimensional analysis. Separation at low pH in low acrylamide concentrations will separate histones on the basis of charge which is a reflection of their amino acid composition. Separations in the presence of SDS will separate histones on the basis of size; although compared with other proteins histones do migrate anomalously. Other separations depend on the fact that non-ionic detergents such as Triton X-100, Triton DF-16 or Lubrol-WX will bind to proteins in proportion to the degree of hydrophobicity of the proteins. This binding greatly reduces protein mobility and separates sub-species of the various histones differently to that obtained in acid-urea or SDS-PAGE.

Histone proteins have been separated very successfully on two-dimensional systems incorporating acid—urea gels in the first dimension with Triton—acid—urea gels in the second as described by Hoffmann and Chalkley (15). Because the basis of separation of the histones in these gel systems is different, good resolution is usually obtained. One problem encountered when using Triton gels is the possibility of artefactual electrophoresis patterns caused by oxidation, particularly of methionine residues, in the histones. Triton has less affinity for oxidized histones and this may lead to artefactual heterogeneity. Therefore it is advantageous to keep histones in a reducing

Table 11. Other two-dimensional gel methods for the separation of ribosomal proteins

Authors	Gel	References
L. J. Mets and L. Borograd	4% acrylamide rod gel in 8 M urea at pH 5.0 in the first dimension followed by separation in a 10% acrylamide slab gel containing SDS.	38
O. H. W. Martini	4% acrylamide rod gel in urea—acetate in the first dimension followed by separation in a 12% acrylamide slab gel containing SDS.	39
W. L. Hoffman and R. M. Dowben	8% acrylamide rod gel at pH 4.5 in urea—acetate containing Triton X-100 followed by separation in a 15% acrylamide slab gel containing SDS.	57

environment, such as 1 mM DTT, to prevent this. The buffers and gel systems for this separation are listed in *Table 12*. The gel solution has a final composition of 0.9 M acetic acid, 2.5 M urea, 15% acrylamide, and 0.1% bisacrylamide. Preparation and use of these gels is described in *Protocol 7*.

Table 12. Recipes for Triton – acid – urea gels

First-dimensional gel
3.0 ml of 40% acrylamide, 0.267% bisacrylamide
1.0 ml of 43.2% glacial acetic acid, 4% TEMED
4.0 ml of 5 M urea, 0.2% ammonium persulphate

Second-dimensional gel
15.0 ml of 40% acrylamide, 0.267% bisacrylamide
5.0 ml of 43.2% glacial acetic acid, 4% TEMED
20.0 ml of 2% Triton X-100, 5 M urea, 0.2% ammonium persulphate

---

Protocol 7. Two-dimensional separation of histones on Triton – acid – urea gels

1. Degas the first-dimensional gel mixture (*Table 12*) and pour it into cylindrical gel tubes 0.3×7 cm. Overlayer with 0.5 ml of cold 0.9 M acetic acid.

2. After polymerization, place the rod gels in the electrophoresis apparatus and pre-electrophorese for 4 h at 130 V, using 0.9 M acetic acid as electrophoresis buffer.

3. Then layer the samples onto the top of the rod gels and electrophorese from the anode to the cathode at 1.56 mA per gel for approx. 2 h at room temperature. A tracking dye of methyl green can be used, the blue component of which moves just ahead of histone H4 in this system.

4. Prepare the second-dimensional gel mixture (*Table 12*) and degas it. Then pour the mixture to form slab gels of dimensions 0.3 cm × 14 cm × 10 cm.

5. Overlayer with 0.9 M acetic acid, 1% Triton X-100, and allow to polymerize.

6. After polymerization, pre-electrophorese the slab gel for 12 h at 20 mA constant current, using 0.9 M acetic acid, 1% Triton X-100, as the electrophoresis buffer.

7. Following pre-electrophoresis, pour off the electrophoresis buffer and seal the first-dimensional gel to the slab gel, using 2 ml of Triton gel mix. Overlayer this in turn with 0.9 M acetic acid, 1% Triton X-100, and allow to polymerize for approx. 1 h.

8. Electrophorese the second-dimensional gel from positive to negative, using 0.9 M acetic acid, 1% Triton X-100, as electrode buffer, for 15 h at 20 mA constant current.

9. After electrophoresis, carefully remove the gel from the gel plates and stain in 0.1% Amido Black in 0.7% acetic acid, 30% ethanol, for 3 h at 80°C or overnight at room temperature, and then destain in 7% acetic acid, 20% ethanol.

Two-dimensional separations of histones and 'histone-like' proteins from lower eukaryotes are routinely carried out in the authors' laboratory by the combination of an acid−urea gel in the first dimension followed by an SDS separation at high pH in the second. The acid−urea system is a modification of Reisfeld *et al.* (41) and the SDS slab gel is based on that of Thomas and Kornberg (42). Recipes for the gels required are given in *Table 13*. All solutions with the exception of the 9 M urea can be stored for several weeks frozen in the dark. A full description of the use of this acid−urea system for histone fractionation is given in *Protocol 8*.

Table 13. Acid−urea gels for the two-dimensional separation of histones

First-dimensional gel

Resolving gel:
4.0 ml of 60% acrylamide, 0.4% bisacrylamide
2.0 ml of 0.48 M KOH, 17.2% glacial acetic acid, 4% TEMED
9.0 ml of 9 M urea
50 μl of 15% ammonium persulphate

Stacking gel:
2.0 ml of 10% acrylamide, 2.5% bisacrylamide
1.0 ml of 0.048 M KOH, 2.87% glacial acetic acid, 0.46% TEMED
4.0 ml of 9 M urea
1.0 ml of 0.4 mg/ml riboflavin

Second-dimensional gel

Resolving gel (50 ml):
30 ml of 30% acrylamide, 0.15% bisacrylamide
0.5 ml of 10% SDS
12.5 ml of 3 M Tris−HCl (pH 8.8)
12.5 μl of TEMED
water to 50 ml final volume

Stacking gel (20 ml):
3.33 ml of 30% acrylamide, 0.8% bisacrylamide
0.2 ml of 10% SDS
2.5 ml of 1 M Tris−HCl (pH 6.8)
10 μl of TEMED
water to 20 ml final volume

Protocol 8. Two-dimensional separation of histones on acid−urea gels

1. Prepare the first-dimensional resolving gel mixture (*Table 13*) and adjust the pH to pH 4.7 by addition of acetic acid prior to the addition of the ammonium persulphate.

2. After degassing the gel mix, initiate polymerization by addition of fresh 15% ammonium persulphate, and then pour the gel mixture into cylindrical gel tubes 14 cm×0.5 cm to the 10-cm mark.

3. Overlayer the gels with water and leave to polymerize for about 1 h.

4. After polymerization, remove the overlay and replace it with approx. 150 μl of stacking gel, made up according to the recipe given in *Table 13*. Again

Protocol 8. *continued*

overlayer with water and allow to polymerize with the aid of a fluorescent light source for 30 min.

5. After polymerization, layer $100-200$ μl of histone sample containing $200-300$ μg of protein in 7 M urea onto the stacking gel and electrophorese from positive to negative at 150 V for 3 h, using 31.2 g of β-alanine and 8.0 ml of glacial acetic acid per litre as electrophoresis buffer. Pyronine G in urea can be used as a marker dye.

6. After electrophoresis, gently extrude the rod gels under pressure and equilibrate in each of the following buffers in turn for 40 min at 37°C:

   (a) 1% SDS, 1 mM DTT, 0.1 M phosphate, 10 mM Tris−HCl (pH 7.0);

   (b) 1% SDS, 1 mM DTT, 0.01 M phosphate, 10 mM Tris−HCl (pH 7.0);

   (c) 0.1% SDS, 1 mM DTT, 0.01 M phosphate buffer, 10 mM Tris−HCl (pH 7.0).

7. To prepare the second-dimensional resolving gel (*Table 13*), degas the gel mixture and initiate polymerization by adding 125 μl of 10% fresh ammonium persulphate. After pouring, overlayer the gel with 0.75 M Tris−HCl (pH 8.8), 0.1% SDS, and allow to polymerize for 1 h. After polymerization, pour the overlay off and wet the top of the resolving gel with some of the stacking gel solution (*Table 13*) without ammonium persulphate. Polymerize the stacking gel by adding 100 μl of 10% fresh ammonium persulphate and overlayer with water.

8. After polymerization of the second-dimensional stacking gel, remove the overlay and seal the equilibrated first-dimensional rod gel to the stacker using 1% agarose in 62.5 mM Tris−HCl (pH 6.8), 2.3% SDS, 1 mM DTT. Bromophenol Blue can be used as a marker dye.

9. Electrophorese the gel at 150 V constant voltage overnight, or until the marker dye runs off the gel, using 0.05 M Tris base, 0.38 M glycine, and 0.1% SDS, as the electrode buffer.

10. After electrophoresis, carefully remove the gel from the gel plates and stain in 0.1% Coomassie Blue in methanol:water:glacial acetic acid (5:5:1 by vol.) for 3 h. Destain by diffusion overnight in the same solvent, without the Coomassie Blue stain. A typical separation of histones is shown in *Figure 6*.

5.4 Chromatin non-histone proteins

Problems caused by protein aggregation prevent the use of ribosomal separation procedures for some proteins, especially for nuclear proteins. Orrick *et al.* (12) have developed a high resolution two-dimensional separating system for extracts of both whole nuclei and nucleoli which can be used as an alternative separation method

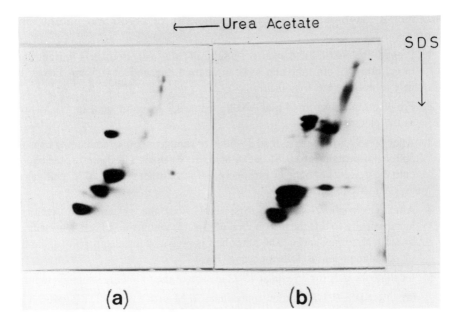

Figure 6. Two-dimensional separation of histones. Purified histones of (a) *Dictyostelium discoideum* and (b) calf thymus were separated by two-dimensional gel electrophoresis as described in the text. (Figure kindly provided by J. Sinclair.)

to that devised by O'Farrell (11) which is usually used. The method consists of a separation using a continuous buffer system at acid pH in the first dimension followed by a separation on the basis of molecular masses in the presence of urea and SDS in the second. The gels are prepared according to the recipes given in *Table 14*. A description of the use of this gel system is given in *Protocol 9*.

The first-dimensional gel separates the proteins on the basis of charge which is related to the amino acid composition of the polypeptides. Even so, as larger polypeptides move more slowly than smaller polypeptides of the same amino acid composition, the polypeptides are also separated on the basis of molecular mass.

Table 14. Gels for the separation of chromatin non-histone proteins

First-dimensional gel
5.0 ml of 40% acrylamide, 1.4% bisacrylamide, in 4 M urea
5.0 ml of 4% TEMED in 2 M urea
5.0 ml of 0.21% ammonium persulphate, 21% glacial acetic acid, in 6 M urea

Second-dimensional gel
18 ml of 20% acrylamide, 0.52% bisacrylamide in 8 M urea
7.5 ml of 0.2% TEMED, 0.4% SDS in 0.4 M sodium phosphate (pH 7.1)
4.5 ml of 0.5% ammonium persulphate in 8 M urea

Protocol 9. Two-dimensional electrophoresis of chromatin non-histone proteins

1. Prepare the first-dimensional gel mixture (*Table 14*). Pour gels 9.5 cm long in gel tubes 12 cm long and with an internal diameter of 0.5 cm. Leave the gels to polymerize for 60 min.

2. Pre-electrophorese for 2 h at 120 V, using 0.9 M acetic acid in 4.5 M urea as the electrode buffer.

3. After pre-electrophoresis, load 2−50 μl of sample mixture, containing approx. 500 μg of protein in 0.9 M acetic acid, 10 M urea, 1% 2-mercaptoethanol, onto the gel and continue electrophoresis for a further 5 h at 120 V, equivalent to 1.5 mA per gel.

4. After electrophoresis in the first dimension, the gel may be sectioned longitudinally so that the distribution of proteins within the gel can be visualized by staining immediately. The other half is prepared for electrophoresis in the second dimension as follows.

5. Incubate the gels for 35 min at 45°C in each of the following solutions in turn:
   (a) 2% SDS, 0.1 M sodium phosphate, 6 M urea, 1 mM DTT (pH 7.1);
   (b) 1% SDS, 10 mM sodium phosphate, 6 M urea 1 mM DTT (pH 7.1);
   (c) 0.1% SDS, 10 mM sodium phosphate, 6 M urea, 1 mM DTT (pH 7.1).

6. Prepare the second-dimensional gel mixture (*Table 14*) and pour 25 ml of the gel mixture to form gel slabs 10 cm×9.5 cm×0.3 cm. Overlayer with water.

7. Gently soak the portion of the rod gel equilibrated with SDS buffer in the sealing gel mixture which is identical to the second-dimensional slab gel (*Table 14*) except that it contains a final concentration of only 10 mM phosphate (pH 7.1).

8. Seal the first-dimensional gel to the second-dimensional gel using about 5 ml of sealing gel.

9. Electrophorese the gels for 16 h at 50 mA per gel slab with 0.1% SDS, 0.1 M phosphate (pH 7.1) as the electrophoresis buffer.

10. After electrophoresis, carefully remove the gel from the gel plates and stain it.

Consequently, the final pattern on the slab gel tends to be essentially diagonal in nature (*Figure 7*) and hence the degree of resolution may be poorer than that obtained using electrofocusing and SDS-PAGE. However, polypeptides whose amino acid composition is markedly different from the other polypeptides in the sample will be located away from the diagonal.

6. Quantifying two-dimensional gel patterns

6.1 Introduction

Experienced researchers are now capable of producing reproducible gels from which data can be collected on several hundred and perhaps over a thousand proteins from

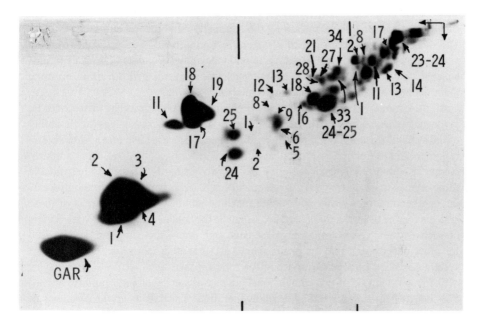

Figure 7. Two-dimensional separation of nucleolar proteins. Acid-soluble nucleolar proteins were extracted from purified rat-liver nucleoli and separated by two-dimensional gel electrophoresis as described in the text. (Reproduced from ref. 12 with permission.)

a single gel of one sample. To handle and analyse these large quantities of data, several groups have found it necessary to automate the analysis of gels and to develop computer databases and processing systems to deal with the data generated (43–49). The methods that are being developed all have their own characteristics and it is beyond the scope of this chapter to describe them individually. Rather we will outline the mechanisms and procedures involved.

The functions required of an automated gel quantifying system, its database and processing system are:

- acquisition of data from a gel
- storage of data for later processing
- processing of data

The last of these includes the removal of extraneous noise (staining background, scratches on film etc.), quantification of data from individual spots, and comparison of two or more gels.

Systems can have additional functions. For example the QUEST system of Garrels (47) includes a number of features that are relevant to management of the laboratory (keeping track of consumable items, budget, exposure schedules for gels and so on) but not directly relevant to the acquisition and manipulation of data from two-dimensional gels. Most other systems, including commercial ones such as those from Bio-Image, appear to concentrate on the acquisition and manipulation of gel data. The most important part of such a system is its ability to store the data from the scanning system and allow its manipulation in the analysis of data. The system must be able to compare and match up patterns from several gels to allow accurate identification of spots for quantitative analysis.

All the commonest systems (43−49) perform these functions. The differences lie in the mathematical methods used (i.e. the algorithms). Criteria such as the flexibility of the system, the computing power required and the ease of use (user friendliness) reflect these differences. A great deal of data collection can be done using a microcomputer with 500 kbytes of memory and 20 Mbytes hard-disc drives are available for such machines to store very large quantities of data. Unfortunately the processing can become very slow with such a machine and because of this most of these systems require at least a minicomputer to handle the data in a reasonable time. Setting up such a system can become very expensive and, unless the computer is to be used for other purposes, it may be preferable to go to another laboratory where such a system has been established and pay for time on that.

Computer analysis of two-dimensional gel patterns is developing rapidly; at least one such system is described as second generation. This brings up the problems of inter-operability and of transferring data from an older system to a newer one that may suit your needs better. This problem is acknowledged by those developing the systems and may be resolved in the not-too-distant future. At the moment the choice seems to stand between using an older, established system such as TYCHO (48), QUEST (47), or GELLAB (49) or becoming involved in a newer system which may be more flexible but still needs development.

The analysis of two-dimensional gels can be either at the level of qualitative analysis, that is finding out if different samples contain the same polypeptides, or quantitative analysis in which case it is necessary to know not only if the same polypeptides are present but also whether they are present in similar amounts.

Comparison of gel patterns is a common requirement of both one- and two-dimensional gel electrophoresis but it is obvious that the analytical methods used for one-dimensional gel electrophoresis cannot be transferred directly to two-dimensional gels. After a two-dimensional separation, polypeptides appear as spots distributed over most of the area of the slab of polyacrylamide and since simple linear scanners are designed to record in one direction only they cannot be used for scanning two-dimensional gels. This means that two-dimensional scans require more elaborate scanners. Before considering these scanners and subsequent treatment of the data it is worth listing the minimal amount of data needed from two-dimensional gel experiments, namely that it must be possible to compare the coordinates of spots from two or more records and it is necessary to estimate the relative amounts of all or selected polypeptides.

In general, it is not worth performing experiments where the result is a single gel pattern followed by further individual experiments because it is difficult to achieve absolute reproducibility of the coordinates when individual gels are cast and run separately. Thus, it is always desirable to prepare and run several gels at a time using the kind of apparatus referred to in Section 2. If one has run 10 or more gels then manual recording and manipulation of data is so tedious that the results are far from reliable. The answer to this problem is to employ automatic procedures. This means that computers must be involved. This section describes computer systems which are either dedicated to the analysis of two-dimensional gels or are general purpose computers having the necessary hardware to make scans of two-dimensional gels. At first glance this problem may seem rather trivial. Most computer magazines carry advertisements for digitizers based on linking a TV camera to a personal computer. However, the problem is that not only does each image require a great deal of memory but also capturing the image is only the first and relatively straight-forward step in the computer analysis of two-dimensional gels. It is the software needed for the subsequent manipulation and comparison of images that is very complex and here we can give the reader only an insight into the methods used and their complexity.

6.2 Primary data from gels

The stained gels are the primary data (or X-ray films if the peptides were labelled) and it is these that must be recorded. However, it is of paramount importance that the method used for visualizing polypeptide spots on the gel is thoroughly characterized in terms of its specificity, sensitivity and linearity with protein concentration. Some staining methods do not give a linear response with the mass of protein and some polypeptides may appear as multiple spots as a result of modification either *in vivo* or as a result of sample preparation. Everything possible should be done to avoid the latter possibility. Photography is the most permanent way to store the gel patterns but, although the positions are faithfully recorded, the relative intensities of the spots are only reliable if great care is taken in the choice of film and its development. Under optimal conditions, a 35 mm film can distinguish 100 μm spots. This is much smaller than the average two-dimensional gel spot. Replacing the film by a digitizing system capable of recording spots from the gel of 1 mm diameter means that the spatial resolution in the presence of noise must be about 0.3 mm. This means recording at least 512×512 intensities (these intensities are called pixels, picture elements) giving a quarter of a megabyte of information (assuming the densities are measured to 1 part in 256). This is the minimum amount of data that is necessary for a complete image although smaller amounts can be collected if only limited regions of the gel are studied (50). This comparison between film and computer illustrates the problem of handling the data. A film is stable but still needs to be measured while the design limits set on the measurements in the computer-based system determine the final resolution; further measurements cannot be made if the gel has been discarded.

6.3 Types of gel scanner

Scanner instruments can be roughly divided into two types:

- The scanning mechanism is built into the photometric measuring device [Vidicon and similar tubes or a two-dimensional array of charged couple devices (CCD) array]
- The gel is moved mechanically either in one direction over a linear array of detectors (as in the case of a linear CCD) or in both directions over a photometer (photomultiplier tube).

A variation on moving the gel is to move the source of light using a cathode ray tube (50). It is possible to get square arrays of CCDs, which means no mechanical scanning is required, but at the present time the resolution is limited and arrays that are guaranteed free from blemishes and record 512 × 512 pixels are expensive. The simplest and most common scanners use variables of Vidicon tubes (TV cameras) so that the image is scanned in its entirety 50 times a second. The image is digitized using an analogue-to-digital converter and the digitized version of the image is stored. The camera measures intensities to an upper precision of about 64 Grey levels (essentially a range of discrete optical densities for a monochrome camera), so although the memory stores 8 bits, only 6 bits are faithful records of the intensities. Intensities must be converted to absorbances by dividing a reference intensity by that of the sample and taking the logarithm of the product (Beer−Lambert law).

These processes can be followed more easily if a schematic system using a TV camera (as shown in *Figure 8*) is described. Here the intensities from the camera (TVC) are first digitized (A/D) and then fed to a linear set of registers called a look-up table (ILU). The connection from A/D to ILU is shown as a switch to illustrate the working of the table; the digitized value of the intensity (say 0−255) is found in the look-up table and the equivalent value in the table is fed to the frame-store (FS). These are special stores provided with the scanner and are the only ones that can receive the TV camera data or display it on the monitor. If the look-up tables are part of the hardware it is possible to convert from digital intensities to a previously loaded set of values, that is the logarithms of 1−256 so that the values are calculated while the data are being collected. This example assumes there are 256 registers in the table (each holding the equivalent value for the 8 bits of the digitized intensity) but this is not always the case with simpler display systems supplied with micro-computers designed mainly for business use.

When using a computer system, a blank gel is first recorded to provide the values for the reference pixels. This is then followed by the test gel. Since the logarithmic values are loaded into the frame-store, the values for the test gel can be subtracted pixel by pixel from the reference values to give a record of absorbances. This reference frame need not be a frame-store; if it is possible to add enough memory to the computer then the contents of the frame-store can be copied to the computer memory and the subtraction done using this stored information. These operations with frame-stores are fast since they use pre-wired hardware with dedicated processors. Their operation can be followed by the researcher if the manufacturer of the scanner supplies software which can be linked with the user's high level program. Scanning each gel several times to reduce the background electronic noise

262

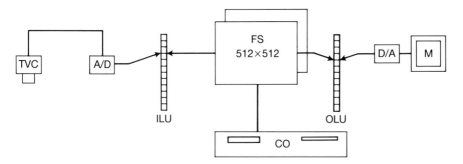

Figure 8. A schematic of a typical scanning system containing a TV camera (TVC), fast analogue-to-digital converter (A/D), input and output look-up tables (ILU and OLU), frame-stores (FS), digital-to-analogue converter (D/A), monitor (M) and computer (CO). The connections to the look-up tables are shown as switches but this is not necessarily how they work.

from the camera (often this equals about 10% of the intensity) is recommended. In general, this is random noise and can be reduced by making repeated scans, summing the optical densities for each scan of the pixel and dividing the sum by the number of scans made. Again this process can be carried out using the software and hardware supplied by the manufacturer of the scanning hardware. Note that the averaging procedure will only remove random noise but not periodic fluctuations coming from the illumination. Spatial variations in light intensity are recorded from the blank gel and should not fluctuate in the interval between recording the blank and test gels.

The contents of the frame-stores can be displayed on a monitor through a second look-up table (OLU, *Figure 8*), which again can be loaded with suitable values to provide shaded contours on the display. If colour displays are to be used, the OLU must have three sets of registers, one for each primary colour. Generating suitable colours by combining these colours is just like painting but the results can be far from satisfying if some rules of combination are not followed. These rules are used in colour TV transmissions and have been summarized in a chromaticity diagram produced by the Commission Internationale de l'Eclairage. Many colour display systems of computers do not use this form of look-up table because the number of colour combinations is limited to 8 or 16. Nevertheless, it is always possible to adjust the combinations to suit the user.

The limits on levels of Grey scales from TV cameras are such that the relationship between absorbance and concentration is only linear to an absorbance of about 1.5.

The computer files used for storing the image must be designed so that in addition to saving the transformed image itself, one must be able to include with each image the title of the experiment, value of maximum absorbance (found after removing the minimum value from the whole image) and space left so that the subsequent history of the image can be added later. This space could be used to store notes about smoothing, boxing of areas and finding peaks (see subsequent sections). If several gels are to be recorded then after each recording, the data from each image must be stored on disc. This can only be done realistically if hard discs are used. It is also worth remembering that every four images fill about a megabyte so the disc must either be very large (300 Mbytes or more) or the data must be transferred regularly to high density floppies or magnetic tape.

6.4 Numerical analysis of gel patterns

The data now stored in the computer can be manipulated using special procedures. In order to visualize the processing of the data it is essential to have a high resolution colour display for viewing the images. These may be displayed simply as a contour map using pseudocolours to define the contours (outlined in Section 6.3). Thus, one could define blue for lowest absorbance going through green, yellow, red and finally black for maximum absorbance. More complicated displays can be made to project pseudo-three-dimensional isometric views, but although these are visually interesting they are less useful for processing the image. Numerical methods must be provided to carry out four sequential operations:

(a) Remove background noise without distorting the image appreciably.

(b) Locate significant spots in the gel.

(c) Reduce the amount of data in each spot to a mathematical model.

(d) Compare the models from different gels for spatial and quantitative differences.

6.4.1 Removal of background noise

With gels stained with Coomassie Blue, the background is not appreciable if care is taken with destaining but with silver-stained gels there is always appreciable background. There are three ways to remove this background.

- It is possible to remove the residual background in the computer using a special digital filter (often called a Sobell filter) which is a 3×3 window in which the eight edge elements have a lower weighting coefficient than the centre element. This window is passed over the whole image in order to average the values within the window.

- A similar procedure is to smooth the image out mathematically. The intensities along the row and column that the pixel is on are fitted to a curve that describes the distribution of the values in space. The type of curve used is an orthogonal Legendre polynomial (cubic with 5 points). The method generates a curve that best fits the intensities recorded on the two axes. If the gel has a high background it is possible to reduce the level by setting a minimal threshold value for the whole image and remove all below this threshold. The simplest way is to clear some of the least significant bits in the display memory (if the display is from the frame-stores). This procedure reduces the quantitative relationships between the spots so it is only used for mapping the contours and not for estimating peak parameters.

- A more versatile method is to transform the whole image to notional frequencies using a Fourier transform. In essence one describes individual rows or columns of pixels as a group of sine waves of varying amplitudes and frequencies at a specific time (time domain). The image can then be filtered to remove the low frequencies, which correspond to the low-level background effects, and then one can reconstruct the original image (the original time domain) with these filtered data. Although this is the most versatile method, it is slow in execution (even with fast Fourier transforms it takes about 30 minutes to process an image having 512×512 pixels) and high pass filters (removing low frequency noise) can be temperamental and will garble the image by showing the results of differentiating the functions of the sine waves if the constants for the filter are wrong.

6.4.2 Finding peaks

Spots extend in two dimensions. It is therefore necessary to explore the image using a two-dimensional window (say 3×3) which the computer slides over the image and compares the rectangular coordinates of each maximum found within the window with those at the centre. To illustrate this procedure a grid is shown in *Figure 9* in which the centre square corresponds with the coordinates of the absorbance being examined. Thus, the routine finds the indices of the maximum absorbances within the 3×3 grid (allowing for noise by subtracting an arbitrary amount, found by experience, from each absorbance). If the x, y indices of the maximum absorbance are the same as those for the centre of the window then a peak has been found and its coordinates noted. The selection becomes better as the window size is increased until a point is reached where fewer peaks will be found. Having found a peak some further criterion or criteria must be defined to see if it is 'real' and the simplest test is to set a threshold level of absorbance. If the maximum absorbance of the peak (centre of the window) is less than this value, it is not a real peak.

Having located the peak, the absorbance within the peak can be abstracted by expanding out along the circumradii of an octagon (*Figure 9b*) having sides which are not necessarily equal, until the absorbance either starts to increase (a point of inflection) or becomes less or equal to the pre-defined background value. This is the point at the bottom of the peak or a cusp between incompletely separated peaks. The absorbance within these eight triangles can be transferred to routines which fit mathematical models to represent the peaks (see ref. 51). Each triangle within the octagon has two sides radiating out from the peak centre. The routine to extract the data must allow for the three cases where the two sides may or may not be equal (*Figure 9c*). The triangulated data can now be used to calculate the sum of absorbances (pixels) in the peak (which approximates to the mass of peptide in the spot providing the conversion constant for changing from intensity to concentration is shown) and for fitting models of the spots. Before making these calculations, remember to check that the data does not contain repeats caused by running down the same boundary twice (that is, the one separating two triangles).

Automatic procedures will often fail to locate a spot because the characteristics of the data do not match the general rules used to define a peak. For this reason

Two-dimensional gel electrophoresis

it is essential to have an interactive procedure for overlaying the image with spots that have been found and then by using a cursor it is possible to point out the missed spots. Similarly it is useful to be able to examine the profiles of the image in confused areas. These profiles should cut across the image in both the x- and y-coordinates so producing two profiles which can be displayed in unique colours at the edges of the whole image. Assuming models have been fitted to the whole image but there still remain peaks which are thought to be significant, then it is possible to force the fitting of the data to the model by selecting centres of spots manually. It is not usually worth extending the automatic search algorithm providing it finds at least 90% of the spots, because the atypical spots in every image may be of greater interest than those following the pre-set rules and should be specially noted.

Displays of silver-stained gels will invariably have a disfiguring background with limited contrast but it is possible to improve the contrast in two ways. The simplest is to remove from the images all pixels below a pre-defined value. This value may vary between images but can be found by examining the images interactively using the cursor and printing values under the cursor. The removal can be done on the whole image either using a separate program or through the look-up tables (this is a more advanced use of the tables and the manufacturer's literature will show how this is done). The second procedure does not change the data in the image but evens out the probabilities for all the absorbances in the displays. This procedure is called histogramming and is described in detail in the literature (52). The principle is to

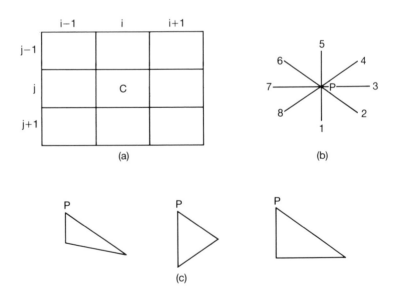

Figure 9. (a) Grid showing the arrangement of the 3 × 3 window for finding spots, C is the pixel under consideration; (b) octagon with circumradii from the centre peak (p) illustrating the directions the search will be made to find the limits of the spot; (c) the three situations that can occur for the sides of each triangle at the end of the octagonal search.

make a histogram of frequencies of optical densities of all the pixels (if possible using software provided by the supplier of the frame-store hardware), then calculate the cumulative distribution function. This function increases monotonically from 0 to 1 and is used as indices for the old ILO and so inserts a new set of values to put in a second ILO (*Figure 8*) which is then used as the control of the displayed image. The original values in the image are not lost but the contrast is markedly increased, making it possible to find spots which were previously masked by general high background. These can then be fitted to the models by manual selection with the cursor.

A further useful feature which can be employed through interaction with the displayed image is to select areas of significance rather than search over the whole display. Often all that is necessary is to produce a rectangle about the area using the cursor and retain the coordinates of the top and bottom diagonal corners ready for peak detection.

All these interactive manipulations can be included in one program. This can be quite large since it must contain the statistical routines for fitting the models to selected peaks, but the selection of the operations can be controlled through a menu displayed on the computer monitor.

6.4.3 Mathematical models

Each image will contain upwards of 250 kbytes of data of which a lot represents space between the spots. It is possible to reduce the stored data considerably if the spots are fitted to models which describe both the rectangular coordinates of the maximum absorbance and sufficient parameters to describe the three-dimensional shape of the peak. A physically correct model which incorporates all the features of a moving electrophoretic boundary is complicated and fortunately not required. In general two models have been used for two-dimensional electrophoresis.

- A double Gaussian (51,53). This assumes that the spot is a normal distribution along the x and y axes. The spot can then be described as a function of its coordinates and the statistical parameters of the normal distributions as measured in the scan.

- An ellipsoid. This is the three-dimensional shape that results when an ellipse is rotated about an axis. Again the spot can be described as a function of its position and the properties of the distribution as measured in the scan.

It is harder to fit the data to a double-Gaussian form than to an ellipsoidal one because the double-Gaussian requires non-linear fitting procedures while fitting an ellipsoid is a linear regression equation and routines to do this are often in the computer library. If these routines are not available, the routines coded in Fortran can be found in the literature (54 – 56). Fitting non-linear mathematical models to experimental data is an iterative procedure and one must supply assumed values of the parameters before starting. If these values are not reasonable, the required convergence may not occur; that is, the problem becomes insoluble. In the analyses discussed here, good starts can be made by always fitting the ellipsoid and using these parameters to calculate Gaussian parameters. The parameters can be readily converted using simple formulae.

Experience in fitting both models to the same data show that they give equivalent estimates the reduced masses in the spot (calculated as volumes from the models since both yield analytical equations for the volumes). It is also worth remembering that although it would be expected that a diffusing spot would be better represented by a Gaussian model, this will only occur for the isoelectric focusing dimension; non-linear concentration dependence of the transporting boundaries will distort the symmetry of the spot to produce a sharpened leading edge. *Figure 10* shows profiles of a spot in a gel which illustrate the distortions as well as profiles calculated using the two models. The profile is relatively symmetrical in the direction of isoelectric focusing but the leading edge is sharpened for the transporting direction.

After fitting the data to the equations, the parameters can be used to decide further if the selected set of data is from a 'real' spot or not. For example if any of the parameters describing the elipsoid do not have the expected sign or the fitting fails for numerical reasons then the data must be rejected. Similarly, if the ratio between the ellipsoidal volume and its standard error (calculated by combining the errors and covariances from the regression) is less than unity the result is doubtful. A further useful test is to compare the volumes calculated from the ellipsoidal parameters with those from the summed intensities within the octagon. If the ratio is <0.05 or >2 it is doubtful that the data are from a unique spot. Employing models to describe the patterns may seem an unnecessary step but if the parameters are scaled correctly they can be stored as integer values occupying two or three bytes for each parameter of which only five are required per spot (x- and y- coordinates of the maximum absorbance plus the three parameters from the equations). Standard errors on the masses can also be calculated and used in future comparisons of gels although care must be taken in interpreting errors arising from the non-linear fitting. The parameter values for one or both models can be stored in separate files and these can be compared later with the original image using the interactive program.

6.4.4 Comparison of two-dimensional gel patterns

In order to compare patterns from several experiments it is necessary to overlay the patterns notionally and note the differences. Thus, the system must allow for unreproducible distortions in the patterns caused by variability in the gels as well as be capable of holding in memory two or more images. In order to adjust the coordinates so that two patterns are superimposed, internal references (also known as landmarks) must be selected. These can be standard proteins added to the sample before electrophoresis or spots known to be common to all samples and present in the tissue to be studied. Then, by comparing the coordinates in the two or more images, the coordinates of all the spots in all but one reference pattern can be adjusted to match the reference pattern. The distortions are not necessarily uniform over the whole pattern so reference spots must be chosen from several positions in the pattern and allowances made in the calculations for the variations over the whole pattern.

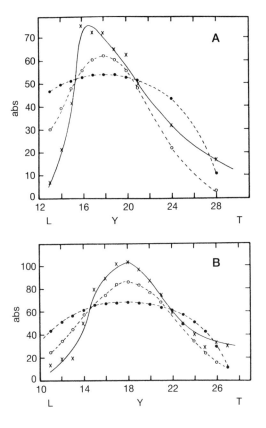

Figure 10. Profiles of boundaries stained with silver and plotted through the maximum absorbance of the spot (a peptide complexed with SDS). L is the leading edge of the moving boundary when travelling towards the anode, T is the trailing edge. (**A**) is the axis parallel to the transport; (**B**) is the axis perpendicular to the transport (isoelectric focusing direction). x----x, experimental data; O----O, profile of double Gaussian; ●----●, profile for ellipsoid.

These compensations are not as easy to make as would appear from visual comparisons of gel patterns. Holding two or more complete images in memory means that most small computer systems are not capable of performing these comparisons unless the images have been modelled. Using the parameters from the models to reconstruct patterns in the display memory and restricting the number of bits for each image makes it possible to superimpose two images in the display. Then, using the cursor, the internal standard spots can be noted and the coordinates of the rest of the spots adjusted. With the modelled image this means adjusting the values of the centres of the peaks.

6.5 Conclusions

It has become increasingly obvious that two-dimensional protein gels can only be satisfactorily analysed if they are digitized and the features are abstracted using a semi-automated system. It is not necessary to invest in mainframe computers to do this since many micro- and minicomputers with large memories (4 Mbytes or more) that may be capable of processing the data in reasonable time are available. Scanners are now available which will capture the data and usually these come with electronic boards which contain display memory and the contents can be shown on high resolution monitors. Manipulation of the data collected by this system requires several programs that are not necessarily available as a package so that one would have to put this together. This requires a degree of computing expertise in addition to the specialized hardware. However, with the increasing realization of the importance of such methods of gel analysis, such practical problems will no doubt be solved.

Acknowledgements

The authors wish to thank John Sinclair for his contributions to Section 5 of this chapter that were originally published in the first edition of this book. J.A.A.C. wishes to thank Jeri Higginbotham for critical reading of the manuscript and help with the literature and Helen Hoeven of the Plant Breeding Library of Pioneer Hi-Bred International Inc for conducting literature searches.

References

1. Kaltschmidt, E. and Wittman, H. G. (1970). *Anal. Biochem.*, **36**, 401.
2. Dean, B. (1979). *Anal. Biochem.*, **99**, 105.
3. Anderson, N. G. and Anderson, N. L. (1978). *Anal. Biochem.*, **85**, 331 and 341.
4. Garrels, J. I. (1979). *J. Biol. Chem.*, **254**, 7961.
5. Jones, M. I., Massingham, W. E., and Spragg, S. P. (1980). *Anal. Biochem.*, **106**, 446.
6. Johns, E. W. (1976). In *Subnuclear Components: Preparation and Fractionation*, (ed. G. D. Birnie), p. 202. Butterworths, London.
7. Kruh, J., Schapira, G., Lareau, J., and Dreyfus, J. C. (1964). *Biochim. Biophys. Acta*, **87**, 669.
8. Adamietz, P. and Hiltz, H. (1976). *Hoppe-Seylers Z.Physiol. Chem.*, **357**, 527.
9. Sinclair, J. H. and Rickwood, D. (1985). *Biochem. J.*, **229**, 771.
10. MacGillivray, A. J. and Rickwood, D. (1974). *Eur. J. Biochem.*, **41**, 181.
11. O'Farrell, P. H. (1975). *J. Biol. Chem.*, **250**, 4007.
12. Orrick, L., Olson, M., and Busch, H. (1973). *Proc. Natl. Acad. Sci. USA*, **70**, 1316.
13. Ames, G. F. L. and Nikaido, K. (1976). *Biochemistry*, **15**, 616.
14. Howard, G. A. and Traut, R. R. (1973). *FEBS Lett.*, **29**, 177.
15. Hoffmann, P. and Chalkley, R. (1976). *Anal. Biochem.*, **76**, 539.
16. Wray, W., Boulikas, T., Wray, V. P., and Hancock, R. (1981). *Anal. Biochem.*, **118**, 197.
17. Chambers, J. A. A., Hinkelammert, K., degli Innocenti, F., and Russo, V. E. A. (1985). *Electrophoresis*, **6**, 339.
18. Poehling, H. M. and Neuhoff, V. (1980). *Electrophoresis*, **1**, 90.
19. Hochstraser, D., Augsburger, V., Funk, M., Appel, R., Pelligrini, C., and Mueller, A. F. (1986). *Electrophoresis*, **7**, 505.
20. Gianazza, E., Astrua-Testori, S., Righetti, P. G., and Bianchi-Bosisio, A. (1986). *Electrophoresis*, **7**, 435.

21. Gianazza, E., Astrua-Testori, S., Giacon, P., and Righetti, P. G. (1985). *Electrophoresis,* **6**, 332.
22. Cuono, C. B. and Chapo, G. A. (1982). *Electrophoresis,* **3**, 65.
23. Mukasa, H., Tsumori, H., and Shimamura, A. (1987). *Electrophoresis,* **8**, 29.
24. Dunn, M. J. and Burghes, A. H. M. (1983). *Electrophoresis,* **4**, 97.
25. Dunbar, B. S. (1987). *Two-dimensional Gel Electrophoresis and Immunological Techniques.* Plenum Press, New York.
26. Burghes, A. H. M., Patel, K., and Dunn, M. J. (1985). *Electrophoresis,* **6**, 453.
27. Wilson, D., Hall, M. E., Stone, G. C., and Rubin, R. W. (1977). *Anal. Biochem.,* **83**, 33.
28. Horst, M. N., Mahaboob, S., Basha, M., Baumbach, G. A., Mansfield, E. H., and Roberts, E. M. (1980). *Anal. Biochem.,* **102**, 399.
29. Schaffner, W. and Weissman, C. (1973). *Anal. Biochem.,* **56**, 502.
30. Ramagli, L. S. and Rodriguez, L. V. (1985). *Electrophoresis,* **6**, 559.
31. Duncan, R. and Hershey, J. E. B. (1984). *Anal. Biochem.,* **138**, 144.
32. Voris, B. P. and Young, D. A. (1980). *Anal. Biochem.,* **104**, 478.
33. Nguyen, N. Y. and Chrambach, A. (1977). *Anal. Biochem.,* **79**, 462.
34. Nguyen, N. Y. and Chrambach, A. (1977). *Anal. Biochem.,* **82**, 54.
35. Nguyen, N. Y. and Chrambach, A. (1977). *Anal. Biochem.,* **82**, 226.
36. O'Farrell, P. Z., Goodman, H. M., and O'Farrell, P. H. (1977). *Cell,* **12**, 1133.
37. Young, D. A., Voris, B. P., Maytin, E. V., and Colbert, R. A. (1983). In *Methods in Enzymology*, (ed. C. H. W. Hirs and S. N. Timasheff), Vol. 91, p. 190. Academic Press, New York.
38. Mets, L. J. and Bogorad, L. (1974). *Anal. Biochem.* **57**, 200.
39. Martini, O. H. W. (1974). *PhD Thesis,* University of London.
40. Chambers, J. A. A., Hinkelammert, K., and Russo, V. E. A. (1985). *EMBO J.,* **4**, 364.
41. Reisfeld, R. A., Lewis, U. J., and Williams, D. E. (1962). *Nature,* **195**, 281.
42. Thomas, J. and Kornberg, R. (1975). *Proc. Natl. Acad. Sci. USA,* **72**, 2626.
43. Miller, M. J., Olson, A. D., and Thorgeirsson, S. S. (1984). *Electrophoresis,* **5**, 297.
44. Olson, A. D. and Miller, M. J. (1988). *Anal. Biochem.,* **169**, 49.
45. Appel, R., Hochstrasser, D., Roch, C., Funk, M., Muller, A. F., and Pellegrini, C. (1988). *Electrophoresis,* **9**, 136.
46. Tarroux, P., Vincennes, P., and Rabilloud, T. (1987). *Electrophoresis,* **8**, 187.
47. Garrels, J. I. (1984). In *Two-dimensional Gel Electrophoresis of Proteins: Methods and Applications.* (ed. I. E. Celis and R. Bravo), p. 38. Academic Press, New York.
48. Anderson, N. L. *et al.* (1981). *Clin. Chem.,* **27**, 1807.
49. Lempkin, P. F. and Lipkin, L. E. (1981). *Computer Biomed. Res.,* **14**, 272.
50. Spragg, S. P., Amess, R., Jones, M. I., and Ramasamy, R. (1985). *Anal. Biochem.,* **147**, 120.
51. Taylor, J., Anderson, A. L., Coulter, B. P., Scandora, A. E., and Anderson, N. G. (1980). In *Electrophoresis 79*, (ed. B. J. Radola), p. 329. de Gruyter, Berlin.
52. Gonzalez, R. C. and Wintz, P. (1977). *Digital Image Processing.* Addison Wesley, London.
53. Lemkin, P. F., Lipkin, L. E., and Lester, E. P. (1982). *Clin. Chem.,* **28**, 840.
54. McCalla, T. R. (1967). *Introduction to Numerical Methods and FORTRAN Programming.* J. Wiley & Sons, New York.
55. Press, W. H., Flannery, B. P., Teukolsky, S. A., and Vetterling, W. T. (1986). *Numerical Recipes.* Cambridge University Press.

56. Spragg, S. P., Amess, R., Jones, M. I., and Ramasamy, R. (1987). In *Two-Dimensional Gel Electrophoresis and Immunological Techniques* (ed. B. S. Dunbar). Plenum Press, New York.
57. Hoffman, W. L. and Dowben, R. M. (1978). *Anal. Biochem.*, **89**, 540.

4

Immunoelectrophoresis

T. C. BØG-HANSEN

1. Introduction

Immunoelectrophoresis is a procedure in which proteins and other antigenic substances are characterized by both their electrophoretic migration in a gel (usually agarose) and their immunological properties. We can distinguish between immuno-electrophoretic analysis [the classic immunoelectrophoresis of P.Graber (1) often referred to as IEA], crossed immunoelectrophoresis [the two-dimensional quantitative immunoelectrophoresis of Clarke and Freeman (2) often referred to as CIE or CRIE], and rocket immunoelectrophoresis [the one-dimensional immunoelectrophoresis of Laurell (3), also called electroimmunoassay, EIA]. Numerous variations of these techniques exist, some of which will be described in this chapter. For further details and references the reader is referred to refs 4−8.

In the classic immunoelectrophoretic analysis, the proteins are separated by zone electrophoresis and subsequently diffuse into the gel. Precipitation lines are formed with corresponding antibodies at positions between the antigen and antibody application sites where the antigen to antibody ratio is optimal. The technique allows the detection and characterization of individual proteins.

Crossed immunoelectrophoresis is analogous to classical immunoelectrophoretic analysis apart from the second step where electrophoresis into an antibody-containing gel substitutes for the diffusion step. This method allows for the detection and characterization of individual proteins and their quantitation. Rocket immuno-electrophoresis is a one-dimensional electrophoresis of antigen into antibody-containing gel, used for measuring the amount of a specific protein. These immunological methods can be used to characterize individual proteins and even variants of individual proteins in complex mixtures such as crude extracts, tissue homogenates and body fluids, all without prior purification. The quantitative immunoelectrophoresis methods may be regarded as specific examples of a general principle known as affinity electrophoresis (9) where interacting components are allowed to react during electrophoresis.

Immunoelectrophoresis contains an element of irreproducibility or at least an element of biological variation. The information derived from any kind of immunoelectrophoresis is very much dependent upon the antisera or antibody used. These can be polyspecific or monospecific depending on the purpose of the analysis.

The animal species in which the antiserum was raised is also an important factor. Horse antibodies produce reversible precipitation lines, that is, they may redissolve in the presence of an excess of antigen or antibody, whereas rabbit antibodies bind much more firmly in the precipitate. A large number of reliable antisera to proteins of human body fluids are now commercially available for medical, clinical, and research purposes. The titres of these antisera vary and so it is advisable to establish the optimum working concentration by preliminary experiments. In some cases, particularly for reasons of economy, home-made antisera may serve as well as commercial products. If commercial antibodies are not available, the animal should be chosen judging from its response to the antigen in question. Descriptions of immunization procedures are given in Appendix 6 and are published elsewhere (6,10).

After bleeding the animal, the serum is either used directly as a crude antiserum or it is fractionated to purify the immunoglobulin or just the specific antibody. The immunoglobulin fraction can be obtained by protein purification procedures such as salting-out, ion-exchange, T-gel chromatography, or protein A chromatography. The specific antibody fraction is often referred to as the 'no-nonsense' fraction and is prepared by immunosorption.

The use of monoclonal antibodies in immunoelectrophoresis is not yet very widespread since they do not precipitate normal antigens in gel precipitation systems. However, their use either as a first-dimensional ligand in crossed immuno-electrophoresis or as a local absorption ligand in the intermediate gel technique is not excluded (see Section 8). Presumably, however, precipitating systems could be constructed by mixing several non-precipitating monoclonal antibodies.

2. Basic techniques and their applications

The electrophoresis step is normally carried out in thin layers of agarose; a gel 1.0–1.8 mm thick is often used. Since the proteins and the antibodies interact with each other, the system is sensitive to changes in temperature. The thin gels required and the temperature sensitivity of the antibody-antigen reaction impose restrictions on the type of equipment which can be used. Preferred equipment is described below. After electrophoresis and reaction with antibodies the gel is washed to remove unprecipitated protein and unreacted antibody and is then dried and stained.

2.1 Equipment

2.1.1 Power supply

Generally a power pack capable of delivering 200–300 V and 200 mA can be used to power several electrophoresis tanks. *Figure 1* shows a typical power supply with four outlets, each of which can be set individually.

2.1.2 Electrophoresis apparatus

The electrophoresis equipment is a tank comprising a cooled surface to support the gel and buffer vessels with electrodes. A wide variety of suitable apparatus is

T. C. Bøg-Hansen

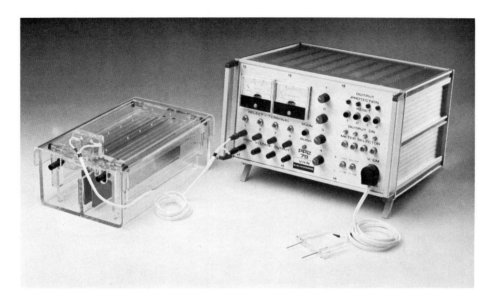

Figure 1. Electrophoresis tank and power pack for immunoelectrophoresis. The electrophoresis tank is especially designed for crossed and rocket immunoelectrophoresis and the power pack has four outlets with individual settings. There is a built-in voltmeter and an ammeter. The voltage in the gel may be measured directly by means of the test probe. (Reproduced with permission from Daela, Denmark.)

commercially available (e.g. LKB, Bio-Rad). *Figures 1* and *2* show a compact model used by the author which may be used for all types of immunoelectrophoresis. The components can be easily taken apart for inspection and cleaning.

Support plate with cooling coil, working surface 120 × 220 mm
The cooling coil is connected to a water bath fitted with a thermostat. For electrophoresis using field strengths greater than 10 V/cm, the temperature of the coolant should be 10−14 °C, depending on the humidity. For field strengths below 10 V/cm, a temperature of 16−20 °C is adequate. For most techniques it may be sufficient to connect the cooling coil to an ordinary water tap. However, circulating thermostatted cooling is easier to regulate. Furthermore, the cooling coil will not become fouled if the system is filled with thermostatted distilled water. Prolonged use of tap water may lead to deposits in the cooling coil. The coil should then be cleaned with 2% nitric acid.

Platinum electrodes
A plastic support incorporating the electrodes is mounted underneath the cooling coil. To protect against mechanical damage the electrode wires run in grooves.

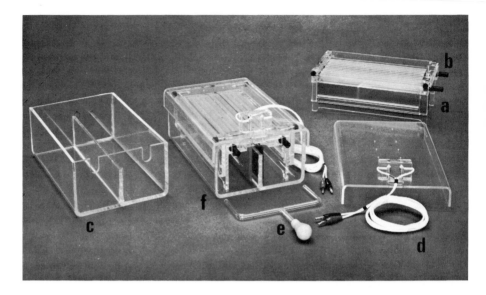

Figure 2. The electrophoresis tank taken apart for inspection. (a) The support plate with cooling coil; (b) inlet and outlet for cooling water; (c) buffer vessels and protective container; (d) lid with connecting cable; (e) level regulator; (f) complete tank assembled for electrophoresis. The tank is connected to the power pack and cooling water circulation is started. Electrode buffer is added in each compartment, and gels may then be placed on the cooling surface. Connection to the buffer is established with filter paper wicks. (Reproduced with permission from Daela, Denmark.)

Buffer reservoirs and protective containers

The buffer reservoirs have a volume of approximately 1200 ml, ensuring sufficient buffer capacity (i.e. constant pH) even during prolonged electrophoretic runs. Only when assembled with the safety-lid in place can electrophoresis occur. Without the lid, no current is applied to the electrodes. When on, the lid rests on the frame of the supporting plate, creating a closed chamber with controlled humidity from the buffer reservoirs. Systematically arranged holes in the lid allow the introduction of test electrodes to measure field strength directly in the gel.

Level regulator

Unequal liquid levels in the two buffer vessels may cause an undesirable flow of buffer in the gel plates. Therefore, it is essential to equalize the buffer levels before starting electrophoresis. The two arms of the level regulator are dipped into the buffer vessels and, by fully squeezing the rubber ball, the liquid is drawn up into the bulb. A three-minute wait is necessary to allow the buffer levels to equalize.

2.1.3 Horizontal table

A level table is necessary for casting the gels on glass plates. We have found it convenient to use a 40 × 60 cm (or larger) glass plate as a working surface on the laboratory bench. The glass plate is carefully levelled by means of a spirit level. Horizontal tables (20 × 30 cm) with built-in spirit levels are available commercially.

2.1.4 Well cutters

Circular wells need to be punched in the agarose gel for sample loading. In order to cut good wells easily in agarose gel, a well cutter incorporating a telescopic suction device is recommended (*Figure 3*). Several commercial sources of these exist. If simple well cutters are used, the gel layers may detach from the glass plate by the simultaneous punching and suction removal of the agar gel plugs. However, the telescopic suction device allows the plug to be removed safely, regardless of the strength of the vacuum. After punching out the well with the larger outer needle, continued gentle pressure will establish the connection between the vacuum source (e.g. a water pump) and the gel surface. The gel plug is sucked out from the well without any risk of lifting the rest of the gel from the glass plate. Cutters are available commercially for well diameters of 2.0, 2.5, 3.0, and 4.0 mm.

2.1.5 Punching template

The Perspex punching template consists of a base plate and a sliding bridge piece which can be locked into position (*Figure 4*). The bridge piece has guide-holes for 28 and 40 linearly-arranged wells, or for 79 wells arranged in a staggered or W-shaped pattern. The gel plate in which the wells are to be cut is inserted into the punching template between the base plate and the bridge. It is held in position against two pins inserted into holes in the base plate.

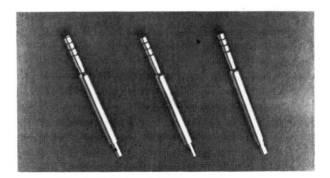

Figure 3. Well cutters, each incorporating a telescopic suction device. During use, the well cutter is connected to a water pump to provide suction. (Reproduced with permission from Daela, Denmark.)

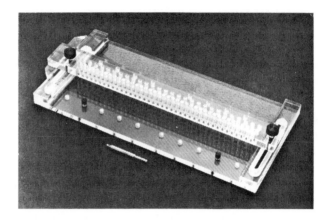

Figure 4. A punching template. A Perspex punching template such as this is especially useful when many wells must be cut as, for instance, in rocket immunoelectrophoresis and for fused rocket immunoelectrophoresis. (Reproduced with permission from Daela, Denmark.)

2.1.6 Glass plates

Glass plates of the following dimensions are available commercially for various electrophoretic techniques: 205 × 100 mm, 110 × 90 mm, 100 × 100 mm, 50 × 50 mm, 70 × 100 mm, 50 × 70 mm, 70 × 70 mm including microscope slides and projector slides. We normally use plates 1 mm thick.

2.1.7 Water bath

When melted, agarose can be kept liquid in a water bath set at 45−55°C, the exact temperature depending upon the type of agarose used. The water bath should be conveniently arranged with vessels and test tubes for agarose handling and antibody mixing as appropriate.

2.1.8 Cutting blades

The 155 mm long cutting blades (razor blades) sharpened on both edges (*Figure 5*) are used in two-dimensional immunoelectrophoresis to cut and separate the gel strips of the samples used in the first-dimensional step. They are also used to transfer gel strips to the glass plate prepared for the second-dimensional step (see Section 5.3). They can be obtained commercially from Daela.

2.1.9 Plate holders

A plate holder (*Figure 6*) provides ease of handling for staining and rinsing the gels or for temporary storage in a moist chamber. Several Perspex plate holders are available commercially (Daela). These typically hold eight glass plates (100 × 100 mm) or, after detaching the central divider, four glass plates (205 × 110 mm), or 32 glass plates (50 × 50 mm).

T. C. Bøg-Hansen

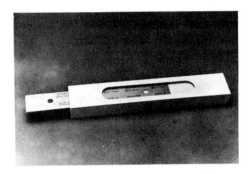

Figure 5. The long cutting blades (155 mm) that are required for handling the first dimensional gels which must be transferred in cross immunoelectrophoresis. (Reproduced with permission from Daela, Denmark.)

Figure 6. Plate holders. One type of Perspex plate holder for staining and rinsing immunoplates is shown here. Other types of holders are also commercially available. (Reproduced with permission from Daela, Denmark.)

2.2 Gel preparation

2.2.1 Preparation of the agarose solution

Typically 1% (w/w) agarose in Tris−barbital (pH 8.6) is used in immunoelectrophoresis. This is prepared as described in *Table 1*.

Table 1. Buffer and gel reagents for immunoelectrophoresis

Electrophoresis buffer (Tris – barbital, pH 8.6)

This is prepared as a stock solution as follows:

5,5′-Diethyl-barbituric acid (Veronal)	2.4 g
Tris (e.g. Sigma 7-9)	44.3 g
Distilled water	to 1000 ml

Dilute 5-fold before use.

1% (w/w) agarose[a]

Agarose (HSA Litex or similar)	1 g
Tris – barbital buffer, pH 8.6	100 ml

Dissolve the agarose by boiling for 5 min. It can be kept molten in a water bath at 50 – 60°C. Once cool and set, the gel can be stored at 4°C, and is ready for use once more after a short period of boiling.

[a] For normal procedures, use agarose with electroendosmosis $M_r = 0.13$.

2.2.2 Coating

Prior to casting of the gel, the glass plate may be 'coated' in order to prevent the gel floating off during washing and handling. This is achieved by spreading a diluted agar solution (0.2 – 0.5%) by means of a brush, a tissue, or by using a slide (as for blood smears).

2.2.3 Casting the gel

Without antibody in the gel

Agarose without antibody is used for immunoelectrophoresis analysis and for first-dimensional electrophoresis in crossed immunoelectrophoresis. The agarose solution (*Table 1*) is melted and kept at 55°C in the thermostatted water bath until use. A measured amount is poured onto a well-cleaned glass plate to give the required thickness (1 – 1.8 mm). After solidification, wells are punched in the gel by means of a well cutter (see *Figure 9*).

With antibody in the gel

Agarose containing antibodies is used for rocket immunoelectrophoresis and the second-dimensional gel in crossed immunoelectrophoresis. The agarose solution (*Table 1*) is kept at 55°C in a thermostatted water bath and a calculated volume of gel is transferred to a test-tube or another vessel of suitable size kept warm in the water bath. Antibody is added to the gel, then mixed and poured immediately on to the previously prepared glass plate.

2.3 Electrophoresis

Samples are loaded onto the gel by gentle pipetting and electrophoresis is performed at 0.8 – 10 V/cm. For first-dimensional electrophoresis, any value in this range which is convenient can be chosen. Often we use 10 V/cm which allows albumin (i.e. human serum albumin) to migrate at $4-5$ cm/h (at pH 8.6). The migration may be checked with a sample of serum or albumin using Bromophenol Blue as a stain. The albumin adsorbs the dye and may be seen clearly in the agarose gel. Free Bromophenol Blue will migrate somewhat faster than albumin at pH 8.6 except in the presence of non-ionic detergents.

The second dimension in crossed electrophoresis and in other electrophoretic separations where protein antigens are to react and precipitate with antibodies, is normally performed at $0.8-1$ V/cm.

2.4 Inspection

On completion of the immunoelectrophoresis or immunodiffusion stage the plate is inspected for the presence and pattern of the precipitation lines. This is best done by using oblique illumination against a dark background. Special viewing boxes are commercially available. The plates may be photographed at this stage but it is much simpler and more satisfactory to use stained immunoelectrophoretic plates, both for interpretation as well as for permanent recording. Incubated immunoelectrophoresis plates should be inspected during the immunodiffusion process because certain precipitation arcs may appear very quickly (e.g. free light chains, Bence Jones proteins) and then disappear.

2.5 Pressing, washing, and drying

It is important to remove most of the non-precipitated proteins in order to obtain low background staining. This is done as described in *Protocol 1*.

Protocol 1. Preparing gels for staining

1. After electrophoresis, place the gel plate onto a sheet of filter paper.
2. Pour some distilled water over the gel so that all the wells are filled.
3. Place five layers of filter paper or absorbing material onto the gel followed by a thick glass plate. Allow this to stand for 5 min. This will 'press' the gel.
4. Renew the upper layers of filter paper and press for another 5 min.
5. Wash the gel for 15 min in 0.1 M NaCl or water, during which the gel swells again.
6. Press the gel for 10 min.
7. The procedure of pressing and washing can be repeated $2-3$ times.
8. Dry the gel in a stream of hot or cold air (cold air if the gel will subsequently be stained for enzyme activity).
9. Stain and destain the gel (see Section 2.6). Finally dry the gel; it may be stored if desired.

2.6 Staining and destaining

2.6.1 Coomassie Blue R-250

Coomassie Brilliant Blue R-250 is a commonly used stain for immunoelectrophoresis. A typical procedure is given in *Protocol 2*.

Protocol 2. Staining agarose gels with Coomassie Brilliant Blue R250

  1. Prepare the staining and destaining solutions as follows.

Staining solution

  • Coomassie Brilliant Blue R-250   5 g
  • Ethanol 96%   450 ml
  • Glacial acetic acid   100 ml
  • Distilled water   450 ml

After mixing the ingredients, leave the solution overnight with gentle mixing. The following day filter the solution.

Destaining solution

  • Ethanol 96%   450 ml
  • Glacial acetic acid   100 ml
  • Distilled water   450 ml

  2. Stand the gel in the stain for 5 min.

  3. Destain the gel in destaining solution 2−3 times, each time for 5 min.

  4. Dry the stained and destained gel in front of a stream of hot or cold air.

2.6.2 Other general protein stains

Ponceau S is an excellent stain for strong immune precipitates (0.2% Ponceau S in 3% trichloroacetic acid). It is an aqueous stain and destaining is carried out in aqueous 5% acetic acid. There is very little background stain left after a relatively short washing period (15 min or less with agitation). The red colour behaves like black to the photographic emulsion and therefore provides very good photographic reproduction. It is important to obtain a good, reliable brand of the dye (e.g. Raymond Lamb).

When a more sensitive dye is required for weak precipitates, Nigrosin has proven to be a useful stain. Rapid staining occurs with an alcoholic solution, details of which are given in *Protocol 3*. Alternatively, a procedure using aqueous Nigrosin solution is slower but cleaner. Aqueous Nigrosin solution is 0.001−0.002% Nigrosin in 5% acetic acid. Destain and wash in 5% acetic acid or tap water.

With the aqueous Nigrosin solution the staining time is not critical and the immuno plates may be left overnight in the stain. Nigrosin solutions should be kept clean and changed frequently; on prolonged staining the alcoholic rapid stain (*Protocol 3*) tends to accumulate a flocculant precipitate.

Protocol 3. Rapid Nigrosin staining

1. Prepare the alcoholic Nigrosin staining solution as follows.

   - Nigrosin   1 g
   - Glacial acetic acid   150 ml
   - Sodium acetate, 0.1 M   400 ml
   - Methanol   100 ml
   - Glycerol   50 ml

2. Incubate the gel with this solution until the precipitates are stained sufficiently.

3. Destain the gel in 20% methanol in 5% acetic acid followed by 5% acetic acid and water.

2.6.3 Special and selective stains

Special staining procedures are used to demonstrate the presence of lipids, glycoproteins, haemoglobin, specific enzymes, and any other substances which can be detected by a suitable combination of reagents. Acid mucopolysaccharide stains cannot be used as agarose itself is an acid mucopolysaccharide. Examples of specific staining applications are given in refs 5−8. Numerous details of the various staining procedures including the substrate formulae and developing reagents for enzyme visualization and identification can be found in ref. 11. These staining techniques apply primarily to electrophoretic separation patterns but can be used equally well for gel precipitation techniques.

Autoradiography or radioimmunoelectrophoresis is a very sensitive technique based on the use of labelled proteins or labelled secondary antibodies, for example the use of anti-IgE in crossed radioimmunoelectrophoresis for the specific detection of allergens. Immunoelectrophoresis is performed in the usual manner but after development of the specific precipitate or after incubation with patient serum and labelled reagent, the dried plate is exposed to a suitable X-ray film. This technique can also be particularly valuable in cases where non-precipitating antibodies must be identified.

3. Image processing

Without doubt, the main drawback of immunoelectrophoresis techniques is the slow evaluation procedure. Traditionally, a great number of precipitates or autoradiographs have to be compared manually. However, the advent of powerful tools for computerized image-processing has spurred the development of systems applicable to this evaluation procedure. The main problem is how to determine the similarities and differences appearing in the patterns using different antibodies or different antigen preparations and how to determine the number of precipitates in crossed immuno-electrophoresis that can be identified as a specific protein or an allergen labelled with second antibody. Applications include more efficient data management (which is of utmost importance in diagnostic applications of immunoelectrophoresis), various research objectives, immunochemical quality control in the manufacture of allergens and other proteins isolated from natural products, and control of batch-to-batch variations in antibody production.

3.1 Hardware and storage

A number of computer programs have been written for the evaluation and comparison of two-dimensional gel electrophoresis patterns other than crossed immuno-electrophoresis. In this regard, the reader should also refer to Section 6 of Chapter 3. However, to our knowledge, CREAM (Kem-En-Tec) is the first computer program designed especially for use with immunoelectrophoresis techniques (12). The hardware requirements have been limited to a video camera with interface and a personal computer with a hard disc. This modest hardware requirement renders this technique within the reach of any analytical biochemical laboratory. The optical part of the system consists of a video camera, rather than the densitometer used in many other evaluation techniques, so that the cost of implementation of the technique is reduced. The visual data are digitized using an ordinary video camera (Panasonic WV 1800/G) which subdivides its field of vision into 256×256 pixels. The light intensity of each pixel is represented by an integer between 0 and 255 (8 bits). This digitized data is the input to the computer system, an IBM-PC or IBM-AT with Professional Graphics Display, a 20-Mb hard disc, and an attached 'mouse' which is used to operate the menu-driven software.

3.2 Image processing and pattern recognition

For crossed immunoelectrophoresis the curve detection is performed in the following way. A simple averaging filter (window 3×3 pixels) for suppression of optical noise, that is, an algorithm which transforms the value of each pixel according to the values of the pixel surrounding it, can be applied initially. After noise suppression, the computer program detects gradients in the grey scale values in order to search for pixels that could be peak or maximum values for each curve. A peak situated on top of a curve that follows the edge of an immunoprecipitate can be identified using a number of feature extractors (FEXs).

In the present context, a FEX is a computer algorithm which assesses the similarity between the vicinity of a pixel and the different characteristics of peak points in general. The primary characteristics are estimates of local curvature depending on first and second derivatives of the curve. The edges of the curves are found by local thresholding of the gradients and the second derivatives of the curves. Once the edge has been determined, a convolution filter with weights resembling curve values around a peak point is applied. The application of this corresponds to a matched filtering, and the magnitude of the output at a given pixel may be used as a measure of the likelihood of that pixel being a peak point. This is done for different training sets. The closeness of fit of a particular point to what is expected for a peak point is then assured by using the classification functions on the FEX values. In this way a peak point candidate is classified as a true or false peak point. Each precipitation curve is then fitted around a peak point by Gaussian curves or by using spline functions. In order to obtain the best visual resolution on the screen, a pseudocolour scale is applied to the images. The linear resolution in the present implementation is 0.2 mm horizontally and 0.33 mm vertically.

3.3 Applications of image processing

Rocket precipitates (Section 6) can easily be measured and transformed to yield the amount of protein per volumetric unit by CREAM. However, the system is not confined to this type of analysis and can also be used for any applications in which evaluation and comparison of a number of immunoelectrophoresis patterns is of value, such as quality control in antigen and antibody production, analysis of daily batch-to-batch variation in allergen preparations, control of natural products such as milk and other food stuffs, and any analysis of antigens against which antibodies can be raised. This computer analysis can also be applied to combinations of different techniques such as the intermediate gel technique (see Section 8). Since the areas under the precipitates can easily be calculated, quantitative analysis of the individual antigens in a complex mixture can be performed. It is also possible to measure the intensity of the autoradiographic images since the colour of each pixel is given a value. Thus, for example, the binding of human antibodies in the crossed radioimmunoelectrophoresis system can be quantitated. In future, it is likely that array counters measuring radioactivity will replace the video camera, since the output data from a radioactivity counter are in digital form and can be measured directly, thus speeding up the generation of the image of the CREAM program.

4. Immunoelectrophoresis

4.1 Purpose

Investigation of a complex mixture of proteins with respect to the presence and characteristics of one or more proteins.

4.2 Principle

The protein samples are placed in circular wells and separated by electrophoresis. Antiserum is then placed in a central slot and individual proteins are precipitated by the antibodies during diffusion (e.g. *Figure 7*).

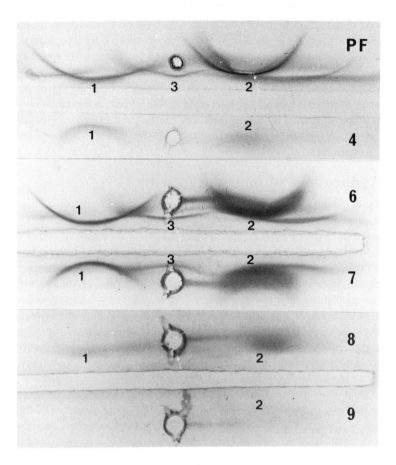

Figure 7. Immunoelectrophoretic analysis of the perivitelline fluid (PF) and embryo extracts from the snail *Biomphalaria glabrata* at days 4, 6, 7, 8, and 9. The antigen extracts were applied in the wells and electrophoresis was performed (anode to the right). After electrophoresis, the troughs were filled with an antiserum against spawn extract. Antigen 1 is a lectin seen to appear on day 4, and to disappear again from the embryo on day 8. Antigen 2 is a major protein of unknown function. Antigen 3 is a galactan. Adapted with permission from H.Bretting (15).

4.3 Procedure

Protocol 4. Immunoelectrophoresis

  1. When the agarose gel has been cast and set, cut circular wells about 10 mm apart with a well cutter (see Section 2.1.4). For weak antigen solutions the sample wells may be larger (*Figure 7*).

Protocol 4. *continued*

2. Apply $3-5$ μl of the sample to the well using a micropipette, microsyringe or capillary. The well should be filled flush with the gel surface in order to obtain an even electric field in the gel. The minimum concentration of antigen for staining with Coomassie Blue is of the order of 1 mg/litre for most proteins when applied in samples of $5-10$ μl.

3. Perform electrophoresis for about 1 h at 10 V/cm.

4. At the end of the electrophoretic run, cut troughs in the gel according to the template (*Figure 7*).

5. Remove the gel from the trough. This is a delicate operation and requires a fair amount of experience and manual dexterity. The gel is scooped up either by suction or using a suitable needle or a narrow instrument.

6. Fill the troughs with antiserum using a suitable pipette with a fine tip. The troughs should be filled flush with the surface as overfilling or spillage will ruin the precipitation pattern.

7. After the antiserum troughs have been filled, place the loaded gel plates in a suitable moist chamber for immunodiffusion and the subsequent immunoprecipitation to take place. Overnight incubation at room temperature is normally adequate, but stronger precipitates may develop after another day of incubation.

8. Stain and destain the precipitates as described in Section 2.6.

4.4 Modifications for identification of specific antigens or antibodies

Identification and characterization of a large number of antigen−antibody reactions in a complex pattern is not possible by observation alone, even though their positions may be known from other experiments. Numerous variations of the immuno-electrophoretic analysis are available for specific purposes, including characterization of the precipitated protein for biological properties (see Section 11 and ref. 13).

5. Crossed immunoelectrophoresis

5.1 Purpose

Investigation of a complex mixture of proteins or antibodies with respect to the presence and characteristics of one or more proteins or antibodies.

5.2 Principle

The sample proteins are separated electrophoretically in an agarose gel. The separated proteins are then electrophoresed, at right angles to the first direction, into a gel containing antibodies against the protein. With a sufficient amount of corresponding antibody, each protein will give rise to a precipitate, a peak or a rocket (*Figure 8*). The area enclosed by the precipitate is proportional to the amount of antigen. The resolving power of this analysis is high, making it possible to quantify 20−30 proteins in one experiment. The precision is 2−18% depending on the antibody-antigen system (4,5). The technique is very versatile and numerous modifications exist (6). The method is useful especially for biomolecular characterization of antigens without having them in a pure state.

5.3 Procedure

The technique is very easy to practise and demands only a limited amount of work (maximally 15 min per plate). The crucial step in the practical performance is the transfer of each lane from the first-dimensional gel after electrophoresis to a separate second-dimensional plate (*Figure 9*). In order to retain the qualitative aspect of this analysis, the transfer should be completely quantitative.

Protocol 5. Crossed immunoelectrophoresis

1. Cut out the agarose corresponding to single lane of the first-dimensional gel with a scalpel or a cutting blade (*Figure 6*).

2. Using a long cutting blade, transfer it to the second-dimensional plate. The transfer should be performed within a reasonable time of the completion of the first-dimensional electrophoresis in order to maximize the reproducibility.

3. Pour the antibody-containing gel onto the second-dimensional plate.

4. To avoid diffusion, start the second-dimensional electrophoresis immediately after the gel has been cast.

5. After electrophoresis, stain and destain the precipitates (Section 2.6).

When the mobility of the antibodies used is zero, both cathodally and anodally moving antigens can be detected and quantitated. A peculiarity quite often seen is an antigen moving anodally in the first dimension but forming a cathodic precipitate in the second dimension, indicating that the complex of IgG and the antigen moves differently from the antigen itself.

The sensitivity of crossed immunoelectrophoresis is comparable to or even better than that of rocket immunoelectrophoresis (see Section 6). When a monospecific antibody is available, however, rocket immunoelectrophoresis is quicker and easier to perform and measure, especially if a large number of samples is to be compared. However, with the new possibilities for video and computer processing of the experimental data (see CREAM, Section 3) one should choose the method which is most adequate for each particular measurement or investigation.

The main advantage of crossed immunoelectrophoresis is its potential for molecular characterization, identification and resolution of complex mixtures of proteins.

T. C. Bøg-Hansen

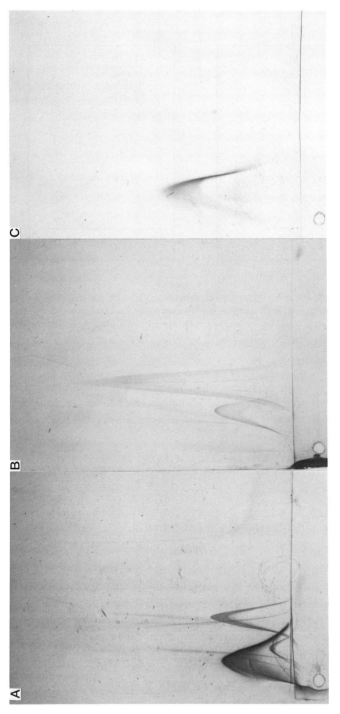

Figure 8. Crossed immunoelectrophoresis. An antigen mixture (rabbit IgG) analysed with a crude antiserum and fractions from a purification of the antiserum by immunosorption. The immunosorption was performed on rabbit IgG (the antigen) immobilized on Mini Leak activated agarose (Kem-En-Tec). (A) crude antiserum against rabbit IgG; (B) material which failed to bind to the rabbit IgG antigen; (C) monospecific IgG eluted from the immunosorbent.

289

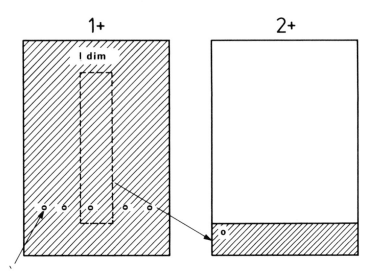

Figure 9. Template for crossed immunoelectrophoresis. The first plate is used for the first dimensional electrophoresis. After electrophoresis, each gel lane is transferred to a separate plate onto which is then cast the antibody-containing gel ready for the second-dimensional electrophoresis.

Accordingly, the technique has been applied to a number of very different problems (4–9). Some of the major modifications will be described below. A comprehensive survey has been published by Axelsen (6).

6. Rocket immunoelectrophoresis, electroimmuno-assay

6.1 Purpose

The measurement of the amount of a specific protein.

6.2 Principle

When electrophoresis of an antigen is performed in an agarose gel containing the corresponding antibody, a long rocket-like immunoprecipitate develops (*Figure 10*). The height of the rocket is linearly correlated with the amount of antigen. Rocket immunoelectrophoresis gives rapid and precise identification and quantitation of even minor amounts of protein; the standard deviation on double estimates is 1–3%; as little as 4 ng of albumin can be detected and the electrophoretic run can be completed in only 2–16 h depending on the electrophoretic mobility of the antigen and the field strength used.

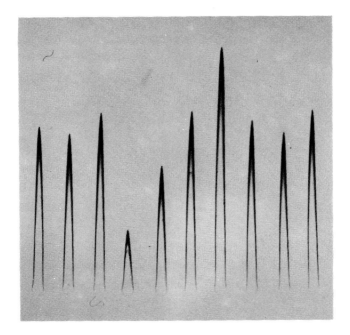

Figure 10. Rocket immunoelectrophoresis which allows the amount of a specific protein to be measured in a crude mixture of proteins. The height of the rocket is linearly correlated with the amount of antigen.

The original procedure of Laurell (14) is still a very useful alternative to single radial immunodiffusion and nephelometry for quantitation of antigens. Only if the antigen in question is of low molecular mass (below 30 kd) or if it exhibits extensive electrophoretic heterogeneity, should single radial immunodiffusion or ELISA (enzyme-linked immunosorbent assay) be considered.

6.3 Procedure

Protocol 6. Rocket immunoelectrophoresis

1. Load the samples in individual wells which have been punched along one side of the agarose gel. Also load a series of standard dilutions of the antigen.
2. Electrophorese overnight at low voltage or for a few hours at higher voltage.
3. Stain and destain the precipitates (Section 2.6).
4. Measure the heights of the rockets and correlate these to the known standards run in adjacent lanes. This may now be performed electronically with video image processing (see CREAM, Section 3).

A series of standard dilutions of the antigen must normally be included on all plates as the calibration curve can change from one plate to the other. The curve is seldom linear because of the non-homogeneous distribution of antibodies which occurs during electrophoresis. In order to obtain correct quantification it is important to ensure that the measured sample and the standard are antigenically identical. It is also important to fill the wells to the same extent in both the sample and the standard wells to ensure a uniform electric field. With strict control of field strength and temperature, the reproducibility is as good as 99% (15).

7. Fused rocket immunoelectrophoresis

7.1 Purpose

To characterize quantitatively the content of fractions from a fractionation procedure for the antigens of interest.

7.2 Principle

This method is a modification of rocket immunoelectrophoresis. Small volumes from, for example, fractions obtained in separation experiments or from extracts from tissue, are applied in a row of sample wells across the plate in an antibody-free gel, where the samples are allowed to diffuse for up to about 1 h.The proteins are then electrophoresed into the antibody-containing gel. The resulting fused precipitate pattern of proteins from the various fractions will be an immunological profile of each individual protein, either the elution profile or a profile showing the appearance and disappearance of each protein in that tissue during an experiment.

7.3 Procedure

Protocol 7. Fused rocket immunoelectrophoresis

1. Prepare a blank gel with sample wells and an antibody-containing gel (see *Figure 11*).
2. Apply aliquots of the samples close together in the antibody-free gel and allow them to diffuse into the gel for some time, typically 0.5−1 h at room temperature.
3. Electrophorese them into the antibody-containing gel.
4. Stain and destain the precipitates (Section 2.6).

The procedure is thus a one-dimensional technique, the electrophoresis taking place overnight at a relatively low voltage. In this way all fractions containing an antigen are delineated by a precipitate showing the presence of this particular antigen in different fractions. It may be said that each precipitate describes the elution profile

T. C. Bøg-Hansen

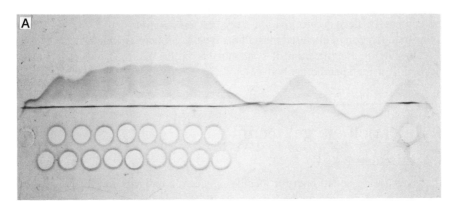

Figure 11. Fused rocket immunoelectrophoresis. Plasminogen was purified by affinity chromatography on two different commercial brands of lysine agarose and the elution profile analysed by fused rocket immunoelectrophoresis. One analysis (**A**) shows that the column is overloaded since plasminogen appears in the initial column run-through fractions, while from the other column (**B**) the plasminogen is eluted homogeneously in a few fractions (at the extreme right).

of a particular protein (*Figure 11*). With multispecific antibodies, the elution profiles for all individual proteins in the sample which was fractionated can be obtained, provided the multispecific antibody preparation contains antibodies against all the protein constituents.

7.4 Applications

This variant of rocket immunoelectrophoresis is especially well suited to following the fate of a protein during fractionation, whether it is separated from the other constituents or whether it is still contaminated. For careful inspection of the fused rocket immunoelectrophoresis experiments and other immunoelectrophoretic analyses it may also be possible to identify the contaminating proteins so that a purification strategy can be designed to give the very best purification.

One modification of the fused rocket immunoelectrophoresis which we have found most useful for identification of contaminants is the intermediate gel technique (Section 8) using specific antibodies.

8. Intermediate gel technique

8.1 Purpose

This technique is suited both for identifying specific antigens and for establishing the presence of specific antibodies against the proteins being investigated.

8.2 Principle

The modification in this technique takes place in the second dimension of the crossed immunoelectrophoresis (or the equivalent gel in other analyses); part of the gel is replaced by a gel containing another antibody. Antigens will then be precipitated in this intermediate gel according to the specificity of the antibody. If a monospecific antibody is placed in an intermediate gel, only one precipitate will be seen, displaced from its usual position with the reference antibodies (*Figure 12*).

If the antibody is of very low titre, the normal precipitation pattern will not be distorted greatly. Axelsen has described the reaction of 'inward feet' with such antibodies as the last stage for the determination of interaction in this system (6).

8.3 Procedure

Electrophoresis is carried out according to the standard procedure for the immuno-electrophoresis method being used but replacing part of the antibody-containing gel with a gel containing another antibody (e.g. *Figure 12*).

8.4 Applications

This variation can often be extended, with great advantage, to the other immuno-electrophoretic techniques. It is often successfully applied to, for example, fused rocket immunoelectrophoresis. The degree of retention is always dependent upon the relationship between the titre of the second-dimensional antibody (the standard antibody or the reference antibody) and the intermediate gel antibody against the common antigens.

With commercial antibodies we often see additional precipitates in the upper gel. Any antigen present in an intermediate gel will cause line precipitates to develop in the second-dimensional gel by their reaction with the corresponding antibody. This is often observed with liquid phase absorbed monospecific antibodies, and so can be used as a test for complete removal of antigens for absorption.

T. C. Bøg-Hansen

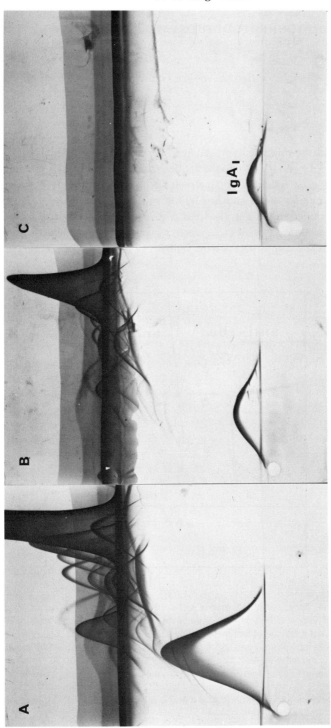

Figure 12. The intermediate gel technique. This figure shows an example of the intermediate gel technique with crossed immunoelectrophoresis. The intermediate gel contained a specific antibody against IgA; the upper gel contained an antiserum against all serum proteins. **(A)** Unfractionated serum (control); **(B)** run-through fraction; and **(C)** eluted fraction from an affinity chromatographic fractionation on immobilized jacalin, a lectin that binds IgA₁ specifically (Kem-En-Tec). One precipitate only is seen in **C** apart from the lines originating from remains of the liquid phase absorption of the monospecific anti-IgA.

9. Line immunoelectrophoresis

9.1 Purpose

To compare and evaluate different antigen samples and different antibody preparations.

9.2 Principle

The antigen samples are mixed into melted agarose, cast as well-defined blocks of agarose, and then electrophoresed into the antibody-containing gel(s). A precipitate line is formed for each individual antigen – antibody system and its position is dependent on the amounts of the antigen and the antibody. When several samples or antibodies are applied in the gel, the lines will fuse from one part of the plate to the other part(s), wherever there is cross-reactivity between antigens and antibodies, thus revealing the relationships between the samples or the antibodies (*Figure 13*).

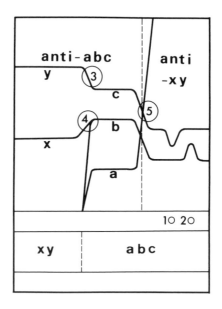

Figure 13. Line immunoelectrophoresis. The diagram shows the use of line immuno-electrophoresis for the comparison of antigens and antibody preparations. The parameters to be compared could be titre, specificity, cross-reactivity, and/or immunological identity. Both the antigens (designated **a, b, c, x, y**) and the antibodies (**anti-abc** and **anti-xy**) are incorporated into the molten agarose and poured into the positions indicated before electrophoresis. Only one electrophoresis is performed. Well 1 contained **anti-y** and well 2 contained antigen **b** or **x**. The fusion of precipitin lines at point 3 indicates immunological identity between antigens **y** and **c** and the fork at position 4 indicates partial identity between antigens **x** and **b**. The cross-over at position 5 indicates that antigens **a** and **c** are immunologically different. With kind permission of J.Krøll (19).

9.3 Procedure

Protocol 8. Line immunoelectrophoresis

1. Prepare a blank agarose gel so as to cover the plate.
2. Cut a rectangular piece from the agarose gel, mix the first antigen sample with melted agarose and pour this into the mould (see *Figure 13*).
3. Repeat step 2 for each antigen, cutting moulds adjacent to each other.
4. Remove a part of the blank gel and cast the antibody-containing gel, leaving some blank gel between the sample gel(s) and the antibody-containing gel.
5. Perform electrophoresis (usually overnight).

9.4 Applications

Line immunoelectrophoresis is mostly used for comparing complex samples qualitatively and quantitatively. It is also most useful for comparing polyvalent antibodies against the same complex antigen, or for comparing consecutive bleedings from an immunized animal.

An interesting application is the preparative aspect. The separation of proteins into clearly positioned and well-spaced precipitates provides an easy means for providing purified antigens for immunization for production of specific antisera. By loading the sample gel maximally with antigen, there will be sufficient material for successful immunization (1, 2). Other applications include a comparison of different sera from patients, characterization of alterations in the protein pattern during a biological process, and determination of normal values for various human serum proteins.

10. Line immunoelectrophoresis with *in situ* absorption

10.1 Purpose

To detect, measure, and evaluate antibodies or interacting substances such as lectins (*Figure 14*).

10.2 Principle

Basically, this analysis is a line immunoelectrophoresis, where deviations of the line ('dips') are introduced by small samples of antibody or lectin which bind to the antigen instead of the antibody in the agarose. This is seen in diagrammatic form on the right-hand side of *Figure 13* where well 1, containing antibody, induces a deviation in the corresponding precipitin line in the top gel. The size of the dip is a quantitative measure of the titre of the antibody or lectin in the sample.

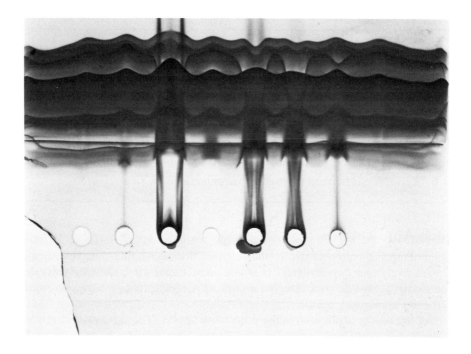

Figure 14. Line immunoelectrophoresis with *in situ* absorption. Deviations in the line precipitates of the serum glycoproteins is caused by lectins in the small aliquots of seed extracts in the circular wells.

10.3 Procedure

Protocol 9. Line immunoelectrophoresis with *in situ* absorption

1. Prepare a blank agarose gel so as to cover the plate.
2. Remove a part of the gel and cast the sample-containing agarose (see *Figure 14*).
3. Remove another part of the blank gel and cast the antibody-containing gel, leaving a large part of the blank gel between the sample gel and the antibody-containing gel.
4. Cut wells for the samples or antibodies to be tested in the blank gel between the sample gel and the antibody-containing gel.
5. Apply the samples and perform electrophoresis (normally overnight).

10.4 Applications

We have used this test to evaluate the specificity and the titre of consecutive bleedings from animals immunized with various antigens. The method gives a direct visual comparison between the bleedings such that the immunization of certain animals can be discontinued on the basis of such comparisons.

The author has also used the method to screen a number of Australian seeds for lectins. The screening is performed by means of electrophoresis and agglutination. Prior to electrophoresis, the dry seeds are crushed in a meat grinder, then homogenized using a mortar and pestle with a small volume of buffer (< 10 ml/g). After centrifugation the supernatant is tested by electrophoresis. Electrophoresis is first performed as 'line immunoelectrophoresis with *in situ* absorption' and is the easiest way to test a large number of lectin extracts for precipitating capacity towards glycoproteins. Several different reactions are seen (*Figure 14*). The specificity of the lectin in each reacting extract is analysed by the very sensitive method of including the lectin in the first dimension in crossed immunoelectrophoresis (16).

11. Detection of biological activity

Probably the most important single advance to patient management to occur with crossed immunoelectrophoresis was the possibility for the identification of specific allergenic components in allergens. This advance is based upon the fact that the immunoprecipitate which forms is very open with room for more antibody molecules to react. This means that patient IgE can be incorporated specifically into individual allergenic precipitates in crossed immunoelectrophoresis. Subsequently, a reaction takes place with labelled anti-IgE. Since the sample proteins are not denatured before immunoelectrophoresis, enzymatic activities are often retained. Numerous staining methods exist for enzymes (see Appendix 2) and so can be applied to detect specific antigens following any of the techniques described above.

Another aspect which can be investigated is glycoprotein microheterogeneity. By introducing lectins into the first dimension and into an intermediate gel, it is possible to characterize the carbohydrate moiety of glycoproteins and to quantify individual subpopulations (17).

References

1. Grabar, P. and Williams, C. A. (1953). *Biochim. Biophys. Acta,* **10**, 193.
2. Clarke, H. G. M. and Freeman, T. (1968). *Clin. Sci.,* **35**, 403.
3. Laurell, C. B. (1966). *Anal. Biochem.,* **15**, 45.
4. Axelsen, N. H., Krøll, J., and Weeke, B. (ed.) (1973). *A Manual of Quantitative Immunoelectrophoresis. Methods and Applications.* Blackwell Scientific Publications, Oxford.
5. Axelsen, N. D. (ed.) (1975). *Quantitative Immunoelectrophoresis. New Developments and Applications.* Blackwell Scientific Publications, Oxford.
6. Axelsen, N. H. (ed.) (1983). *Handbook of Immunoprecipitation-in-Gel Techniques.* Blackwell Scientific Publications, Oxford; also *Scand. J. Immunol.,* **17**, Suppl. 10.
7. Bjerrum, O. J. (ed.) (1983). *Electroimmunochemical Analysis of Membrane Proteins.* Elsevier, Amsterdam.

8. Heegaard, P. M. H. and Bøg-Hansen, T. C. (1986). In *Gel Electrophoresis of Proteins* (ed. N. J. Dunn), p. 262. Wright, Bristol.

9. Bøg-Hansen, T. C. (1973). *Anal. Biochem.*, **56**, 480.

10. Jurd, R. D. (1981). In *Gel Electrophoresis of Proteins: A Practical Approach* (ed. B. D. Hames and D. Rickwood), 1st edn. IRL Press, Oxford.

11. Smith, I. (ed.) (1976). *Chromatographic and Electrophoretic Techniques,* Vol. 2., 4th edn. William Heinemann, London.

12. Søndergaard, I., Poulsen, L. K., Hagerup, M., and Conradsen, K. (1987). *Anal. Biochem.*, **165**, 384.

13. Feinstein, A. (1976). In *Chromatographic and Electrophoretic Techniques.* (ed. I. Smith), p. 138. William Heinemann Medical Books, London.

14. Laurell, C. B. (1966). *Anal. Biochem.*, **15**, 45.

15. Axelsen, N. H., Bock, E., Larsen, P., Blirup-Jensen, S., Svendsen, P. J., Pluzek, K. J., Bjerrum, O. J., Bøg-Hansen, T. C., and Ramlau, J. (1983). *Scand. J. Immunol.*, **17**, Suppl. 10.

16. Bøg-Hansen, T. C., Bjerrum, O. J., and Ramlau, J. (1975). *Scand. J. Immunol.*, **4**, Suppl. 2, p.141.

17. Bøg-Hansen, T. C. (1983). In *Solid Phase Biochemistry. Analytical and Synthetic Aspects* (ed. W. H. Scouten), p. 223. Wiley, New York.

18. Bretting, H. (1988). In *Lectins, Biology, Biochemistry, Clinical Biochemistry.* (ed. T. C. Bøg-Hansen and D. L. J. Freed), Vol. 6, p 205. Sigma Library, St. Louis.

19. Krøll, J. (1981). In *Methods in Enzymology.* (ed. J. J. Langane and H. van Vunakis), Vol. 73B, p. 370. Academic Press, New York.

5

Peptide mapping

ANTHONY T. ANDREWS

1. Introduction

In recent years electrophoretic methods have gained an impressive and well-deserved place in the armoury of scientists investigating the composition and structure of nucleic acids, proteins, and peptides. Current versions of techniques such as polyacrylamide gel electrophoresis (PAGE), polyacrylamide gel electrophoresis in the presence of detergents such as sodium dodecyl sulphate (SDS-PAGE), immunoelectrophoresis (IE) and isoelectric focusing in polyacrylamide gels (IEF) have very high resolution, but at the same time are relatively inexpensive and simple to set up and use, are reproducible, reliable and rapid. At the cost of slightly greater complexity, the combination of two of the methods into a two-dimensional separation gives unparalleled resolution but the number of samples which can be examined in a given time is much reduced.

In spite of their capabilities, however, there are situations in protein analysis where neither one- nor two-dimensional methods can give an unequivocal answer. In essence, most of these devolve into a question of 'relatedness' between two or more proteins. Identifying an unknown protein or following changes in protein components typically involves comparison with known proteins or standards. In most one-dimensional electrophoretic systems it is usual to run at least one standard sample together with a series of unknowns on a single gel slab under the same conditions of time, voltage, current, temperature, gel composition, etc. SDS-PAGE separates proteins and peptides primarily on the basis of size differences, whereas PAGE separations are determined by a combination of both size and charge differences. Obviously two unrelated proteins can be similar in size (and hence fail to separate by SDS-PAGE) or in molecular charge (and hence have similar pIs, so separate poorly or not at all by isoelectric focusing), but perhaps less obviously, in PAGE for example, a large highly charged protein may have a similar mobility to a small molecule with a low net molecular charge. Traditional physical, chemical, and biochemical methods of analysis can be applied to solving such problems but they usually require substantial amounts of material (on a milligram scale) and there are many occasions where this is not available or cannot be spared. A major advantage of electrophoretic methods of analysis is the very small amounts of material needed (microgram range or even less), and there are indeed electrophoretic methods specifically devised to deal with this problem of relatedness.

Of course, all the one-dimensional methods provide some evidence on similarities and differences between proteins, so the more that are applied to a sample the more rigorous this evidence becomes. For example, in PAGE, separations in concentrated gels (high values of %*T*) are determined largely by molecular sieving (size differences), while charge differences predominate in low %*T* gels. Thus running samples on a series of gel slabs of different %*T* can be used to demonstrate homogeneity (all molecules closely related) in terms of both size and charge, and incidentally enables Ferguson plots (1−3) to be constructed and molecular masses measured (see Chapter 1, Sections 4.6 and 4.7), but this approach can be rather time-consuming.

Unfortunately even this is not sufficient because unrelated proteins can still, on rare occasions, fail to be distinguished while related ones may run very differently. This is often the case with protein−precursor relationships such as enzymes and their zymogens, for example, or when proteins are subject to some form of post-synthetic modification (e.g. glycosylation, phosphorylation). Degradation during sample preparation is another potential source of heterogeneity because virtually all biological materials contain at least traces of proteinase activity.

The best approach to dealing with such problems will often be peptide mapping (3−6). This consists of breaking down the unknown protein into a number of peptide fragments in a specific and controlled manner, separating the peptide mixture and then comparing the pattern of separated zones with those of one or more standard proteins treated in the same way.

With such a general definition of peptide mapping it is obvious that there are a great many possible ways of doing it. Procedures can be varied at the peptide production stage as well as by using different peptide separation methods. Initially, two decisions have to be made, namely what fragmentation method is to be used and should it be performed before or after the sample protein has been applied to the electrophoresis gel.

If one is confident that the sample protein is relatively pure, it will usually be preferable to generate the peptide mixture *in vitro* before applying the peptide separation method, because it is then easier to define and control the hydrolysis reaction. Either chemical (7,8) or enzymic methods (9,10) can be used to hydrolyse the protein. A high degree of specificity in peptide bond cleavage is perhaps less easy to achieve with chemical methods of hydrolysis than with enzymic methods, but with suitable care, some reasonably selective methods are now available (*Table 1*). Almost any proteinase could, in theory, be used for enzymic cleavage of polypeptide chains, but the most popular ones tend to be *Staphylococcus aureus* V8 proteinase, trypsin, chymotrypsin, papain, and pepsin (*Table 2*). Enzymic hydrolysis has the advantages that it is quick, more specific in most cases, and the buffers used are often compatible with the subsequent electrophoretic separation of peptides, so digests may be applied directly to the gels for analysis. Unfortunately, the enzymes themselves may contribute peptides to the mixture so a blank enzyme-only digest must always also be analysed. If chemical methods are used, no spurious peptides are added, of course, but it may be necessary to remove reagents and/or reduce

Anthony T. Andrews

Table 1. Chemical methods for selective peptide bond cleavage in proteins

Reagent	Bonds cleaved	References
Cyanogen bromide	$-Met-X-$	7,11–14
Hydroxylamine	$-Asn-Gly-$	7,8,11,14,15
BNPS-skatole	$-Trp-X-$	7,14,16
N-Chlorosuccinimide (N-Bromosuccinimide)	$-Trp-X-$	7,17
Partial acid hydrolysis	$-Asp-Pro-$	7,8,11,14,18
Partial basic cleavage	$-Ser-X-$	7,19
Heat (110°C, pH 6.8, 1–2 h)	$-Asp-Pro-$	20
2-Nitro-S-thiocyanobenzoate	$-X-Cys-$	8,14

the ionic strength and change the pH before analysis. Dialysis will not usually be practical for this if small peptides are present, so other techniques, such as column desalting, precipitation, ion-exchange treatment and so on will be needed. These have their own drawbacks and are, at the very least, time-consuming. In such cases it may be best to use isotachophoresis (25, 26) or other non-electrophoretic means of separating peptides, for which excellent methods exist, such as reversed-phase HPLC (26–28) or hydrophobic interaction chromatography by HPLC or FPLC.

In many cases, however, sample proteins are not readily available in a purified form. Then, instead of some preliminary purification scheme, it is often convenient to apply the samples directly to an electrophoresis gel and, once the components have been separated into their respective zones, to perform the hydrolysis to peptides and map the peptides in a second electrophoresis stage. Again, there are many possible permutations as to how this is done. For example, various methods (e.g. PAGE, SDS-PAGE, IEF, etc.) can be used for the initial separation of the protein components in the sample. If a gel matrix is present, as is usually the case, are the protein zones to be individually cut out and the protein eluted or not before hydrolysis; is a chemical or enzymic hydrolysis stage to be used; what method of mapping the resulting peptides is intended, etc.? Faced with such a large range of possibilities it is perhaps surprising to find a remarkable degree of uniformity in peptide mapping experiments in the literature. A very large majority of such experiments employ SDS-PAGE in the final stage to separate the peptides into a one-dimensional map and usually the peptides have been generated from protein zones obtained by subjecting the sample to PAGE or SDS-PAGE as a first step. *In situ* hydrolysis of the protein zones in the gel without extraction is most common and is performed enzymically with proteinases.

Although there are many good and very successful earlier peptide mapping experiments, the most quoted modern method is that described by Cleveland *et al.* (21).

2. Apparatus

The apparatus required for peptide mapping is the same as that required for PAGE, SDS-PAGE, IEF, etc. (see Chapter 1, Section 5; Chapter 2, Section 2.2). Since

303

Table 2. Proteinases used for selective peptide bond hydrolysis of proteins for peptide mapping

Proteinase	pH optimum	Bond specificity	References
Staphylococcus aureus V8 proteinase	4–8	Glu–X, Asp–X	9,11,21–23
α-Chymotrypsin	7–9	Trp–X, Tyr–X, Phe–X, Leu–X	10,11,21–23
Trypsin	8–9	Arg–X, Lys–X	10,11,22,23
Pepsin	2–3	N- and C-sides of Leu, aromatic residues, Asp and Glu	10
Thermolysin	~8	Ile–X, Leu–X, Val–X, Phe–X, Ala–X, Met–X, Tyr–X	10,23
Subtilisin	7–8	Broad specificity	21,23
Pronase (*Streptomyces griseus* proteinase)	7–8	Broad specificity	21
Ficin	7–8	Lys–X, Arg–X, Leu–X, Gly–X, Tyr–X	21,23
Elastase	7–8	C-side of non-aromatic neutral amino acids	21,23
Papain	7–8	Lys–X, Arg–X, (Leu–X), (Gly–X), (Phe–X)	23,24
Clostripain	7–8	Arg–X	10,23

the peptide map from a sample protein is compared to the pattern of peptides obtained from one or more standard proteins, the use of the slab gel format is virtually obligatory.

3. Methodology

3.1 The standard technique

If the samples contain a number of component proteins which require separation from each other before protein mapping can be applied, this can easily be done by any of the one- or two-dimensional methods described in Chapters 1 and 3 respectively. If the proteins are radioactively labelled (see Appendix 3), the separated protein zones are detected autoradiographically or fluorographically (see Chapter 1, Section 7.5.1) in the usual way, the gel slabs being dried down before measurement if desired. The dried gels are then rehydrated and, using the autoradiographic plate as a guide, the areas of gel containing the zones of protein to be mapped are cut out. With non-labelled proteins, gel slabs are stained with a dye such as Coomassie Blue R-250 or G-250 and destained by the usual methods (Chapter 1, Section 7.2). It is a good idea to photograph the gel at this stage to make a record of the protein separation, before cutting out the protein zones of interest (with radiolabelled samples the autoradiograph provides a suitable record). Cleveland *et al.* (21) recommended that both staining [30 min in 0.1% Coomassie Blue R-250 in methanol:acetic acid:water (5:1:4 by vol.)] and diffusion destaining (60 min in aqueous 5% methanol, 10% acetic acid) should be kept as brief as conveniently possible. While others (29) have maintained that this is not important and have advocated much more thorough staining and destaining, it is probably sound advice not to prolong these steps unnecessarily. It is possible, for example, that the acid labile Asp−Pro bond, if present, may be cleaved to some extent under acid staining and destaining conditions, but this should not be of any practical importance as long as the standard protein or proteins have been run on the same gel slab as the sample proteins (emphasizing the importance of treating sample and standards in an identical fashion throughout) because it should be partially hydrolysed in the same way.

In the original protocol (Cleveland *et al.*, ref. 21) purified proteins are made up as 0.5 mg/ml samples in 0.125 M Tris−HCl buffer (pH 6.8) containing 0.5% SDS, 10% (v/v) glycerol and 0.001% Bromophenol Blue and heated at 100°C for 2 min to inhibit extraneous proteinase activity and to unfold and denature the proteins to ensure optimum SDS binding. Known amounts of proteinase are then added and the samples incubated at 37°C. Since proteinases vary greatly in activity depending upon their specificity and purity (commercially available proteinases such as many of those in *Table 2* are usually employed in this type of work) and also upon the identity of the substrate protein/proteins as well as temperature, pH, etc., it is difficult to give universally ideal conditions for generating peptide digests. It is worth emphasizing that complete digestion to the smallest peptides that can be achieved with the proteinase in question (a so-called limit digest) is not required or even desirable. What is needed is a partial hydrolysate with a maximal number of different peptides and preferably a considerable proportion of large peptides, since many small 'limit peptides' are likely to be poorly fixed and stained in the gel and may be washed out and contribute nothing to the peptide map. Typical amounts of proteinase required with incubation times of about $10-60$ min may be of the order of $1-100$ μg/ml. In order to establish good hydrolysis conditions it may be useful to do a preliminary experiment in which the substrate is incubated with proteinase and samples withdrawn at different time intervals, boiled briefly to end the hydrolysis and examined by PAGE or SDS-PAGE. An example of this is shown in *Figure 1*. It can be seen that the qualitative appearance of the peptide band pattern is surprisingly not very sensitive to incubation time. Most of the peptides that can be seen after 10 min of hydrolysis are also observable after 90 min or even 3 h, although naturally there are marked quantitative differences as the initial large peptide products are subsequently broken down to smaller peptides. In this example one might choose a 30-min hydrolysis time to give a good spread of peptides for mapping. A similar trial experiment could be conducted by choosing a fixed incubation time and varying the amount of proteinase added, and again it will be found that the qualitative pattern of peptide zones is not very sensitive to changes in proteinase level. These findings reflect the relative insensitivity of protein digestion and of the subsequent peptide mapping to the hydrolysis conditions. This is a very useful feature of the method when dealing with proteins of unknown susceptibility to the proteinase used, and also means that it is often not necessary to define hydrolysis conditions precisely.

When a suitable incubation time has elapsed, proteolysis is stopped by adding 2-mercaptoethanol (to a final concentration of 10%) and SDS (to 2%) to the samples and then boiling them for about 2 min. The samples ($20-30$ μl containing about $10-15$ μg of peptide) are then loaded onto the SDS-PAGE slab for peptide mapping. Cleveland *et al*. (21) used a conventional SDS-PAGE slab gel arrangement for this with a $T = 15\%$, $C = 2.6\%$ gel made up in the Laemmli (30) buffer system (see Chapter 1, Section 6) which was then run, stained and destained or subjected to autoradiography or fluorography in the usual way to reveal the zones of separated peptides. Sharper peptide bands and greater resolution is often obtained if instead of a uniform concentration gel of $T = 15\%$, a concentration gradient gel is used. *Figure 2* shows that this improvement in band sharpness can be very pronounced. The preparation of concentration gradient gels has been described earlier (Chapter 1, Section 9.3.3) and depending upon the molecular mass of the original protein and size of the peptide fragments produced, gels with a linear $T = 5-20\%$ or $10-25\%$ gradient are often suitable for peptide mapping work.

On many occasions, however, it will not be convenient or possible to prepare solutions of pure sample proteins for digestion in this way. In these cases, the sample proteins are best separated by a preliminary gel electrophoresis step using virtually any of the usual methods. After this separation, the gels are briefly stained and destained as indicated above and placed on a transparent plastic sheet over a light box. Individual protein zones are then cut out with a scalpel or razor blade and, if necessary, trimmed to a size small enough to fit easily into the sample wells of

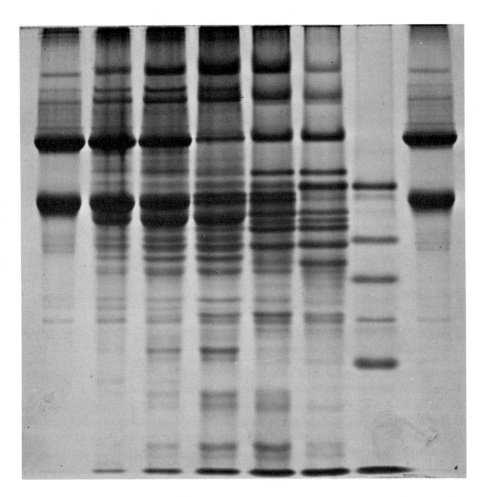

Figure 1. A typical preliminary experiment in peptide mapping in which the substrate protein and a proteinase are incubated for various times to establish conditions for optimum hydrolysis. In this example a 1% solution of total casein (Hammarsten casein, BDH Ltd) in 0.1 M sodium phosphate buffer (pH 7.5) was incubated at 37°C with 3 μg ml^{-1} of trypsin (Sigma type XIII, TPCK-treated). After the desired incubation time samples were withdrawn, heated at 100°C for 2 min to inactivate the trypsin and diluted 5-fold with stacking gel buffer containing 5% sorbitol and 0.001% Bromophenol Blue. The digests were then applied to a T = 12.5%, C = 4% PAGE gel for analysis. Samples **left to right** were: control (0 min incubation), 3 min, 10 min, 30 min, 90 min, 3 h, 24 h, and control (0 min).

the SDS-PAGE slab to be used for the peptide mapping stage. Before doing this, however, the trimmed pieces of gel are soaked for 30 min in about 10 ml of 0.125 M Tris−HCl (pH 6.8) containing 0.1% SDS, and in the original method (21) also containing 1 mM EDTA. This can inhibit some proteinases, such as *S.aureus* V8 proteinase for example, so its addition may be inadvisable as a routine addition.

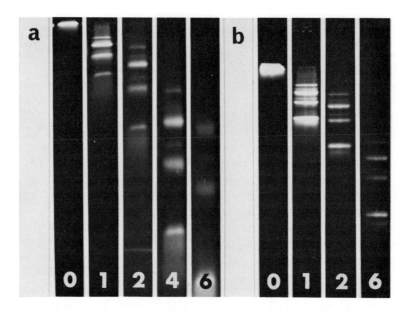

Figure 2. Increased peptide band sharpness using gradient polyacrylamide gels. Digestion products of p49 were separated on either (a) a uniform acrylamide gel (15% acrylamide, 2.66% bisacrylamide) or (b) a gradient gel (10-20% acrylamide, 5% bisacrylamide). The numbers indicate the electrophoresis time in hours after penetration of the p49 digest into the resolving gel. (Reproduced from ref. 8 with permission.)

Likewise there have been reports that excessive losses of protein can occur during the 30-min incubation in the buffer. This is probably not important for most 'average' proteins, but very small proteins, especially if they are highly glycosylated or have extreme isoelectric points, may be more susceptible to leaching out of the gel. If the final peptide maps are very faint, it may be worth reducing this washing time substantially (e.g. to 5−10 min) to see if it helps. The washing buffer is identical to the stacking gel buffer in the Laemmli system (30), so once equilibrated by this washing stage, the gel pieces are simply pushed into place in the sample wells of the slab gel. The spaces around and immediately above and below the gel pieces are filled by adding a few microlitres of this same stacking gel buffer to which 20% glycerol has been added. Finally, the given amount of the proteinase to be used is dissolved in this same buffer but containing 10% glycerol and 10 μl is added to each sample well overlaying the sample protein gel piece. The SDS-PAGE mapping gel itself is the same as that used above with soluble protein digests (i.e. $T = 15\%$, $C = 2.6\%$ or a gradient gel), the only difference being that a rather longer than normal stacking gel phase is preferred. Electrophoretic conditions are likewise the same as usual except that when the Bromophenol Blue tracking dye has reached a position close to the bottom of the stacking gel, the electrical current is switched off for 20−30 min. At this point both the sample proteins and the overlaid proteinase will have migrated down to the Kohlrausch boundary or 'stack' and will be in close proximity to each other, so switching off the current allows time for the hydrolysis reaction to occur before separation of the resulting peptides is resumed. Having a longer than normal (e.g. 5 cm as opposed to 2.5 cm) stacking gel ensures that complete stacking together of the sample and proteinase has time to occur. Since digestion only occurs when the substrate and proteinase are together in the stack, the extent of hydrolysis naturally depends upon the length of time for which the electric current is switched off, as well as upon the type and activity of the proteinase used, the ratio of proteinase to substrate and also upon the length of time the reactants are comigrating in the stack itself even when the power is on. Indeed, some research workers do not switch the power off at all, but merely migrate the components slowly through the stacking gel at a low constant current and then increase it to full power once the Bromophenol Blue reaches the end of the stacking gel.

After the peptide separation, radiolabelled peptides can be detected by autoradiography or fluorography (Chapter 1, Section 7.5.1). Unlabelled peptides are usually stained with Coomassie Blue R-250 but nowadays very often by silver staining (Chapter 1, Section 7.3), because the former is often not sufficiently sensitive to reveal a good map in many cases. Depending upon the number of peptide zones formed, 10 μg of a starting protein per sample well is really about the minimum needed to give a peptide map of reasonable intensity, but silver staining can reduce this requirement by a factor of about 100. Since its introduction in 1979 (31) for gel staining and application shortly afterwards to peptide mapping (29), many variations of silver staining have been described. By comparison with earlier methods, recent versions (e.g., ref. 32) represent both reductions in the amount of silver required (and hence in cost) and improvements in terms of sensitivity.

A typical result (*Figure 3*) shows that different proteins treated with the same proteinase give very different patterns of peptide bands. Likewise, of course, since different proteinases have different bond specificities, a series of enzymes applied to a single substrate protein also gives different peptide patterns.

3.2 Variations of the standard technique

Variations to the original methodology in which chemical cleavage as an alternative to hydrolysis with proteinases, and in which *in vitro* hydrolysis before applying samples to the mapping gel as opposed to *in situ* hydrolysis in the stacking gel itself, have already been discussed. When an SDS-PAGE slab gel is used as the mapping gel, only size differences between the peptides determine their separation. Chemical modifications that alter peptide molecular charge are without effect on the peptide band pattern as long as they do not disturb the extent of SDS binding. This means that when *in vitro* chemical or enzymic hydrolysis is used to generate the peptide mixture a pre-labelling technique can be employed. Most dye or fluorescent labelling reagents react with terminal $\alpha-NH_2$ groups and side chain $\epsilon-NH_2$ groups of lysine residues and alter the net charge on the peptide molecules, so this approach cannot be used if detergent-free PAGE or IEF is used in the subsequent separation. Most

Peptide mapping

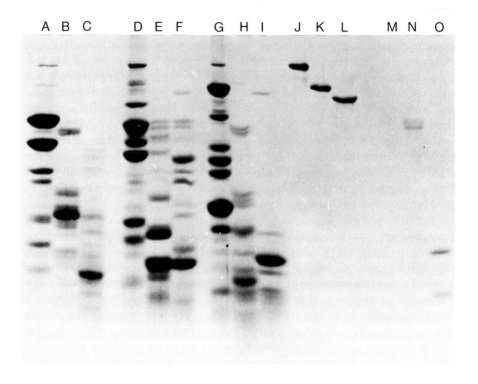

Figure 3. Peptide maps of albumin, tubulin, and alkaline phosphatase using several different proteases; proteolytic digestion allowed to occur in solution. Albumin, tubulin, and alkaline phosphatase, each at 0.67 mg/ml in sample buffer, were incubated at 37 °C for 30 min with the following proteases and then 30 μl of each sample loaded onto a 20% polyacrylamide slab gel. A to C; peptide maps of albumin, tubulin, and alkaline phosphatase after digestion with 33 μg/ml, 3.3 μg/ml, 33 μg/ml, respectively, of papain (final concentrations). D to F; peptide maps of albumin, tubulin, and alkaline phosphatase after digestion with 133 μg/ml, 67 μg/ml, 67 μg/ml, respectively, of *S. aureus* V8 protease. G to I; peptide maps of albumin, tubulin, and alkaline phosphatase after digestion with 133 μg/ml, 67 μg/ml, 67 μg/ml, respectively, of chymotrypsin. J to L; 2.5 μg each of undigested albumin, tubulin, and alkaline phosphatase, respectively. M to O; papain, *S. aureus* V8 protease, and chymotrypsin, respectively, at the highest amounts used in the digestion. (Reproduced from ref. 5 with permission.)

dye pre-labelling methods with Remazol BBR (33) or Uniblue A (34) are less sensitive than conventional post-separation staining, but if a 10 mM solution in acetone of 4-dimethylamino-azobenzene-4'-sulphonyl chloride (dabsyl chloride) is mixed with an equal volume of a 20 mg/ml peptide solution in borate buffer of pH 9 containing 5% SDS and heated at about 60°C for 5 min (35), the peptides are labelled with a strong orange colour. If this mixture is applied directly to the gel, excess reagent and byproducts can act as a suitable tracking dye and no Bromophenol Blue need be added.

Rather more sensitive than dye labelling, however, is fluorescent labelling with reagents such as dansyl chloride (1-dimethylaminonaphthalene-5-sulphonyl chloride), fluorescamine {4-phenylspiro[furan-2(3H),1′-phthalan]-3,3′-dione} or *o*-phthaldi-aldehyde (36−38). These procedures are very simple. Fluorescamine labelling, for example, is achieved by the usual heating (e.g. 100°C, 5 min) of peptide with 5% SDS in 0.15 M phosphate buffer pH 8.5, followed by cooling and the addition of a solution of fluorescamine (1 mg/ml) in acetone, using 5 μl for every 100 μl of peptide solution. After shaking briefly, a small amount of Bromophenol Blue track-ing dye (e.g. 5 μl, 5 mg/ml) is added and the sample applied directly to the sample well in the mapping gel. A very similar procedure is used with *o*-phthaldialdehyde except that a 1% solution in methanol is employed, and samples kept in the dark for 2 h before applying to the gel. With *o*-phthaldialdehyde, 2-mercaptoethanol is added to samples at the heating stage, but with fluorescamine labelling it is best added after the addition of fluorescamine.

The advantages of using a pre-labelling method are that there are no fixing and staining steps after the separation, which saves time and can be most beneficial if there are a number of small or very soluble peptides present which are poorly fixed and liable to be washed out of the gel during staining and destaining and hence may escape detection. Another major advantage is that the progress of the peptide separation can be followed during the run, visually if dye pre-labelling is used or by examining the gel with a UV lamp if fluorescent pre-labelling has been employed. The main disadvantages are that it tends to be less sensitive than post-separation staining, especially if dye pre-labelling is used, and that it can only be employed when SDS-PAGE is used to separate the peptides into a map and not with any other method.

Of much greater sensitivity is to use radioactively-labelled substrate proteins. There are many different ways of radioactive labelling (3). Proteins can be labelled *in vivo* using labelled amino acid precursors such as [³H]leucine or [³⁵S]methionine, but this often gives low specific radioactivities. Labelling of either purified proteins or of mixtures in solution is also possible using a range of procedures (see Appendix 3). Indeed, proteins can even be radiolabelled after PAGE separation without even being eluted from the gel matrix (39) but this procedure can result in high backgrounds.

It is, of course, also possible to separate unlabelled peptides into a map and then to transfer the whole pattern by capillary blotting (40) or electroblotting (41) to an immobilizing membrane for subsequent detection (Chapter 1, Section 7.10) by the usual general protein detection methods. Due to the large number of different peptides present it is not usually practical to use immunodetection methods. Nitrocellulose membranes are the least expensive and will often be satisfactory but it is worth remembering that small peptides are not strongly bound to it and may either pass through the membrane or be lost during staining. The more strongly binding nylon membranes are likely to be preferable in these cases, although methods for peptide detection are then more limited. An alternative is to transfer the intact proteins after the preliminary separation to the immobilizing nitrocellulose membrane and then to digest the bound protein while still on the membrane. The usual blotting procedures (40, 41) often involve blocking residual binding sites on the membrane with excess protein (e.g. gelatin, serum albumin) before detection of the transferred zones (Chapter 1, Section 7.10.3). These would naturally interfere with the peptide mapping, but by using radiolabelled proteins and autoradiographic detection of peptides on the final map this can be overcome. Carrey and Hardie (42) successfully used Western blotting (electroblotting) to nitrocellulose membranes, trypsin treatment of the transferred proteins on strips of membrane and IEF mapping of the peptides released (poorly bound to nitrocellulose).

A number of gel electrophoretic procedures specifically intended for peptide analysis have been introduced recently and have clear advantages over some of the earlier methods, although they have not yet been widely applied to peptide mapping studies. They include improved procedures for PAGE (43), SDS-PAGE (44, 45) and IEF in immobilized pH gradients (46).

3.3 Peptide mapping of protein mixtures

Cutting zones out of a gel in which a preliminary separation has been done and then applying them to individual wells in a peptide mapping gel slab is, in essence, almost the same as a more conventional two-dimensional scheme, with a hydrolysis of proteins to peptides in between the two stages. In fact Bordier and Crettol-Järvinen (47) and Saris *et al.* (15) carried this concept further and applied peptide mapping to a number of proteins in a whole strip of gel without isolation of the individual protein zones. First a number of proteins in a heterogeneous sample were separated in the usual way on a normal slab gel. They used SDS-PAGE for this but of course there is no reason why other techniques such as PAGE, IEF, agarose gel electrophoresis, etc., should not be used just as successfully. Once this separation has been completed, a strip of gel containing a complete lane of the separated components is cut out and the rest of the gel stained or autoradiographed in the usual way. The gel strip is then equilibrated for 30 min in about 50 ml of the same buffer as that used in the stacking gel of the peptide mapping gel (i.e. 0.125 M Tris−HCl, pH 6.8, containing 0.1% SDS in the usual Cleveland *et al.* method). This mapping gel can be either a homogeneous (e.g. $T = 15\%$) slab gel or one with an acrylamide concentration gradient, as above. The stacking gel phase is poured without any well-forming sample comb, however, and the top 2−3 cm of the gel mould should be left empty to accommodate the gel strip. The stacking gel mixture is overlaid with water, stacking gel buffer or 10% ethanol before polymerization in order to obtain a flat surface.

When the stacking gel has polymerized, the overlaying liquid is removed and the gel strip of sample proteins added across the top of the stacking gel. This is easier to do if the peptide mapping gel is slightly thicker than the strip of sample gel (e.g. 1.8−2.0 mm and 1.5 mm respectively). The former authors (47) then sealed the gel strip in place with melted 1% agarose in stacking gel buffer, taking care not to trap air bubbles. It may be easier to do this if the slab gel, mould and agarose solutions are all kept warm (e.g. $55-60°C$) to keep the agarose molten and if a little agarose is added above the polyacrylamide stacking gel before adding the sample gel strip. The sample gel strip should be covered with $2-3$ mm of the agarose solution and the whole assembly allowed to cool, upon which the agarose sets to a gel. The gel assembly is then mounted in the electrophoresis apparatus which is then filled with the appropriate reservoir buffer. Finally, 1 ml of the protease solution, as used in the usual Cleveland *et al.* method (21) in 0.125 M Tris−HCl, pH 6.8, containing 0.1% SDS, 10% glycerol and 0.001% Bromophenol Blue (i.e. stacking gel buffer plus tracking dye and glycerol to increase the density of the solution), is injected down through the buffer on to the top surface of the gel. As in the original method (21) the electrical voltage is applied and the sample and proteinase either allowed to migrate slowly through the stacking gel or, after the sample proteins and pro-teinase have stacked together and the Bromophenol Blue is close to the bottom of the stacking gel, the power is switched off for $20-30$ min. In both approaches there is time for the sample proteins to be hydrolysed to a mixture of peptides which, when electrophoresis is resumed at full power, are then separated into a series of zones characteristic of both the protein substrate and of the proteinase used. The gel running conditions, staining and destaining are performed as above.

Typical results are shown in *Figure 4* in which erythrocyte membrane components are separated into about 30 zones in the preliminary SDS-PAGE separation, most of which give lanes of separated peptide zones in the second dimension. Even with the relatively high sample loadings used here, the pattern of zones revealed by Coomassie Blue staining is faint, and silver staining could probably have been used with advantage. The poorly-resolved curve of material extending from top left to bottom right largely represents unhydrolysed sample proteins and a more extensive hydrolysis (larger proportion of proteinase or longer time in the stacking gel stage) might have been beneficial. The method does have the advantage that hitherto unsuspected homologies between different protein components in the sample may be revealed by similarities in peptide spot patterns (e.g. the two components arrowed in *Figure 4*).

3.4 Characterization of proteinases

When portions of a substrate protein solution are incubated *in vitro* with a number of different proteinases, the enzymes will only hydrolyse the protein polypeptide chain at particular points depending upon the specificity of the enzyme concerned (e.g. see *Table 2*). The resulting mixture of peptides is therefore highly reproducible and characteristic of that particular enzyme, although of course in all cases some bonds are more susceptible to attack than others and the large peptides formed initially are broken down to smaller peptides as time progresses. All these features are apparent when the resulting peptide mixtures are analysed by electrophoretic techniques to give peptide maps (e.g. *Figure 5*). Peptide maps of this type can be constructed using any one- or two-dimensional electrophoretic separation technique, but SDS-PAGE and PAGE are probably the most popular. The method is essentially the same

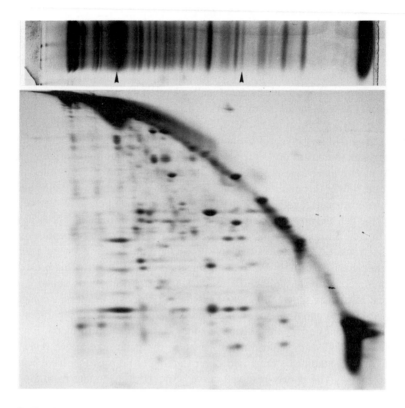

Figure 4. Peptide mapping of heterogeneous protein samples. Human erythrocyte membranes (100 μg protein) were electrophoresed on a linear 4-13% gradient slab gel and then the gel strip containing the sample was transferred to a uniform acrylamide gel (15% acrylamide) and overlayered with 1 μg of *S. aureus* V8 protease in sample buffer. The stained electrophoretogram of the same sample (50 μg) after one-dimensional electrophoresis is shown at the top of the two-dimensional gel. (Reproduced from ref. 9 with permission.)

as that described earlier for the peptide mapping of unknown purified proteins treated *in vitro* with added proteinases of known identity to generate the peptide mixture, but in this case the unknown is the proteinase and the substrate is the purified known component. The difference of course is that in the former case the peptide map is related to the amino acid sequence of the unknown substrate protein while here the peptide map is made up of peptides derived from a known protein, often with known amino acid sequence, and therefore reflects the enzymatic activity of the proteinase (i.e. variations reflect differences in bond specificity).

The activity of the unknown proteinase can also be explored by peptide mapping using what could be considered as the inverse of the Cleveland *et al.* (21) procedure. In this a proteinase is separated from other components in a preliminary electrophoretic run, for example by PAGE, IEF, etc. After the zones have separated,

Anthony T. Andrews

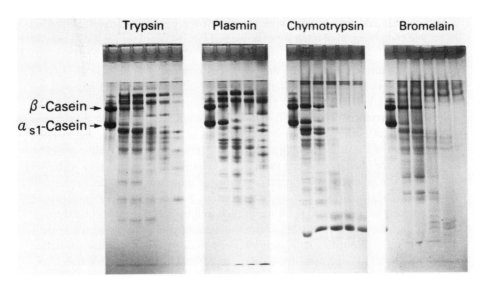

Figure 5. PAGE analysis on a T = 12.5%, C = 4% slab gel containing 4.5 M urea of the hydrolysis of total casein by a number of different proteinases. Total casein (3.0 mg/ml) precipitated isoelectrically from bovine milk was dissolved in 0.1 M sodium phosphate buffer, pH 7.0 (pH 5.0 for bromelain) and trypsin (2 μg/ml), plasmin (20 μg/ml), chymotrypsin (3 μg/ml) or bromelain (20 μg/ml) added. Incubation times, **left** to **right** were 0, 2.5, 5, 20, 60, and 240 min for trypsin; 0, 10, 60, 80 and 1200 min for plasmin; 0, 5, 12, 40, 120 and 360 min for chymotrypsin and 0, 2.5, 5, 12, 40, and 240 min for bromelain. At the end of each incubation samples were heated at 100°C for 2 min to inactivate the proteinases and urea added to 8 M and 2-mercaptoethanol to 0.1 M. Portions of 25 μl were applied to the slab gel. Note that the peptide band patterns are very different for each enzyme and that there are even marked differences between the peptide maps given by enzymes as similar as trypsin and plasmin which have very similar bond specificity (Lys – ,Arg –) but differ only in the rates at which individual bonds are cleaved. (Reproduced from ref. 5 with permission.)

the gel can sometimes be stained briefly and under non-denaturing conditions by immersion for a few seconds or minutes in 0.1% Coomassie Blue G-250 in aqueous 10% methanol, followed by transfer into water or the stacking gel buffer to be used in the peptide mapping gel. When the preliminary run has been performed at an acid pH (e.g. acid-gel PAGE or the acidic regions of an IEF gel) a 30-sec staining time is sufficient but with basic gels slightly longer is needed (e.g. 5 – 10 min). When high loadings of sample are used, as is usually required in this mapping technique, the zones are generally visible immediately so the band of interest can be easily cut out. The advantage of using mild non-denaturing conditions for both the preliminary separation and the staining step is that enzyme activity is preserved to the maximum extent. For this reason SDS-PAGE is often unsuitable for the preliminary separation, SDS being a potent denaturing agent. As an alternative to a rapid staining of this type, the guide strip method of locating enzyme bands can be used. Strips of gel

315

are cut off each side of the slab gel and stained in the conventional way (Chapter 1, Section 8.1) and then aligned with the rest of the gel so that zones containing the sample component(s) of interest can be located and cut out of the central, unstained piece of gel.

If it has not already been done, the gel pieces containing proteinase are soaked for about 30 min in the same buffer as that used in the stacking gel phase of the peptide mapping gel. The mapping gel is the same as that described above and used in the standard protocol (21) (i.e. a $T = 15\%$ gel or a gradient gel). When equilibrated with stacking gel buffer, the gel pieces are cut to an appropriate size and inserted into the sample wells of the peptide mapping gel. Any gaps around, above or below the gel pieces are filled by covering them with a few microlitres of stacking gel buffer containing 20% glycerol and the gel assembly is mounted in the electrophoretic chamber which is then filled with apparatus buffer. Portions (15 μl) of the substrate protein solution are then added to the top of each well by injecting them down through the buffer into the wells. A suitable substrate solution for most purposes will consist of $3-5$ mg ml^{-1} of purified α_{s1}-casein or β-casein dissolved in stacking gel buffer containing about 10% glycerol or sorbitol, and 0.001% Bromophenol Blue.

Interpretation of the final peptide maps is usually easier if the substrate protein is purified and itself migrates as a single band during electrophoresis, but as long as control samples receiving no proteinase treatment are run on the same slab gel for comparison this is not an essential requirement. In order to keep costs to a minimum, if care is taken a 5 mg ml^{-1} solution of unfractionated total casein (with added glycerol and Bromophenol Blue, of course) may prove to be an acceptable alternative as substrate to purified individual caseins.

Electrophoretic running conditions, staining and destaining are all performed in the usual way (see Section 3.3), with the power again being switched off for $20-30$ min when the Bromophenol Blue is close to the bottom of the gel to allow time for the proteinase(s) in the sample to digest the casein substrate to peptides which are then separated into a map.

4. Interpretation of results

It has been implicit in this discussion so far that similar band patterns produced by peptide digests of an unknown sample and a standard known protein, or between two unknowns indicates a degree of identity. Obviously if the two band patterns are identical this is strong evidence that the two proteins are identical, especially if the same result is obtained when the digests are examined on more than one peptide mapping gel run under different conditions or using different techniques. But what of the situation when some bands appear to coincide and others do not? In this case perhaps the most likely explanation is that the two proteins are related but not identical, as may occur, for example, in a precursor−product relationship or between a number of isozymes or when there are varying degrees of post-synthetic modification (e.g. glycosylation, phosphorylation, etc.). The more zones there are in common, the more closely related the two proteins are. However, it is sometimes possible for two related proteins sharing close sequence homology to give rise to very different peptide maps on peptide hydrolysis, particularly if they differ slightly in molecular weight (11). In the example shown diagrammatically in *Figure 6*, protein A is slightly larger than protein B and points X and Y represent preferred major sites of cleavage by the proteinase while 1−7 are secondary cleavage sites. If the principal cleavage occurs at X or at X and Y, then very similar peptide mixtures are formed, but if for some reason there is little or no cleavage at X while cleavage at the other sites continues, then many of the peptides formed will not coincide (e.g. 0−1, 0−2, 0−3, etc., from A but 0′−1, 0′−2, 0′−3, etc. from B). Although there are still likely to be a number of peptides that do coincide (e.g. 1−2, 1−3, 2−3, etc.) these require two cleavages at secondary sites and may be relatively minor products, so the peptide maps may look very different. Lam and Kasper (11) did in fact report that cleavage of two major nuclear envelope polypeptides with cyanogen bromide, hydroxylamine and chymotrypsin gave peptide mixtures that led to similar maps for the two proteins, but cleavage with *S. aureus* proteinase gave two very different peptide maps. Thus it may be necessary to use several methods for generating peptides to establish a sequence homology unequivocally between two proteins.

If two lanes of zones on a peptide mapping gel have some zones of apparently identical mobility and some different, the degree of similarity can be expressed quantitatively (48). For example, if the resolving power of the gel is such that bands 0.5 mm apart can be distinguished, then a gel could be considered as made up of a number of elements N, each of which may or may not contain a peptide band.

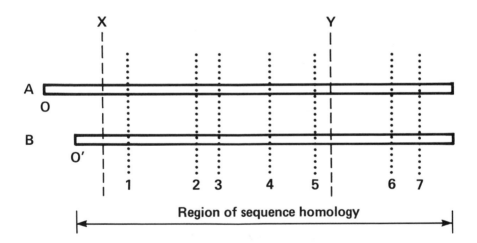

Figure 6. Hypothetical peptide mixtures that might be derived from two polypeptides, A and B, which have extensive primary amino acid sequence homology. A has a higher molecular weight than B, with an N-terminal extension to the sequence (as might occur with a signal sequence or a product − precursor relationship). If X and Y are major cleavage sites and 1 − 7 are minor sites and hydrolysis occurs at all of them, then most of the peptides generated from A will match those from B, but if little cleavage at X occurs, then many will be different.

When comparing two lanes on a peptide map containing m and n bands distributed over the N elements, the probability of a number of coincidences, x, is given by:

$$P(x) = \frac{m!n!\ (N-m)!\ (N-n)!}{N!x!\ (m-x)!\ (n-x)!\ (N-m-n+x)!}.$$

Thus, if x is large, $P(x)$ becomes very small, which means that it is very improbable that such an extent of coincidence could arise purely by chance, which suggests that the two protein sequences are homologous.

In the example (48) of two spectrin subunits digested to peptides with papain, 7-cm-long gels were used for mapping the peptides so $N = 140$ (7 cm gel; 0.5 mm resolution), $m = 23$, $n = 20$, $x = 11$ (the number of bands appearing to be identical in both maps) and therefore

$$P(x) = 1 \times 10^{-5}.$$

The usual criterion for a valid homology is $P(x) < 0.01$, so the value obtained excludes the possibility that this level of similarity in the maps occurred by chance and the two subunits are therefore at least partially homologous in amino acid sequence.

References

1. Ferguson, K. A. (1964). *Metabolism,* **13**, 985.
2. Hedrick, J. L. and Smith, A. J. (1968). *Arch. Biochem. Biophys.,* **126**, 155.
3. Andrews, A. T. (1986). *Electrophoresis: Theory, Techniques and Biochemical and Clinical Applications,* 2nd edn. Oxford University Press, Oxford.
4. James, G. T. (1980). In *Methods of Biochemical Analysis* (ed. D. Glick), Vol. 26, p. 165. J. Wiley, New York.
5. Andrews, A. T. (1984). In *Methods of Enzymatic Analysis.* (ed. H. U. Bergmeyer, J. Bergmeyer, and M. Grassl), Vol. V, p. 277. Verlag Chemie, GbmH, Weinheim.
6. Gooderham, K. (1987). *Science Tools,* **34**, 4.
7. Croft, L. R. (1980). *Handbook of Protein Sequence Analysis,* 2nd edn, p. 19. J. Wiley, Chichester, UK.
8. Stark, G. R. (1977). In *Methods in Enzymology* (ed. C. H. W. Hirs and S. N. Timasheff), Vol. 47 p. 129. Academic Press, New York.
9. Mitchell, W. M. (1977). In *Methods in Enzymology* (ed. C. H. W. Hirs and S. N. Timasheff), Vol. 47 p. 165. Academic Press, New York.
10. Croft, L. R. (1980). *Handbook of Protein Sequence Analysis,* 2nd edn, p. 9. J. Wiley, Chichester, UK.
11. Lam, K. S. and Kasper, C. B. (1980). *Anal. Biochem.,* **108**, 220.
12. Lonsdale-Eccles, J. D., Lynley, A. M., and Dale, B. A. (1981). *Biochem. J.,* **197**, 591.
13. Zingde, S. M., Shirsat, N. V., and Gothoskar, B. P. (1986). *Anal. Biochem.,* **155**, 10.
14. Mahboub, S., Richard, C., Delacourte, A., and Han, K.-K. (1986). *Anal. Biochem.,* **154**, 171.

15. Saris, C. J. M., van Eenbergen, J., Jenks, B. G., and Bloemers, H. P. J. (1983). *Anal. Biochem.*, **132**, 54.
16. Detke, S. and Keller, J. M. (1982). *J. Biol. Chem.*, **257**, 3905.
17. Lischwe, M. A. and Ochs, D. (1982). *Anal. Biochem.*, **127**, 453.
18. Sonderegger, P., Jaussi, R., Gehring, H., Brunschweiler, K., and Christen, P. (1982). *Anal. Biochem.*, **122**, 298.
19. Macleod, A. R., Wong, N. C. W., and Dixon, G. H. (1977). *Eur. J. Biochem.*, **78**, 281.
20. Rittenhouse, J. and Marcus, F. (1984). *Anal. Biochem.*, **138**, 442.
21. Cleveland, D. W., Fischer, S. G., Kirschner, M. W., and Laemmli, U. K. (1977). *J. Biol. Chem.*, **252**, 1102.
22. Herman, H., Pytela, R., Dalton, J. M., and Wiche, G. (1984). *J. Biol. Chem.*, **259**, 612.
23. Mattick, J. S., Tsukamoto, Y., Nickless, J., and Wakil, S. J. (1983). *J. Biol. Chem.*, **258**, 15291.
24. Tijssen, P. and Kurstak, E. (1983). *Anal. Biochem.*, **128**, 26.
25. Verheggen, T. P. E. M., Everaerts, F. M., and Reijenga, J. C. (1985). *J. Chromatogr.*, **320**, 99.
26. Hermann, P., Jannasch, R., and Lebl, M. (1986). *J. Chromatogr.*, **351**, 283.
27. Hancock, W. S. (1984). *CRC Handbook of HPLC for the Separation of Amino Acids, Peptides and Proteins.* Vols. I and II. CRC Press, Boca Raton, USA.
28. Hartman, P. A., Stodola, J. D., Harbour, G. C., and Hoogerheide, J. G. (1986). *J. Chromatogr.*, **360**, 385.
29. Spiker, S. (1980). *J. Chromatogr.*, **198**, 169.
30. Laemmli, U. K. (1970). *Nature*, **227**, 680.
31. Switzer, R. C., Merril, C., and Shifrin, S. (1979) *Anal. Biochem.*, **98**, 231.
32. Daverval, C., le Guilloux, M., Blaisonneau, J., and de Vienne, D. (1987). *Electrophoresis*, **8**, 158.
33. Griffith, I. P. (1972). *Anal. Biochem.*, **46**, 402.
34. Bosshard, H. F. and Datyner, A. (1977). *Anal. Biochem.*, **82**, 327.
35. Tzeng, M. C. (1983). *Anal. Biochem.*, **128**, 412.
36. Strottmann, J. M., Robinson, J. B., and Stellwagen, N. C. (1983). *Anal. Biochem.*, **132**, 334.
37. Tijssen, P. and Kurstak, E. (1979) *Anal. Biochem.*, **99**, 97.
38. Weiderkamm, E., Wallach, D. F. H., and Flückiger, R. (1973). *Anal. Biochem.*, **54**, 102.
39. Elder, J. H., Pickett, R. A., Hampton, J., and Lerner, R. A. (1977). *J. Biol. Chem.*, **252**, 6510.
40. Southern, E. M. (1975). *J. Mol. Biol.*, **98**, 503.
41. Towbin, H., Staehelin, T., and Gordon, J. (1979). *Proc. Natl. Acad. Sci. USA*, **76**, 4350.
42. Carrey, E. A. and Hardie, D. G. (1986). *Anal. Biochem.*, **158**, 431.
43. West, M. H. P., Wu, R. S., and Bonner, W. M. (1984). *Electrophoresis*, **5**, 133.
44. Hashimoto, F., Horigome, T., Kanbayashi, M., Yoshida, K., and Sugano, H. (1983). *Anal. Biochem.*, **129**, 192.
45. De Wald, D. B., Adams, L. D., and Pearson, J. D. (1986). *Anal. Biochem.*, **154**, 502.
46. Gianazza, E., Chillemi, F., Duranti, M., and Righetti, P. G. (1983). *J. Biochem. Biophys. Meth.*, **8**, 339.
47. Bordier, C. and Crettol-Järvinen, A. (1979). *J. Biol. Chem.*, **254**, 2565.
48. Calvert, R. and Gratzer, W. B. (1978). *FEBS Lett.*, **86**, 247.

A1

Suppliers of specialist items for electrophoresis

B. D. HAMES

Many of the larger companies have subsidiaries in other countries whilst most of the smaller companies market their products through agents. The name of a local supplier is most easily obtained by writing to the relevant address listed here.

Aldrich Chemical Company, The Old Brickyard, New Road, Gillingham, Dorset SP8 4JL, UK; 940 West Saint Paul Avenue, Milwaukee, WI 53233, USA.

Amersham International, Lincoln Place, Green End, Aylesbury, Bucks HP20 2TP, UK; 2636 South Clearbook Drive, Arlington Heights, IL 60005, USA.

Amicon Corporation, Upper Mill, Stonehouse, Gloucs GL0 2BJ, UK; 17 Cherry Hill Drive, Danvers, MA 01923, USA.

Anderman & Co. Ltd, 145 London Road, Kingston-upon-Thames, Surrey KT2 6NH, UK.

W & R Balston Ltd, Springfield Mill, Maidstone, Kent, UK.

BDH Chemicals Ltd, Broom Road, Poole, Dorset BH12 4NN, UK.

Bio-Rad Laboratories, Caxton Way, Watford Business Park, Watford, Herts WD1 9RP, UK; 1414 Harbor Way South, Richmond, CA 93804, USA.

Bethesda Research Laboratories (BRL), see Gibco-BRL.

Biotech Instruments Ltd, 183 Camford Way, Luton, Beds LU3 4AN, UK.

Boehringer Mannheim Biochemicals, Boehringer Mannheim House, Bell Lane, Lewes, East Sussex BN7 1LG, UK; PO Box 50816, Indianapolis, IN 46250−0816, USA; PO Box 310120, 6800 Mannheim 31, FRG.

Buchler Scientific Instruments, 1327 Sixteenth Street, Fort Lee, NJ 07024, USA.

Calbiochem Corporation, Novabiochem, UK; PO Box 12087, San Diego, CA 92112−1480, USA.

Cambridge Research Biochemicals, Button End, Harston, Cambridge CB2 5NX, UK; Suite 202, 10 East Merrick Road, Valley Stream, NY 11580, USA.

CP Laboratories, PO Box 22, Bishops Stortford, Herts CM23 3DH, UK.

Daela, DK-7171, Uldum, Denmark.

Difco Laboratories Ltd, PO Box 14B, Central Avenue, East Moseley, Surrey KT8 0SE, UK.

E-C Apparatus Corporation, 3831 Tyrone Boulevard, North St, Petersburg, FL 33709, USA.

Enzymes Systems Products, PO Box 2033, Livermore, California, CA 94550, USA.

Fisher Scientific, 52 Fadem Road, Springfield, NJ 07081, USA.

Fluka Chemical Corp., 980 South Second Street, Ronkonkoma, NY 11779, USA; Industriestrasse 25, CH-9470, Buchs, Switzerland.

FMC Corp., PO Box 308, 5 Maple Street, Rockland, ME 04841, USA.

Fuji Photo Film Co. Ltd, Fuji Photo Film (UK) Ltd, Cresta House, 125 Finchley Road, London NW3 6JH, UK; Chemical Products Dept. 26−30, Nishiazabu, 2-chome, Minato-ku, Tokyo 106, Japan.

Gelman Sciences Inc., 12 Peter Road, Lancing, Sussex, UK; 600 South Wagner Road, Ann Arbor, MI 48106, USA.

Genzyme Corporation, 75 Kneeland Street, Boston, MA 02111, USA.

Gibco-BRL, PO Box 35, 3 Washington Road, Sandyford Industrial Estate, Paisley, Renfrewshire PA3 4EP, UK; 8400 Helgerman Court, Gaithersburg, MD 20877, USA.

V.A.Howe & Co. Ltd, 12−14 St Ann's Crescent, London SW18 2LS, UK.

Hamilton, V.A.Howe & Co. Ltd, UK; PO Box 10030, Reno, NV 89510, USA.

Hamilton Bonaduz AG, PO Box 26, CH-7402 Bonaduz, Switzerland.

Hanimex (UK), Faraday Road, Dorcan, Swindon, Wiltshire SN3 5HW, UK.

Hoefer Scientific Instruments, PO Box 351 Newcastle-under-Lyme, Staffs ST5 0TW, UK; 65 Minnesota Street, PO Box 77387, San Francisco, CA 94107, USA.

ICN Biomedicals Ltd, Freepress House, Castle Street, High Wycombe, Bucks HP13 6RN, UK.

Ilford Ltd, Town Lane, Mobberley, Knutsford, Cheshire WA16 7HA, UK.

Instrumentation Specialities Co. (ISCO), Life Science Laboratories Ltd, UK; PO Box 5347, Lincoln, NE 68505, USA.

Jannsen-Pharmaceutical, ICN Biomedicals Ltd, UK; Jannsen Biotech NV, Lammerdries 55, B-2430 Olen, Belgium.

Joyce-Loebl Ltd, 48 Princes Way, Team Valley, Gateshead NE11 0UJ, UK.

Kem-En-Tec, 2100 Copenhagen, Denmark.

Koch Light Ltd, Rookwood Way, Haverhill, Suffolk CB9 8PB, UK.

Kodak, Kodak Ltd, PO Box 33, Swallowdale Lane, Hemel Hempstead, Herts HP2 7EU, UK; Eastman Kodak Co., 343 State Street, Rochester, NY 14650, USA.

Life Science Laboratories, Sarum Road, Luton, Beds LU3 2RA,UK.

LKB, *see* Pharmacia-LKB Biotechnology.

Marine Colloids Division, FMC Corporation, PO Box 308, 5 Maple Street, Rockland, MA 04841, USA.

E. Merck, Frankfurter Strasse 250, D-6100 Darmstadt 1, FRG.

Miles Research Products Division, Miles Laboratories Ltd, ICN Biomedicals Ltd, UK; PO Box 70, Elkhart, IN 46514, USA.

Millipore Corporation, 11–15 Peterborough Road, Harrow, Middlesex HA1 2YH, UK; 80 Ashby Road, Bedford, MA 01730, USA.

National Technical Information Service (NTIS), Springfield, VA 22161, USA.

New England Biolabs, Inc., CP Laboratories, UK; 32 Tozer Road, Beverly, MA 01915–5510, USA; Postfach 2750, 6231 Schwalbach bei Frankfurt, FRG.

New England Nuclear (NEN) Research Products, NEN Chemicals GmBH, Postfach 401240, 6072 Dreieich, FRG; NEN Corporation, 549 Albany Street, Boston, MA 02118, USA.

Nordic Immunological Laboratories, Dairy House, Moneyrow Green, Holyport, Maidenhead, Berkshire, UK.

Novabiochem, 3 Heathcoat Building, Highfields Science Park, University Boulevard, Nottingham NG7 2QJ, UK.

Oxoid Ltd, Wade Road, Basingstoke, Hants RG24 0PW, UK.

Pierce, Pierce Europe BV, PO Box 1512, 3260 BA Oud-Beijerland, The Netherlands; PO Box 117, Rockford, IL 61105, USA.

Pharmacia LKB Biotechnology, Pharmacia House, Midsummer Boulevard, Milton Keynes MK9 3HP, UK; 800 Centennial Avenue, Piscataway, NJ 08854, USA; S-75182 Uppsala, Sweden.

PL Biochemical, *see* Pharmacia-LKB Biotechnology.

Polaroid, Ashley Road, St Albans, Herts AL1 5PR, UK.

Polysciences, 24 Low Farm Place, Moulton Park, Northampton NN3 1HY, UK; Paul Valley Industrial Park, Warrington, PA 18976, USA.

Sartorius GmbH, Weender Landstrasse 94/108, PO Box 3243, D-3400 Göttingen, FRG.

Schleicher & Schuell, Inc., Anderman & Co. Ltd, UK.

Schleicher & Schüll GmbH, 10 Optical Avenue, Keene, NH 03431,USA; D-3354 Dassel, FRG.

Schwartz Mann Division, Becton Dickinson Corp., Mountain View, CA 940213, USA.

Serva Fine Biochemicals, Cambridge Bioscience, 42 Devonshire Road, Cambridge CB1 2BL, UK; PO Box A, 18 Villa Place, Garden City Park, NY 11040, USA.

Serva Fenbiochemica GmbH, PO Box 105260, Carl-Benz-Strasse 7, D-6900 Heidelberg, FRG.

Shandon Southern, 93–96 Chadwick Road, Astmoor Industrial Estate, Runcorn, Cheshire WA7 1PR, UK.

Sigma Chemical Corporation, Fancy Road, Poole, Dorset BH17 7NH, UK; PO Box 14508, 3500 DeKalb St, St Louis, MO 63178, USA.

Union Carbide, Union Carbide Ltd, Peter House, Oxford Street, Manchester 1, UK; Union Carbide Corporation, 6733 West 65th Street, Chicago, IL 60632, USA.

Universal Scientific Ltd, 9 The Broadway, Woodford Green, Essex 1G8 0H2, UK.

Wellcome Reagents Ltd, 303 Hithergreen Lane, London SE13 6TL, UK.

Whatman Biosystems Ltd, Springfield Mill, Maidstone, Kent ME14 2LE, UK; 9 Bridewell Place, Clifton, NJ 07014, USA.

Bibliography of polypeptide detection methods

B. D. HAMES

1. Detection of proteins *in situ* in gels

1.1 General polypeptide detection methods

1.1.1 Organic dyes

Comparison of a range of organic dyes:

Wilson, C. M. (1983). In *Methods in Enzymology,* (ed. C. H. W. Hirs and S. N. Timasheff), Vol. 91, p. 236. Academic Press, New York.

Merril, C. R., Harasewych, M. G., and Harrington, M. G. (1986). In *Gel Electrophoresis of Proteins,* (ed. M. J. Dunn), Wright, Bristol.

Wilson, C. M. (1979). *Anal. Biochem.,* **96**, 263.

Coomassie Blue R-250 in methanol-acetic acid or methanol TCA:

Righetti, P. G. (1983). In *Laboratory Techniques in Biochemistry and Molecular Biology,* (ed. T. S. Work and E. Work), Vol. 11, p. 148. North-Holland, Amsterdam.

Wilson, C. M. (1979). *Anal. Biochem.,* **96**, 263.

Weber, K. and Osborn, M. (1969). *J. Biol. Chem.,* **244**, 4406.

Coomassie Blue R-250 in isopropanol-acetic acid:

Fairbanks, G., Steck, T. L. and Wallach, D. F. H. (1971). *Biochemistry,* **10**, 2606.

Rapid acid-based Coomassie Blue stains requiring no destaining:

Reisner, A. H. (1984). In *Methods in Enzymology,* (ed. W. B. Jakoby), Vol. 104, p. 439. Academic Press, New York.

Andrews, A. T. (1986). *Electrophoresis: Theory, Techniques and Biochemical and Clinical Applications,* p. 30. Clarendon Press, Oxford.

Wilson, C. M. (1983). In *Methods in Enzymology,* (ed. C. H. W. Hirs and S. N. Timasheff), Vol. 91, p. 236. Academic Press, New York.

Blakesley, R. W. and Boezi, J. A. (1977). *Anal. Biochem.,* **82**, 580.

Chrambach, A., Reisfeld, R. A., Wyckoff, M., and Zaccari, J. (1967). *Anal. Biochem.,* **20**, 150.

Diezel, W., Kopperschläger, G., and Hoffman, E. (1972). *Anal. Biochem.,* **48**, 617.

Malik, N. and Berrie, A. (1972). *Anal. Biochem.,* **49**, 173.

Reisner, A. H., Nemes, P., and Bucholtz, C. (1975). *Anal. Biochem.,* **64**, 509.

Appendix 2

Staining proteins using Fast green:

Allen, R. E., Masak, K. C., and McAllister, P. K. (1980). *Anal. Biochem.*, **104**, 494.

Wilson, C. M. (1983). In *Methods in Enzymology*, (ed. C. H. W. Hirs and S. N. Timasheff), Vol. 91, p. 236. Academic Press, New York.

Staining proteins with Amido Black:

Wilson, C. M. (1979). *Anal. Biochem.*, **96**, 263.

Wilson, C. M. (1983). In *Methods in Enzymology*, (ed. C. H. W. Hirs and S. N. Timasheff), Vol. 91, p. 236. Academic Press, New York.

McMaster-Kaye, R. and Kaye, J. S. (1974). *Anal. Biochem.*, **61**, 120.

Staining proteins prior to electrophoresis:

Bosshard, H. F. and Datyner, A. (1977). *Anal. Biochem.*, **82**, 327.

Sun, S. M. and Hall, T. C. (1974). *Anal. Biochem.*, **61**, 237.

Schägger H, Aquila, H., and Van Jagow, G (1988). *Anal. Biochem.*, **173**, 201.

1.1.2 Silver stains

(see also Chapter 1, Section 7.3)

Reviews:

Dunn, M. J. and Burghes, H. M. (1983). *Electrophoresis*, **4**, 173.

Merril, C. R., Goldman, D., and Van Keuren, M. L. (1984). In *Methods in Enzymology*, (ed. W. B. Jakoby), Vol. 104, p. 441. Academic Press, New York.

A selection of methods:

Oakley, B. R., Kirsch, D. R., and Morris, N. R. (1980). *Anal. Biochem.*, **105**, 361.

Switzer, R. C., Merril, C. R., and Shifrin, S. (1979). *Anal. Biochem.*, **98**, 231.

Sammons, D. W., Adams, L. D., and Nishizawa, E. E. (1981). *Electrophoresis*, **2**, 135.

Morrissey, J. H. (1981). *Anal. Biochem.*, **117**, 307.

Wray, W., Bonlikas, T., Wray, V. P., and Hancock, R. (1981). *Anal. Biochem.*, **118**, 197.

Ohsawa, K. and Ebata, N. (1983). *Anal. Biochem.*, **98**, 231.

Potential artefacts:

Marshall, T. and Williams, K. M. (1983). *Anal. Biochem.*, **139**, 502.

Ochs, D. (1983). *Anal. Biochem.*, **135**, 470.

Merril, C. R., Switzer, R. C., and Van Keuren, M. C. (1979). *Proc. Natl. Acad. Sci. USA*, **76**, 4335.

Hallinan, F. U. (1983). *Electrophoresis*, **4**, 265.

1.1.3 Labelling with fluorophore prior to electrophoresis

Dansyl chloride:

Schetters, H. and McLeod, B. (1979). *Anal. Biochem.*, **98**, 329.

Stephens, R. E. (1975). *Anal. Biochem.*, **65**, 369.

Tjissen, P. and Kurstak, E. (1979). *Anal. Biochem.*, **99**, 97.

Fluorescamine:

Douglas, S. A., La Marca, M. E., and Mets, L. J. (1978). In *Electrophoresis '78*, (ed. N. Catsimpoolas), Vol. 2, p. 155. Elsevier/North-Holland, Amsterdam.

Eng, P. R. and Parker, C. O. (1974). *Anal. Biochem.*, **59**, 323.

Ragland, W. L., Benton, T. L., Pace, J. L., Beach, F. G., and Wade, A. E. (1978). In *Electrophoresis '78,* (ed. N. Catsimpoolas), Vol. 2, p. 217. Elsevier/North-Holland, Amsterdam.

Ragland, W. L., Pace, J. L., and Kemper, D. L. (1974). *Anal. Biochem.,* **59**, 24.

MDPF [2-methoxy-2, 4-diphenyl-3(2H)-furanone]:

Barger, B. O., White, F. C., Pace, J. L., Kemper, D. L., and Ragland, W. L. (1976). *Anal. Biochem.,* **70**, 327.

Douglas, S. A. *et al.* (1978). In *Electrophoresis '78,* (ed. N. Catsimpoolas), Vol. 2, p. 155. Elsevier/North-Holland, Amsterdam.

Ragland, W. L. *et al.* (1978). In *Electrophoresis '78,* (ed. N. Catsimpoolas), Vol. 2, p. 155. Elsevier/North-Holland, Amsterdam.

DACM [*N*-(7-dimethylamino-4-methylcoumarinyl) maleimide]:

Yamamoto, K., Okamoto, Y., and Sekine, T. (1978). *Anal. Biochem.,* **84**, 313.

o-**Phthaldialdehyde:**

Weidekamm, E., Wallach, D. F. H., and Flückiger, R. (1973). *Anal. Biochem.,* **54**, 102.

1.1.4 Labelling with fluorophore after electrophoresis

Anilinonaphthalene sulphone (ANS):

Hartman, B. K. and Udenfriend, S. (1969). *Anal. Biochem.,* **30**, 391.

Bis-ANS:

Harowitz, P. M. and Bowman, S. (1987). *Anal. Biochem.,* **165**, 430.

Fluorescamine:

Jackowski, G. and Liew, C. C. (1980). *Anal. Biochem.,* **102**, 34.

p-**Hydrazinoacridine:**

Carson, S. D. (1977). *Anal. Biochem.,* **78**, 428.

o-**Phthaldialdehyde:**

Liebowitz, M. J. and Wang, R. W. (1984). *Anal. Biochem.,* **137**, 161.

Andrews, A. T. (1986). *Electrophoresis: Theory, Techniques and Biochemical and Clinical Applications,* p. 29. Clarendon Press, Oxford.

1.1.5 Direct detection methods

Via protein phosphorescence:

Mardian, J. K. W. and Isenberg, I. (1978). *Anal. Biochem.,* **91**, 1.

Detection of SDS−polypeptides by chilling:

Wallace, R. W., Yu, P. H., Dieckart, J. P., and Dieckart, J. W. (1974). *Anal. Biochem.,* **61**, 86.

Detection of SDS−polypeptides by precipitation with K⁺ ions:

Nelles, L. P. and Bamburg, J. R. (1976). *Anal. Biochem.,* **73**, 522.

Detection of SDS−polypeptides using sodium acetate:

Higgins, R. C. and Dahmus, M. E. (1979). *Anal. Biochem.,* **93**, 257.

Detection of SDS−polypeptides by reaction with cationic surfactant:

Tagi, T., Kubo, K., and Isemura, T. (1977). *Anal. Biochem.,* **79**, 104.

Appendix 2

Pre-labelling with fluorescent molecules: see Section 1.1.3.

Pre-staining: see Section 1.1.1.

1.2 Detection of radioactive proteins

Radiolabelling proteins *in vivo* prior to electrophoresis:
Dunbar, B. S. (1987). *Two Dimensional Electrophoresis and Immunological Techniques,*
  p. 103. Plenum Press, New York.
Latter, G. I., Burbeck, S., Fleming, S., and Leavitt, J. (1984). *Clin. Chem.*, **30**, 1925.

Radiolabelling proteins *in vitro* prior to electrophoresis:
Dunbar, B. S. (1987). *Two Dimensional Electrophoresis and Immunological Techniques,*
  p. 103. Plenum Press, New York.
Also see Appendix 3.

Radiolabelling proteins after gel electrophoresis:
Christopher, A. R., Nagpal, M. L., Carrol, A. R., and Brown, J. C. (1978). *Anal. Biochem.*,
  85, 404.
Elder, J. H., Pickelt, R. A., Hampton, J., and Lerner, R. A. (1977). *J. Biol. Chem.*, **252**, 6510.
Zapolski, E. J., Gersten, D. M., and Ledley, R. S. (1982). *Anal. Biochem.*, **123**, 325.

Use of radiolabelled probes:
Radiolabelled antibody or protein A; see Section 1.3.
Radiolabelled lectins; see Section 1.4.1.

Indirect autoradiography (^{125}I, ^{32}P) using an X-ray intensifying screen:
Laskey,R.A. (1980). In *Methods in Enzymology,* (ed. L. Grossman and K. Moldave), Vol.
  65, p. 363. Academic Press, New York.
Laskey, R. A. and Mills, A. D. (1977). *FEBS Lett.*, **82**, 314.
Bonner, W. M. (1983). In *Methods in Enzymology,* (ed. S. Fleischer and B. Fleischer), Vol.
  96, p. 215. Academic Press, New York.

Fluorography (^{35}S, ^{14}C, ^{3}H) using PPO in DMSO:
Bonner, W. M. and Laskey, R. A. (1974). *Eur. J. Biochem.*, **46**, 83.
Laskey, R. A. (1980). In *Methods in Enzymology,* (ed. L. Grossman and K. Moldave), Vol.
  65, p. 363. Academic Press, New York.
Laskey, R. A. and Mills, A. D. (1975). *Eur. J. Biochem.*, **56**, 335.
Bonner, W. M. (1984). In *Methods in Enzymology,* (ed. W. B. Jakoby), Vol. 104, p. 461.
  Academic Press, New York.

Fluorography using PPO in glacial acetic acid:
Skinner, K. and Griswold, M. D. (1983). *Biochem. J.*, **209**, 281.

Fluorography using sodium salicylate in polyacrylamide gels:
Chamberlain, J. P. (1979). *Anal. Biochem.*, **98**, 132.
Bonner, W. M. (1984). In *Methods in Enzymology,* (ed. W. B. Jakoby), Vol. 104, p. 461.
  Academic Press, New York.

Fluorography using sodium salicylate in agarose gels:
Heegard, N. H. H., Hebsgaard, K. P., and Bjerrum, O. J. (1984). *Electrophoresis,* **5**, 230.

Fluorography using commercial reagents:
Roberts, P. L. (1985). *Anal. Biochem.,* **147**, 521.
McConkey, E. H. and Anderson, C. (1984). *Electrophoresis,* **5**, 230.

Quenching of radiolabelled proteins by gel conditions:
Harding, C. R. and Scott, I. R. (1983). *Anal. Biochem.,* **129**, 371.
Van Keuran, M. L., Goldman, D., and Merril, C. R. (1981). *Anal. Biochem.,* **116**, 248.

Image intensification:
Rigby, P. J. W. (1981). *Amersham Research News No. 13.*
Laskey, R. A. (1981). *Amersham Research News No. 23.*
Also see Chapter 1, Section 7.5.1.

Double-label detection using X-ray film:
Gruenstein, E. I. and Pollard, A. L. (1976). *Anal. Biochem.,* **76**, 452.
Kronenberg, L. H. (1979). *Anal. Biochem.,* **93**, 189.
McConkey, E. H. (1979). *Anal. Biochem.,* **96**, 39.
Walton, K. E., Styer, D., and Gruenstein, E. (1979). *J. Biol. Chem.,* **254**, 795.
Cooper, P. C. and Burgess, A. W. (1982). *Anal. Biochem.,* **126**, 301.

Electronic data capture:
Davidson, J. B. and Case, A. (1982). *Science,* **215**, 1398.
Burbeck, S. (1983). *Electrophoresis,* **4**, 127.

Gel slicing and counting methods:
See Chapter 1, Section 7.5.2.

1.3 Immunological methods

The most common approach at present for the analysis of protein antigens separated by gel electrophoresis is to transfer these to a filter matrix by electroblotting and then probe with appropriate antibodies ('immunoblotting'). This procedure is described in Chapter 1, Section 7.10 with additional references in this Appendix, Section 2. Only methods for the direct detection of antigens in gels are listed below.

Incubation of the gel with radiolabelled antibody:
Burridge, K. (1978). In *Methods in Enzymology,* (ed. V. Ginsburg), Vol. 50, p. 54. Academic Press, New York.
Kasamatsu, H. and Flory, P. J. (1978). *Virology,* **86**, 344.

Incubation of the gel with unlabelled antibody then with [^{125}I]-protein A:
Burridge, K. (1978). In *Methods in Enzymology,* (ed. V. Ginsburg), Vol. 50, p. 54. Academic Press, New York.
Adair, W. S., Jurivich, D., and Goodenough, U. W. (1978). *J. Cell Biol.,* **79**, 281.
Bigelis, R. and Burridge, K. (1978). *Biochem. Biophys. Res. Commun.,* **82**, 322.
Saltzgaber-Müller, J. and Schatz, G. (1978). *J. Biol. Chem.,* **253**, 305.

Incubation of the gel with fluorescein-labelled antibody:
Groschel-Stewart, U., Schreiber, J., Mahlmeister, C., and Weber, K. (1976). *Histochemistry,* **46**, 229.
Stumph, W. E., Elgin, S. C. R., and Hood, L. (1974). *J. Immunol.,* **113**, 1752.

Incubation of the gel with antibody coupled to peroxidase followed by localization with 3,3'-diaminobenzidine:
Olden, K. and Yamada, K. M. (1977). *Anal. Biochem.,* **78**, 483.

Parish, R. W., Schmidlin, S., and Parish, C. R. (1978). *FEBS Lett.*, **95**, 366.
Van Raamsdonk, W., Pool, C. W., and Heyting, C. (1977). *J. Immunol. Methods*, **17**, 337.

Use of an agarose overlay containing antiserum to produce an immunoreplica:
Showe, M. K., Isobe, E., and Onorato, L. (1976). *J. Mol. Biol.*, **107**, 55.

1.4 Detection of specific classes of proteins

1.4.1 Glycoproteins

Periodic Acid–Schiff (PAS):

Using dansyl hydrazine on polyacrylamide gels:
Eckhardt, A. E., Hayes, C. E., and Goldstein, I. E. (1976). *Anal. Biochem.*, **73**, 192.
Gander, J. E. (1984). In *Methods in Enzymology*, (ed. W. B. Jakoby), Vol. 104, p. 447. Academic Press, New York.

Using dansyl hydrazine on agarose gels:
Furlan, M., Perret, B. A., and Beck, E. A. (1979). *Anal. Biochem.*, **96**, 208.

Using fuchsin:
Fairbanks, G., Steck, T. L., and Wallach, D. L. H. (1971). *Biochemistry*, **10**, 2026.
Zaccharias, R. J., Zell, T. E., Morrison, J. H., and Woodlock, J. J. (1969). *Anal. Biochem.*, **31**, 148.

Using Alcian Blue:
Wardi, A. H. and Michos, G. A. (1972). *Anal. Biochem.*, **49**, 607.

Thymol-sulphuric acid method:
Rauchsen, D. (1979). *Anal. Biochem.*, **99**, 474.
Gander, J. E. (1984). In *Methods in Enzymology*, (ed. W. B. Jakoby), Vol. 104, p. 447. Academic Press, New York.

Periodic acid–silver stain (most sensitive procedure):
Dubray, G. and Bezard, G. (1982). *Anal. Biochem.*, **119**, 325.

P-**Hydrazino-acridine** (fluorescent dye):
Carson, S. D. (1977). *Anal. Biochem.*, **78**, 428.

Stains-all for sialoglycoproteins:
Green, M. R. and Pastewka, J. V. (1975). *Anal. Biochem.*, **65**, 66.
King, L. E. and Morrison, M. (1976). *Anal. Biochem.*, **71**, 223.

Fluorescent lectins:
Furlan, M., Perret, B. A., and Beck, E. A. (1979). *Anal. Biochem.*, **96**, 208.
Cotrufo, R., Monsurro, M. R., Delfino, G., and Geraci, G. (1983). *Anal. Biochem.*, **134**, 313.
Gander, J. E. (1984). In *Methods in Enzymology*, (ed. W. B. Jakoby), Vol. 104, p. 447. Academic Press, New York.

Lectins with covalently-bound enzymes:
Avigad, G. (1978). *Anal. Biochem.*, **86**, 443.
Wood, J. G. and Sarinana, F. O. (1971). *Anal. Biochem.*, **69**, 320.
Moroi, M. and Jung, S. M. (1984). *Biochem. Biophys. Acta*, **798**, 295.

Radiolabelled lectins:
Burridge, K. (1978). In *Methods in Enzymology*, (ed. V. Ginsburg), Vol. 50, p. 54. Academic Press, New York.

Rostas, J. A. P., Kelley, P. T., and Cotman, C. W. (1977). *Anal. Biochem.*, **80**, 366.
Koch, G. L. E. and Smith, M. J. (1982). *Eur. J. Biochem.*, **128**, 107.
Gershoni, J. M. and Palade, G. (1982). *Anal. Biochem.*, **124**, 396.
Dupuis, G. and Doucet, J. P. (1981). *Biochim. Biophys. Acta*, **669**, 171.

Crossed lectin electrophoresis:
West, C. M. and McMahon, D. (1977). *J. Cell Biol.*, **74**, 264.

Labelling of cell surface glycoproteins using galactose oxidase:
Gahmberg, C. G. (1978). In *Methods in Enzymology*, (ed. V. Ginsburg), Vol. 50, p. 204. Academic Press, New York.

Labelling of glycoproteins containing terminal *N*-acetylglucosamine using galactosyl transferase:
Wallenfels, B. (1979). *Proc. Natl. Acad. Sci. USA*, **76**, 3223.

Labelling glycoproteins *in vivo* using radiolabelled sugars:
[^{3}H]glucosamine;
e.g. Taylor, T. and Weintraub, B. D. (1985). *Endocrinology*, **116**, 1968.
[^{3}H]fucose;
e.g. Gregg, J. H. and Karp, G. C. (1978). *Exp. Cell Res.*, **112**, 31.
[^{3}H]mannose;
e.g. Bradshaw, J. P. and White, P. A. (1985). *Biosci. Rep.*, **5**, 229.

1.4.2 Phosphoproteins

Entrapment of liberated phosphate (ELP) method using methyl green:
Cutting, J. A. and Roth, T. F. (1973). *Anal. Biochem.*, **54**, 386.
Cutting, J. A. (1984). In *Methods in Enzymology*, (ed. W. B. Jakoby), Vol. 104, p. 451. Academic Press, New York.

ELP method using rhodamine B:
Debruyne, I. (1983). *Anal. Biochem.*, **133**, 110.

Stains-all:
Green, M. R., Pastewka, J. V., and Peacock, A. C. (1973). *Anal. Biochem.*, **56**, 43.

Silver stain:
Satoh, K. and Busch, H. (1981). *Cell Biol. Int. Rep.*, **5**, 857.

Using electroblotting:
Cantor, L., Lamy, F., and Lecocq,R.E. (1987). *Anal. Biochem.*, **160**, 414.

Radiolabelling: see Section 7.7.

Trivalent metal chelation for acidic phosphoproteins (phosvitins):
Hagenauer, J., Ripley, L., and Nace, G. (1977). *Anal. Biochem.*, **78**, 308.
Cutting, J. A. (1984). In *Methods in Enzymology*, (ed. W. B. Jakoby), Vol. 104, p. 451. Academic Press, New York.

1.4.3 Lipoproteins

Staining before electrophoresis:
Ressler, N., Springgate, R., and Kaufman, J. (1961). *J. Chromatogr.*, **6**, 409.

Staining after electrophoresis:

Prat, J. P., Lamy, J. N., and Weill, J. D. (1969). *Bull. Soc. Chim. Biol.*, **51**, 1367.

Silver-staining:

Tsai, C. M. and Frasch, C. E. (1982). *Anal. Biochem.*, **119**, 115.

Goldman, D., Merril, C. R., and Ebert, M. H. (1980). *Clin. Chem.*, **26**, 1317.

1.4.4 Miscellaneous proteins

Proteins with available thiol groups:

DACM [N-(7-dimethylamino-4-methylcoumarinyl)maleimide]:

Yamamoto, K., Okamoto, Y., and Sekine, T. (1978). *Anal. Biochem.*, **84**, 313.

DTNB [5,5'-dithiobis(2-nitrobenzoic acid)]:

Zelazowski, A. J. (1980). *Anal. Biochem.*, **103**, 307.

Cadmium-containing proteins using dipyridyl-ferrous iodide:

Zelazowski, A. J. (1980). *Anal. Biochem.*, **103**, 307.

1.5 Detection of specific enzymes

A large bibliography of enzyme detection methods is presented in Section 3 of this Appendix.

2. Detection of polypeptides on electroblots

Only selected references can be cited here. The interested reader is referred to the reviews listed for a more comprehensive literature and Chapter 1, Section 7.10.3.

2.1 Reviews

Langone, J. J. (1982). *J. Immunol. Methods*, **55**, 277.

Gershoni, J. M. and Palade, G. E. (1983). *Anal. Biochem.*, **131**, 1.

Towbin, H. and Gordon, J. (1984). *J. Immunol. Methods*, **72**, 313.

Symington, J. (1984). In *Two Dimensional Electrophoresis of Proteins: Methods and Applications*, (ed. J. E. Celis and R. Bravo), p. 127. Academic Press, New York.

Bers, G. and Garfin, D. (1985). *Bio Techniques*, **3**, 276.

Merril, C. R., Harasewych, M. G. and Harrington, M. G. (1986). In *Gel Electrophoresis of Proteins*, (ed. M. J. Dunn), p. 313. Wright, Bristol.

Soutar, A. K. and Wade, D. P. (1989). In *Protein Function: A Practical Approach*, (ed. T. E. Creighton), p. 55. Oxford University Press, Oxford.

2.2 General protein detection methods

Amido Black:

Towbin, H., Staehelin, T., and Gordon, J. (1979). *Proc. Natl. Acad. Sci. USA*, **76**, 4350.

Gershoni, J. M. and Palade, G. E. (1983). *Anal. Biochem.*, **131**, 1.

Soutar, A. K. and Wade, D. P. (1989). In *Protein Function: A Practical Approach*. (ed. T. E. Creighton), p. 55. Oxford University Press, Oxford.

Coomassie Blue:

Burnette, W. N. (1981). *Anal. Biochem.*, **112**, 195.

Fast Green:

Reinhart, M. P. and Malmud, D. (1982). *Anal. Biochem.*, **123**, 229.

Indian ink:

Hancock, K. and Tsang, V. C. W. (1983). *Anal. Biochem.*, **133**, 157.

Silver staining:

Yuen, K. C. C., Johnson, T. K., Dennell, R. E., and Consigli, R. A. (1982). *Anal. Biochem.*, **126**, 398.

Brada, D. and Roth, J. (1984). *Anal. Biochem.*, **142**, 79.

Merril, C. R., Harrington, M., and Alley, V. (1984). *Electrophoresis*, **5**, 289.

Using colloidal gold:

Rohringer, R. and Holden, D. W. (1985). *Anal. Biochem.*, **144**, 118.

Using colloidal iron:

Moeremans, M., Daneels, G., De Raeymaeker, M., and De May, J. (1986). *Electrophoresis '86,* (ed. M. J. Dunn), p. 328. VCH Verlagsgesellschaft mbH.

Protein tagging with dinitrophenol (DNP):

Wotjkowiak, Z., Briggs, R. C. and Hnilica, L. S. (1983). *Anal. Biochem.*, **129**, 486.

Protein tagging with pyridoxal phosphate:

Kittler, J. M., Meisler, N. T., Vicepts-Madore, D., Cidlowski, J. A., and Thomassi, J. W. (1984). *Anal. Biochem.*, **137**, 210.

Protein tagging with biotin:

Bio Radiations (1985). from Bio-Rad Laboratories Ltd, No. 56 EG.

2.3 Detection of radioactive proteins

Roberts, P. L. (1985). *Anal. Biochem.*, **147**, 521.

Towbin, H., Staehelin, T., and Gordon, J. (1979). *Proc. Natl. Acad. Sci. USA,* **76**, 4350.

2.4 Detection of glycoproteins

General procedures:

Gershoni, J. M., Bayer, E. A., and Wilchek, M. (1985). *Anal. Biochem.*, **146**, 59.

Keren, Z., Berke, G., and Gershoni, J. M. (1986). *Anal. Biochem.*, **155**, 182.

Using radiolabelled lectins:

Gershoni, J. M. and Palade, G. E. (1982). *Anal. Biochem.*, **124**, 396.

Dunbar, B. S. (1987). *Two Dimensional Electrophoresis and Immunological Techniques,* p. 345. Plenum Press, New York.

Using peroxidase-labelled lectins:

Moroi, M. and Jung, S. M. (1984). *Biochim. Biophys. Acta.,* **798**, 295.

Using anti-lectin antibody:

Glass, W. F., Briggs, R. C., and Hnilica, L. S. (1981). *Anal. Biochem.*, **115**, 219.

2.5 Detection of lipoproteins

e.g. Bradbury, W. C., Mills, S. D., Preston, M. A., Barton, L. J., and Penner, J. L. (1984). *Anal. Biochem.*, **137**, 129.

2.6 Immunological detection methods

Using radiolabelled antibody:

Towbin, H., Staehelin, T., and Gordon, J. (1979). *Proc. Natl. Acad. Sci. USA,* **76**, 4350.

Dunbar, B. S. (1987). *Two Dimensional Electrophoresis and Immunological Techniques,* p. 341. Plenum Press, New York.

Using fluorescent labelled antibody:

FITC:

Towbin, H., Staehelin, T., and Gordon, J. (1979). *Proc. Natl. Acad. Sci. USA,* **76**, 4350.

The, T. H. and Feltkamp, T. E. W. (1970). *Immunology,* **18**, 865.

MDPF:

Weigele, M., De Bernado, S., Leimgruber, W., Cleeland, R., and Grunber, E. (1973). *Biochem. Biophys. Res. Commun.,* **54**, 899.

Rhodamine:

Towbin, H., Staehelin, T., and Gordon, J. (1979). *Proc. Natl. Acad. Sci. USA,* **76**, 4350.

Using enzyme-conjugated antibody:

Blake, M. S., Johnson, K. H., Russel-Jones, G. J., and Gotschlich, E. C. (1984). *Anal. Biochem.,* **136**, 175.

Towbin, H. and Gordon, J. (1984). *J. Immunol. Methods,* **72**, 313.

Knecht, D. A. and Dimond, R. L. (1984). *Anal. Biochem.,* **136**, 180.

Using biotin-conjugated antibody:

Dunbar, B. S. (1987). *Two Dimensional Electrophoresis and Immunological Techniques,* p. 343. Plenum Press, New York.

Hsu, S. M., Raine, L., and Fanger, H. (1981). *Am. J. Clin. Path.,* **75**, 816.

Using labelled *S.aureus* protein A:

Dunbar, B. S. (1987). *Two Dimensional Electrophoresis and Immunological Techniques,* p. 341. Plenum Press, New York.

Renart, J. and Sandoval, I. V. (1984). In *Methods in Enzymology,* (ed. W. B. Jakoby), Vol. 104, p. 455. Academic Press, New York.

Burnette, W. N. (1981). *Anal. Biochem.,* **112**, 195.

Brada, D. and Roth, J. (1984). *Anal. Biochem.,* **142**, 79.

Renart, J., Reiser, J., and Stark, G. R. (1979). *Proc. Natl. Acad. Sci. USA,* **76**, 3116.

'Immunogold' staining:

Hsu, Y. (1984). *Anal. Biochem.,* **142**, 221.

Surek, B. and Latzko, E. (1984). *Biochem. Biophys., Res. Commun.,* **121**, 284.

Tagging proteins with haptens to detect all polypeptides: *see* Section 2.2.

Immunodetection of pre-stained proteins:

Coomassie Blue:

Jackson, P. and Thompson, R. J. (1984). *Electrophoresis,* **5**, 35.

Silver stain:

Yuen, K. C. C., Johnson, T. K., Dennell, R. E., and Consigli, R. A. (1982). *Anal. Biochem.,* **126**, 398.

Detection of ligand-binding proteins:

Reviews:

Towbin, H. and Gordon, J. (1984). *J. Immunol. Methods,* **72**, 313.

Soutar, A. K. and Wade, D. P. (1989). In *Protein Function: A Practical Approach,* (ed. T. E. Creighton), p. 55. Oxford University Press, Oxford.

Gershoni, J. M. and Palade, G. E. (1983). *Anal. Biochem.,* **131**, 1.

Anderson, N. C., Giometti, C. S., Gemmel, M. A., Nance, S. C., and Anderson, N. G. (1982). *Clin. Chem.,* **28**, 1084.

Symington, J. (1984). *Two Dimensional Electrophoresis of Proteins: Methods and Applications,* (ed. J. E. Celis and R. Bravo), pp. 127–168. Academic Press, New York.

2.7 Detection of specific enzymes

See this Appendix, Section 3.

3. Enzyme detection methods

Three publications by Shaw and Prasad (1), Siciliano and Shaw (2), and Harris and Hopkinson (3a,b,c) describe the preparation and use of reagents to detect a wide variety of enzymes. Often several different approaches to detect a particular enzyme are possible. Harris and Hopkinson (3a,b,c) give an extensive bibliography in each case. The majority of methods in these publications relate to starch gel but most can also be used with polyacrylamide gel. Gabriel (4), Rothe and Maurer (158) and Heeb and Gabriel (159) have reviewed enzyme detection methods which have been applied to polyacrylamide gel. Listed below are the enzyme activities which have been detected on gels, in each case with an indication of the particular compilation which can be consulted for experimental information plus other selected references. The general approaches to enzyme detection on gels, together with some cautionary notes, have been discussed earlier in this book (Chapter 1, Section 7.9) and elsewhere (e.g. refs 159, 160).

Enzyme	*References*
Acid phosphatase	1, 2, 3a, 4, 5, 147, 161.
Aconitase	1, 2, 3a, 6.
Adenine phosphoribosyl transferase (AMP pyrophosphorylase)	3a, 7, 8.
Adenosine deaminase	2, 3a, 9, 21.
Adenosine kinase	3c, 10.
Adenylate kinase	1, 2, 3a, 11.
ADP-glycogen transferase	4, 12.
Alanine aminotransferase	2, 3a, 13.
Alcohol dehydrogenase	1, 2, 3a, 4, 14, 15.
Aldolase	1, 2, 3a, 16, 249.
Alkaline phosphatase	1, 3a, 4, 17, 147, 161.
Amine oxidase	4, 18.
Amino acid oxidase	3b, 4, 19–21.
Aminoacyl tRNA synthetase	154.
5-Aminolaevulinate synthase	162, 163.
Aminopeptidase	164.

AMP deaminase	3c, 21, 22.
Amylase	3a, 4, 23, 24, 223, 235, 237, 238.
Anthranilate phosphoribosyl transferase	165.
Anthranilate synthetase	25.
Arginase	3c, 21, 26.
Arginosuccinase	3c, 21.
Aromatic amino acid decarboxylase	27.
Aromatic amino acid transaminase	1.
Arylamidases	4.
Arylsulphatases	3c, 28, 29, 96, 234.
Aspartate aminotransferase	1, 2, 3a, 4, 30, 209, 210.
Aspartate carbamoyl transferase	3a, 108, 206.
Aspartate oxidase	21.
ATP pyrophosphatase	166, 167.
Carbonic anhydrase	1, 3a, 31, 32.
Carboxypeptidase	243.
Catalase	1, 2, 3a (starch gels only), 4, 202.
Catechol oxidase	168.
Cathepsin B	149, 242.
Cellobiose phosphorylase	4.
Cellulases	33, 148, 155.
Cholinesterase	4, 34.
Chorismate synthase	161.
Chymotrypsin	35, 125, 170.
Citrate synthase	3a, 36.
Creatine kinase	1, 3a, 4, 37, 216–218.
3'5' Cyclic AMP phosphodiesterase	3a, 4, 38.
3'5' Cyclic nucleotide phosphodiesterase	171.
Cystathione β synthase	172.
Cytidine deaminase	3a, 40, 41, 188, 189.
DAHP synthase	161.
Deoxyribonuclease	173, 224, 228–230.
Dextranase	235.
Dextran sucrose	152.
Dihydrouracil dehydrogenase	43.
Dipeptidase	42 (see also peptidases).
DNase	173, 224, 228–230.
DNA nucleotidyl transferase	174.
DNA polymerase	4, 44.
Elastase and pro-elastase	45, 245.
Enolase	2, 3a, 46.

3-Enol pyruvoylshikimate-5-phosphate synthase	161.
Esterases	1, 2, 3a, 4, 5, 47, 220–222.
Ferrisidophore reductase	150.
Folate reductase	4.
β-D-Fructofuranosidase	2, 178, 179, 235, 239.
Fructose-1,6-bisphosphate	1, 161, 175, 176.
Fructose-5-dehydrogenase	177.
Fructosyl transferase	48.
L-Fucose dehydrogenase	49.
α-Fucosidase	3a, 4, 50.
Fumarase	1, 2, 3c, 51.
Fumarate hydratase	250.
Galactokinase	3a, 4, 52.
Galactose-6-phosphate dehydrogenase	1.
Galactose-1-phosphate uridyltransferase	3a, 4, 53 (see also galactosyl transferases).
α-Galactosidase	2, 3a, 54, 55.
β-Galactosidase	3a, 4, 56–58.
Galactosyl transferases	59.
1,4 α-D-Glucan branching enzyme	180, 208.
Glucose oxidase	4.
Glucose-6-phosphate dehydrogenase	1, 2, 3a, 60.
Glucose phosphate isomerase	2.
Glucose-1-phosphate uridylyltransferase	3a, 4, 144, 161, 219.
α-Glucosidase	3a, 4, 61.
β-Glucosidase	3a, 4, 62.
Glucosyltransferase	213.
β-Glucuronidase	1, 2, 3b, 63, 64.
Glutamate dehydrogenase	1, 3b, 4, 65, 181.
Glutamate-oxaloacetate transaminase	See aspartate aminotransferase.
Glutamate-pyruvate transaminase	See alanine aminotransferase.
Glutaminase	66.
Glutamine synthetase	182, 183.
γ-Glutamyl-transferase	207.
Glutathione peroxidase	3a, 67.
Glutathione reductase	1, 3a, 68.
Glutathione S transferase	69, 185.
Glyceraldehyde-3-phosphate dehydrogenase	1, 2, 3a, 4, 70, 161.
Glycerol kinase	4, 8.
Glycerol-3-phosphate dehydrogenase	1, 2, 3a, 4, 70, 196.
Glycogen phosphorylase	4.
Glycogen synthase	153.

Sulphite oxidase	134.
Superoxide dismutase	3a, 135, 136, 192.
Tetrahydrofolate dehydrogenase	193.
Thermolysin	45.
Threonine deaminase	137.
Thymidine kinase	1.
Transamidases	211, 212.
Trehalase	42, 138, 139.
Triose-phosphate isomerase	1, 2, 3a, 139, 140.
Tripeptide aminopeptidase	141.
Trypsin	125, 142, 242, 244.
Tyrosinase	205.
Tyrosine aminotransferase	143, 194.
UDPG dehydrogenase	4.
UDPG pyrophosphorylase	3a, 4, 144, 161, 219.
UMP kinase	3a, 145.
Urease	4, 146.
Urokinase	246, 247.
Xanthine dehydrogenase	1.
β-Xylanase	155.

References

1. Shaw, C. R. and Prasad, R. (1980). *Biochem. Genet.,* **4**, 297.
2. Siciliano, M. J. and Shaw, C. R. (1976). In *Chromatographic and Electrophoretic Techniques,* (ed. I. Smith), Vol. 2, p. 185. William Heinemann Medical Books Ltd, London.
3. (a) Harris, H. and Hopkinson, D. A. (1976). *Handbook of Enzyme Electrophoresis in Human Genetics.* North-Holland, Amsterdam.
    (b) Supplement (1977).
    (c) Supplement (1978).
4. Gabriel, O. (1971). In *Methods in Enzymology,* (ed. S. P. Colowick and N. O. Kaplan), Vol. 22, p. 578. Academic Press, New York.
5. Cullis, C. A. and Kolodynska, K. (1975). *Biochem. Genet.,* **13**, 687.
6. Slaughter, C. A., Hopkinson, D. A., and Harris, H. (1975). *Ann. Hum. Genet.,* **39**, 193.
7. Mowbray, S., Watson, B., and Harris, H. (1972). *Ann. Hum. Genet.,* **36**, 153.
8. Tischfield, J. A., Bernhard, H. P., and Ruddle, F. H. (1973). *Anal. Biochem.,* **53**, 545.
9. Spencer, N., Hopkinson, D. A., and Harris, H. (1968). *Ann. Hum. Genet.,* **32**, 9.
10. Klobucher, L. A., Nichols, E. A., Kucherlapati, R. S., and Ruddle, F. H. (1976). In *Birth Defects; Original Article Series,* Vol. 12, p. 171. The National Foundation, March of Dimes, New York.
11. Wilson, D. E., Povey, S., and Harris, H. (1976). *Ann. Hum. Genet.,* **39**, 305.
12. Frederick, J. F. (1968). *Ann. NY Acad. Sci.,* **151**, 413.
13. Chen, S. H., Giblett, E. R., Anderson, J. E., and Fossum, B. L. G. (1972). *Ann. Hum. Genet.,* **35**, 401.
14. Sofer, W. and Ursprung, H. (1968). *J. Biol. Chem.,* **243**, 3110.

15. Smith, H., Hopkinson, D. A., and Harris, H. (1971). *Ann. Hum. Genet.*, **34**, 251.
16. Lewinski, N. D. and Dekker, E. E. (1978). *Anal. Biochem.*, **87**, 56.
17. Sussman, H. H., Small, P. A., and Cotlove, E. (1968). *J. Biol. Chem.*, **243**, 160.
18. Ma Lin, A. W. and Castell, D. O. (1975). *Anal. Biochem.*, **69**, 637.
19. Barker, R. F. and Hopkinson, D. A. (1977). *Ann. Hum. Genet.*, **41**, 27.
20. Hayes, M. B. and Wellner, D. (1969). *J. Biol. Chem.*, **244**, 6636.
21. Nelson, R. L., Povey, S., Hopkinson, D. A., and Harris, H. (1977). *Biochem. Genet.*, **15**, 1023.
22. Anderson, J. E., Teng, Y. S., and Liblett, E. R. (1975). In *Birth Defects; Original Article Series*, Vol. 11, p. 295. The National Foundation, March of Dimes, New York.
23. Heller, H. and Kulka, R. G. (1968). *Biochim. Biophys. Acta*, **165**, 393.
24. Boettcher, D. and De La Lande, F. A. (1969). *Anal. Biochem.*, **28**, 510.
25. Grove, T. H. and Levy, H. R. (1975). *Anal. Biochem.*, **65**, 458.
26. Farron, F. (1973). *Anal. Biochem.*, **53**, 264.
27. Landon, M. (1977). *Anal. Biochem.*, **77**, 293.
28. Shapira, E., De Gregorio, R. R., Matalon, R., and Nadler, H. R. (1975). *Biochem. Biophys. Res. Commun.*, **62**, 448.
29. Chang, P. L., Ballantyne, S. R., and Davidson, R. G. (1979). *Anal. Biochem.*, **97**, 36.
30. Davidson, R. G., Cortner, J. A., Rattazi, M. C., Ruddle, F. H., and Lubs, H. A. (1970). *Science*, **169**, 391.
31. Hopkinson, D. A., Coppock, J. S., Mühlemann, M. F., and Edwards, Y. H. (1974). *Ann. Hum. Genet.*, **38**, 155.
32. Drescher, D. G. (1978). *Anal. Biochem.*, **90**, 349.
33. Erickson, K. E. and Petterson, B. (1973). *Anal. Biochem.*, **56**, 618.
34. Harris, H., Hopkinson, D. A., and Robson, E. B. (1962). *Nature*, **196**, 1296.
35. Gertler, A., Trencer, Y., and Tinman, G. (1973). *Anal. Biochem.*, **54**, 270.
36. Craig, I. (1973). *Biochem. Genet.*, **9**, 351.
37. Yue, R. H., Jacobs, H. K., Okabe, K., Keutel, H. J., and Kuby, S. A. (1968). *Biochemistry*, **7**, 4291.
38. Monn, E. and Christiansen, R. O. (1971). *Science*, **173**, 540.
39. Teng, Y. S., Anderson, J. E., and Giblett, E. R. (1975). *Am. J. Hum. Genet.*, **27**, 492.
40. Williams, L. and Hopkinson, D. A. (1975). *Hum. Hered.*, **25**, 567.
41. Kaplan, J. C. and Beutler, E. (1967). *Biochem. Biophys. Res. Commun.*, **29**, 605.
42. Suguira, M., Ito, Y., Hirano, K., and Sawaki, S. (1977). *Anal. Biochem.*, **81**, 481.
43. Hallock, R. O. and Yamada, E. W. (1973). *Anal. Biochem.*, **56**, 84.
44. Jovin, T. M., Englund, P. T., and Bertsch, L. L. (1969). *J. Biol. Chem.*, **244**, 2996.
45. Dijkhof, J. and Poort, C. (1977). *Anal. Biochem.*, **83**, 315.
46. Sharma, H. K. and Rothstein, M. (1979). *Anal. Biochem.*, **98**, 226.
47. Coates, P. M., Mestriner, M. A., and Hopkinson, D. A. (1975). *Ann. Hum. Genet.*, **39**, 1.
48. Russell, R. R. B. (1979). *Anal. Biochem.*, **97**, 173.
49. Schachter, H., Sarney, J., McGuire, E. J., and Roseman, S. (1969). *J. Biol. Chem.*, **244**, 4785.
50. Turner, B. M., Beratis, N. G., Turner, V. S., and Hirschhorn, K. (1974). *Clin. Chim. Acta*, **57**, 29.
51. Edwards, Y. H. and Hopkinson, D. A. (1978). *Ann. Hum. Genet.*, **42**, 303.
52. Nicholls, E. A., Elsevier, S. M., and Ruddle, F. H. (1974). *Cytogenet. Cell Genet.*, **13**, 275.

53. Ng, W. G., Bergren, W. R., Fields, M., and Donnell, G. N. (1969). *Biochem. Biophys. Res. Commun.*, **37**, 354.
54. Beutler, E. and Kuhl, W. (1972). *J. Biol. Chem.*, **247**, 7195.
55. Beutler, E., Guinto, E., and Kuhl, W. (1973). *Am. J. Hum. Genet.*, **25**, 42.
56. Norden, A. G. W. and O'Brien, J. S. (1975). *Proc. Natl. Acad. Sci. USA*, **72**, 240.
57. Alpers, D. H. (1969). *J. Biol. Chem.*, **244**, 1238.
58. Alpers, D. H., Steers, E., Shifrin, S., and Tomkins, G. (1968). *Ann. NY Acad. Sci.*, **151**, 545.
59. Pierce, M., Cummings, R. D., and Roth, S. (1980). *Anal. Biochem.*, **102**, 441.
60. Criss, W. E. and McKerns, K. W. (1968). *Biochemistry*, **7**, 125.
61. Swallow, D. M., Corney, G., Harris, H., and Hirschhorn, R. (1975). *Ann. Hum. Genet.*, **38**, 391.
62. Beutler, E., Kuhl, W., Trinidad, F., Teplitz, R., and Nadler, H. (1971). *Am. J. Hum. Genet.*, **23**, 62.
63. Fondo, E. Y. and Bartalos, M. (1969). *Biochem. Genet.*, **3**, 591.
64. Franke, U. (1976). *Am. J. Hum. Genet.*, **28**, 357.
65. Nelson, R. L., Povey, M. S., Hopkinson, D. A., and Harris, H. (1977). *Biochem. Genet.*, **15**, 87.
66. Davis, J. N. and Prusiner, S. (1973). *Anal. Biochem.*, **54**, 272.
67. Beutler, E. and West, C. (1974). *Am. J. Hum. Genet.*, **26**, 255.
68. Kaplan, J. C. and Beutler, E. (1968). *Nature*, **217**, 256.
69. Board, P. G. (1980). *Anal. Biochem.*, **105**, 147.
70. Charlesworth, D. (1972). *Ann. Hum. Genet.*, **35**, 477.
71. Duley, J. and Holmes, R. S. (1974). *Genetics*, **76**, 93.
72. Price, R. G. and Dance, N. (1967). *Biochem. J.*, **105**, 877.
73. Gabriel, O. and Wang, S. F. (1969). *Anal. Biochem.*, **27**, 545.
74. Kompf, J., Bissbort, S., Gussman, S., and Ritter, H. (1975). *Humangenetik*, **27**, 141.
75. Parr, C. W., Bagster, I. A., and Welch, S. G. (1977). *Biochem. Genet.*, **15**, 109.
76. Jamil, T., Fisher, R. A., and Harris, H. (1976). *Hum. Hered.*, **25**, 402.
77. Katzen, H. M. and Schimke, R. T. (1965). *Proc. Natl. Acad. Sci. USA*, **54**, 1218.
78. Rogers, P. A., Fisher, R. A., and Harris, H. (1975). *Biochem. Genet.*, **13**, 857.
79. Ogilvie, J. W., Sightler, J. H., and Clark, R. B. (1969). *Biochemistry*, **8**, 3557.
80. Craig, I., Tolley, E., and Borrow, M. (1976). In *Birth Defects; Original Article Series*, Vol. 12, p. 114. The National Foundation, March of Dimes, New York.
81. Vasquez, B. and Bieber, A. L. (1978). *Anal. Biochem.*, **84**, 504.
82. Fischer, R. A., Turner, B. M., Dorkin, H. L., and Harris, H. (1974). *Ann. Hum. Genet.*, **37**, 341.
83. Babczinski, P. (1980). *Anal. Biochem.*, **105**, 328.
84. Henderson, N. S. (1968). *Ann. NY Acad. Sci.*, **151**, 429.
85. Turner, B. M., Fisher, R. A., and Harris, H. (1974). *Ann. Hum. Genet.*, **37**, 455.
86. Reeves, H. C. and Volk, M. J. (1972). *Anal. Biochem.*, **48**, 437.
87. Allen, J. M. (1961). *Ann. NY Acad. Sci.*, **94**, 937.
88. Werthamer, S., Freiberg, A., and Amaral, L. (1973). *Clin. Chim. Acta*, **45**, 5.
89. Strogin, A. Y., Azarenkova, N. M., Vaganova, T. I., Levin, A. D., and Stepanov, V. M. (1976). *Anal. Biochem.*, **74**, 597.
90. Nachlase, M. M., Morris, B., Rosenblatt, D., and Seligman, A. M. (1960). *J. Biophys. Biochem. Cytol.*, **7**, 261.
91. Millard, S. A., Kubose, A., and Gal, E. M. (1969). *J. Biol. Chem.*, **244**, 2511.

92. Allen, S. L. (1968). *Ann. NY Acad. Sci.*, **151**, 190.
93. Volk, M. J., Trelease, R. N., and Reeves, H. C. (1974). *Anal. Biochem.*, **58**, 315.
94. Cohen, P. T. W. and Omenn, G. S. (1972). *Biochem. Genet.*, **7**, 303.
95. Povey, S., Wilson, D. E., Harris, H., Gormley, I. P., Perry, P., and Buckton, K. E. (1975). *Ann. Hum. Genet.*, **39**, 203.
96. Nichols, E. A., Chapman, V. M., and Ruddle, F. H. (1973). *Biochem. Genet.*, **8**, 47.
97. Poenaru, L. and Dreyfus, J. C. (1973). *Biochim. Biophys. Acta*, **303**, 171.
98. Okada, S., Veath, M. L., Lerov, J., and O'Brien, J. S. (1971). *Am. J. Hum. Genet.*, **23**, 55.
99. Flechner, L., Hirschhorn, S., and Bekjerkunst, A. (1968). *Life Sci.*, **7**, 1327.
100. Ravazzolo, R., Bruzzone, G., Garrè, C., and Ajmar,F. (1976). *Biochem. Genet.*, **14**, 877.
101. Ingle, J. (1968). *Biochem. J.*, **108**, 715.
102. Hucklesby, D. P. and Hageman, R. H. (1973). *Anal. Biochem.*, **56**, 591.
103. Lacks, S. A., Springhorn, S. S., and Rosenthal. A. L. (1979). *Anal. Biochem.*, **100**, 357.
104. Rosenthal, A. L. and Lacks, S. A. (1977). *Anal. Biochem.*, **80**, 76.
105. Spencer, N., Hopkinson, D. A., and Harris, H. (968). *Ann. Hum. Genet.*, **32**, 9.
106. Edwards, Y. H., Hopkinson, D. A., and Harris, H. (1971). *Ann. Hum. Genet.*, **34**, 395.
107. Dvorak, H. F. and Heppel, L. A. (1968). *J. Biol. Chem.*, **243**, 2647.
108. Baron, D. N. and Buttery, J. E. (1972). *J. Clin. Pathol.*, **25**, 415.
109. Feinstein, R. N. and Lindahl, R. (1973). *Anal. Biochem.*, **56**, 353.
110. Scrutton, M. C. and Fatebene, F. (1975). *Anal. Biochem.*, **69**, 247.
111. Lewis, W. H. P. and Harris, H. (1967). *Nature*, **215**, 315.
112. Rapley, S., Lewis, W. H. P., and Harris, H. (1971). *Ann. Hum. Genet.*, **34**, 307.
113 Brock, D. J. H. (1969). *Biochem. J.*, **113**, 235.
114. Niessner, N. and Beutler, E. (1974). *Biochem. Med.*, **8**, 73.
115. De Lorenzo, R. J. and Ruddle, F. H. (1969). *Biochem. Genet.*, **3**, 151.
116. Kühn, P., Schmidtmann, U., and Spielmann, W. (1977). *Hum. Genet.*, **35**, 219.
117. Fildes, R. A. and Parr, C. W. (1963). *Nature*, **200**, 890.
118. Beutler, E. (1969). *Biochem. Genet.*, **3**, 189.
119. Omenn, G. S. and Cheung, S. C. Y. (1974). *Am. J. Hum. Genet.*, **26**, 393.
120. Chen, S. H., Anderson, J., Giblett, E. R., and Lewis, M. (1974). *Am. J. Hum. Genet.*, **26**, 73.
121. Barker, R. F. and Hopkinson, D. A. (1978). *Ann. Hum. Genet.*, **42**, 1.
122. Shier, W. T. and Troffer, J. T. (1978). *Anal. Biochem.*, **87**, 604.
123. Heussen, E. and Dowdle, E. B. (1980). *Anal. Biochem.*, **102**, 196.
124. Klee, C. B. (1969). *J. Biol. Chem.*, **244**, 2558.
125. Andary, T. J. and Dabich, D. (1974). *Anal. Biochem.*, **57**, 457.
126. North, M. J. and Harwood, J. M. (1979). *Biochim. Biophys. Acta*, **566**, 222.
127. Filho, J. X. and De Azevedo Moreira, R. (1978). *Anal. Biochem.*, **84**, 296.
128. Hirsch, A. and Rosen, M. (1974). *Anal. Biochem.*, **60**, 389.
129. Gagelman, M., Pyerin, W., Kübler, D., and Kinzel, V. (1979). *Anal. Biochem.*, **93**, 52.
130. Chern, C. J. and Beutler, E. (1976). *Ann. Hum. Genet.*, **28**, 9.
131. Imamura, K. and Tanaka, T. (1972). *J. Biochem.*, **71**, 1043.
132. Biswas, S. and Hollander, V. P. (1969). *J. Biol. Chem.*, **244**, 4185.
133. Lyublinskaya, L. A., Belyaev, S. V., Strongin, A. Ya., Matyash, L. F., and Stepanov, V. M. (1974). *Anal. Biochem.*, **62**, 371.
134. Cohen, H. J. (1973). *Anal. Biochem.*, **53**, 208.

Appendix 2

135. Beauchamp, C. and Fridovitch, I. (1971). *Anal. Biochem.*, **44**, 276.
136. Beckman, G., Lundgren, A., and Tarnvik, A. (1973). *Hum. Hered.*, **23**, 338.
137. Hatfield, G. W. and Umbarger, H. E. (1980). *J. Biol. Chem.*, **245**, 1736.
138. Killick, K. A. and Wang, L. W. (1980). *Anal. Biochem.*, **106**, 367.
139. Kaplan, J. C., Teeple, L., Shore, N., and Beutler, E. (1968). *Biochem. Biophys. Res. Commun.*, **31**, 768.
140. Scopes, R. K. (1964). *Nature*, **201**, 924.
141. Suguira, M., Ho, Y., Hirano, K., and Sawaki, S. (1977). *Anal. Biochem.*, **81**, 481.
142. Gertler, A., Tencer, Y., and Tinman, G. (1973). *Anal. Biochem.*, **54**, 270.
143. Walker, D. G. and Khan, H. H. (1968). *Biochem. J.*, **108**, 169.
144. Manrow, R. F. and Dottin, R. P. (1980). *Proc. Natl. Acad. Sci. USA*, **77**, 730.
145. Giblett, E. R., Anderson, J. A., Chen, S. H., Teng, Y. S., and Cohen, F. (1974). *Am. J. Hum. Genet.*, **26**, 627.
146. Shaik-M, M. B., Guy, A. L., and Pancholy, S. K. (1980). *Anal. Biochem.*, **103**, 140.
147. Debruyne, I. (1983). *Anal. Biochem.*, **133**, 110.
148. Beguin, P. (1983). *Anal. Biochem.*, **131**, 333.
149. Mort, J. S. and Leduc, M. (1982). *Anal. Biochem.*, **119**, 148.
150. Moody, M. D. and Dailey, H. A. (1983). *Anal. Biochem.*, **134**, 235.
151. Zehender, H., Trescher, D., and Ullrich, J. (1983). *Anal. Biochem.*, **135**, 16.
152. Miller, A. W. and Robyt, J. F. (1986). *Anal. Biochem.*, **156**, 357.
153. Krisman, C. R. and Blumenfeld, M. L. (1986). *Anal. Biochem.*, **154**, 409.
154. Chang, G.-G., Denq, R.-Y., and Pan, F. (1985). *Anal. Biochem.*, **149**, 474.
155. Biely, P., Markovic, O., and Mislovicová, D. (1985). *Anal. Biochem.*, **144**, 147.
156. Kelleher, P. J. and Juliano, R. C. (1984). *Anal. Biochem.*, **136**, 470.
157. Yang, S. S. and Coleman, D. (1987). *Anal. Biochem.*, **160**, 480.
158. Rothe, G. M. and Maurer, W. D. (1986). In *Gel Electrophoresis of Proteins*, (ed. M. J. Dunn), pp. 37–140. Wright, Bristol.
159. Heeb, M. J. and Gabriel, O. (1984). In *Methods in Enzymology*, (ed. W. B. Jakoby), Vol. 104, p. 416. Academic Press, New York.
160. Merri, C. R., Harasewych, M. G., and Harrington, M. G. (1986). In *Gel Electrophoresis of Proteins*, (ed. M. J. Dunn), p. 323. Wright, Bristol.
161. Nimmo, H. G. and Nimmo, G. A. (1982). *Anal. Biochem.*, **121**, 17.
162. Davies, R. C. and Neuberger, A. (1973). *Biochem. J.*, **133**, 471.
163. Davies, R. C. and Neuberger, A. (1979). *Biochem. J.*, **177**, 649.
164. Strongin, A. Y., Azarenkova, N. M., Vaganov, T. I., Levin, E. D., and Stepanov, V. M. (1976). *Anal. Biochem.*, **74**, 597.
165. Henderson, C. J., Nagano, H., Zalkin, H., and Hurang, L. H. (1970). *J. Biol. Chem.*, **245**, 1416.
166. Risi, S., Höckel, M., Hulla, F. W., and Dose, K. (1977). *Eur. J. Biochem.*, **81**, 103.
167. Schäfer, H. J., Scheurich, P., and Rathgeber, G. (1978). *Hoppe-Seyler's Z. Physiol. Chem.*, **359**, 1441.
168. Hanold, G. R. and Stahmann, N. (1968). *Cereal Chem.*, **45**, 99.
169. Gannon, F. and Jones, K. M. (1977). *Anal. Biochem.*, **79**, 594.
170. Solomon, S. S., Palazzolo, M., and King, L. E. (1977). *Diabetes*, **26**, 967.
171. Tsou, K. C., Lo, K. W., and Yip, K. F. (1974). *FEBS Lett.*, **45**, 47.
172. Willhardt, I. and Wiedernanders, B. (1975). *Anal. Biochem.*, **63**, 263.
173. Porter, A. L. G. (1971). *Anal. Biochem.*, **117**, 28.
174. Zöllner, E. J., Müller, W. E. G., and Zahn, R. K. (1973). *Z. Naturforsch*, **28C**, 376.

175. Colombo, G. and Marcus, F. (1973). *J. Biol. Chem.*, **248**, 2743.
176. Hubert, E. and Marcus, F. (1974). *FEBS Lett.*, **40**, 37.
177. Ameyama, M., Shinigawa, E., Matsushita, K., and Adachi, O. (1981). *J. Bacteriol.*, **145**, 184.
178. Faye, L. (1981). *Anal. Biochem.*, **112**, 90.
179. Satoh, K., Imai, F., and Sato, K. (1978). *FEBS Lett.*, **95**, 239.
180. Yonezawa, S. and Hori, S. H. (1975). *J. Histochem. Cytochem.*, **23**, 745.
181. Kimura, K., Miyakawa, A., Imai, T., and Sasakawa, T. (1977). *J. Biochem.*, **81**, 467.
182. Miller, R. E., Shelton, E., and Stadtman, E. R. (1974). *Arch. Biochem. Biophys.*, **163**, 155.
183. Meyer, J. M. and Stadtman, E. R. (1981). *J. Bacteriol.*, **146**, 705.
184. Bieber, A. C. (1974). *Anal. Biochem.*, **60**, 206.
185. Kenney, W. C. and Boyer, T. D. (1981). *Anal. Biochem.*, **116**, 344.
186. Zink, M. W. and Katz, J. S. (1973). *Can. J. Microbiol.*, **19**, 1187.
187. Veng, S. T. H., Hartanowicz, P., Lewandoski, C., Keller, J., Holick, M., and McGuiness, T. (1976). *Biochemistry,* **15**, 1743.
188. Fan, L. L. and Masters, B. S. S. (1974). *Arch. Biochem. Biophys.*, **165**, 665.
189. Ichihara, K., Kusunose, E., and Kusunose, M. (1973). *Eur. J. Biochem.*, **38**, 463.
190. Balakrishnan, C. V., Ravindranath, S. D., and Appaji, R. N. (1974). *Arch. Biochem. Biophys.*, **164**, 156.
191. Kavanolas, H. J., Baedecker, M. L., and Engel, L. L. (1970). *J. Biol. Chem.*, **245**, 4948.
192. De Rosa, G., Duncan, D. S., Keen, C. L., and Hurley, L. S. (1979). *Biochim. Biophys. Acta,* **566**, 32.
193. Nixon, P. F. and Blakely, R. C. (1968). *J. Biol. Chem.*, **243**, 4722.
194. Shuler, J. K. and Tryfiates, G. P. (1977). *Enzyme,* **22**, 262.
195. O'Conner, J. L. (1977). *Anal. Biochem.*, **78**, 205.
196. Leibenguth, F. (1975). *Biochem. Genet.*, **13**, 263.
197. Carda-Abella, P., Perez-Cuadrada, S., Laru-Baruque, S., Gil-Grande, L., and Nunez-Puertas, A. (1982). *Cancer,* **49**, 80.
198. Simon, K., Chaplin, E. R., and Diamond, I. (1977). *Anal. Biochem.*, **79**, 571.
199. Springell, P. H. and Lynch, T. A. (1976). *Anal. Biochem.*, **74**, 251.
200. Povey, S., Wilson, D. E., Harris, H., Gormley, I. P., Perry, P., and Buckton, K. E. (1975). *Ann. Hum. Genet.*, **39**, 203.
201. Peterson, A. C. (1974). *Nature,* **248**, 561.
202. Gregory, E. M. and Fridovich, E. M. (1974). *Anal. Biochem.*, **58**, 57.
203. Liu, E. H. and Gibson, D. M. (1977). *Anal. Biochem.*, **79**, 597.
204. DeLumen, B. O. and Kazeniac, S. J. (1976). *Anal. Biochem.*, **72**, 428.
205. Thomas, P., Delincee, H., and Diehl, J. F. (1978). *Anal. Biochem.*, **88**, 138.
206. Grayson, J. E., Yon, R. P., and Buterworth, P. J. (1979). *Biochem. J.*, **183**, 239.
207. Izumi, M. and Taketa, K. (1981). In *Electrophoresis '81,* (ed. R. C. Allen and P. Arnaud), p. 709. de Gruyter, Berlin.
208. Satoh, K. and Sato, K. (1980). *Anal. Biochem.*, **108**, 16.
209. Scandalios, J. G., Sorenson, J. C., and Ott, L. A. (1975). *Biochem. Genet.*, **13**, 759.
210. Yagi, T., Kagamiyama, H., and Nozaki, M. (1971). *Anal. Biochem.*, **110**, 146.
211. Stenberg, P. and Stenflo, J. (1979). *Anal. Biochem.*, **93**, 445.
212. Lorand, L., Seifring, G. E., Tong, Y. S., Bruner-Lorand, J., and Gray, A. J. (1979). *Anal. Biochem.*, **93**, 453.
213. Mukasa, H., Shimamura, A., and Tsumori, H. (1982). *Anal. Biochem.*, **123**, 276.

214. Phan-Dinh-Tuy, F., Weber, A., Henry, J., Cotterau, D., and Kahn, A. (1982). *Anal. Biochem.*, **127**, 73.
215. Melendez-Heria, C., Corzo, J., and Perez, J. (1981). In *Electrophoresis '81*, (ed. R. C. Allen and P. Arnaud), p. 693. de Gruyter, Berlin.
216. Graeber, G. M., Reardon, M. J., Fleming, A. W., Head, H. D., Zajtchuk, R., Brott, W. H., and Foster, J. H. (1981). *Ann. Thor. Surg.*, **32**, 230.
217. Yamashita, T. and Utoh, H. (1978). *Anal. Biochem.*, **84**, 304.
218. Hall, N. and DeLuca, M. (1976). *Anal. Biochem.*, **76**, 561.
219. Manrow, R. E. and Dottin, R. P. (1982). *Anal. Biochem.*, **120**, 181.
220. Yourno, J. and Mastropaolo, W. (1981). *Blood*, **58**, 939.
221. Rosenberg, V., Roegner, V., and Becker, F. F. (1975). *Anal. Biochem.*, **66**, 206.
222. Herd, J. K. and Tschida, J. (1975). *Anal. Biochem.*, **68**, 218.
223. Hofelmann, M., Kittsteiner-Eberle, R., and Schreier, P. (1983). *Anal. Biochem.*, **128**, 217.
224. Hodes, M. E., Crisp, M., and Gelb, E. (1977). *Anal. Biochem.*, **80**, 239.
225. Hawley, D. M., Tsou, K. C., and Hodes, M. E. (1981). *Anal. Biochem.*, **117**, 18.
226. Hodes, M. E. and Retz, J. E. (1981). *Anal. Biochem.*, **110**, 150.
227. Lo, K. W., Aoyagi, S., and Tsou, K. C. (1981). *Anal. Biochem.*, **117**, 24.
228. Kim, H. S. and Liao, T. H. (1982). *Anal. Biochem.*, **119**, 62.
229. Blank, A., Suigiyama, R. H., and Dekker, C. A. (1982). *Anal. Biochem.*, **120**, 267.
230. Karpetsky, T., Brown, G. E., McFarland, E., Rahman, A., Rictro, K., Roth, W., Haroth, M. B., Ansher, A., Duffey, P., and Levy, C. (1981). In *Electrophoresis '81*, (ed. R. C. Allen and P. Arnaud), p. 674. de Gruyter, Berlin.
231. Smyth, C. J. and Wadstrom, T. (1975). *Anal. Biochem.*, **65**, 137.
232. Karn, R. C., Crisp, M., Yount, E. A., and Hodes, M. E. (1979). *Anal. Biochem.*, **96**, 464.
233. Huet, J., Sentenac, A., and Fromageot, P. (1978). *FEBS Lett.*, **94**, 28.
234. Mannowitz, P., Goldstein, L., and Bellomo, F. (1978). *Anal. Biochem.*, **89**, 423.
235. Musaka, H., Shimura, A., and Tsumori, H. (1982). *Anal. Biochem.*, **123**, 276.
236. Cruickshank, R. H. and Wade, G. C. (1980). *Anal. Biochem.*, **107**, 177.
237. Lacks, S. A. and Springhorn, S. S. (1980). *J. Biol. Chem.*, **255**, 7467.
238. Brown, T. L., Yet, M. G., and Wold, F. (1982). *Anal. Biochem.*, **122**, 164.
239. Babczinski, P. (1980). *Anal. Biochem.*, **105**, 328.
240. Burdett, P. E., Kipps, A. E., and Whitehead, P. H. (1976). *Anal. Biochem.*, **72**, 315.
241. Every, D. (1981). *Anal. Biochem.*, **116**, 519.
242. Lynn, K. R. and Clevette-Radford, N. A. (1981). *Anal. Biochem.*, **117**, 280.
243. Pacaud, M. and Uriel, J. (1971). *Eur. J. Biochem.*, **23**, 435.
244. Ward, C. W. (1976). *Anal. Biochem.*, **74**, 242.
245. Westergaard, J. C. and Roberts, R. C. (1981). In *Electrophoresis '81*, (ed. R. C. Allen and P. Arnaud), p. 677. de Gruyter, Berlin.
246. Granelli-Piperano, A. and Reich, E. (1978). *J. Exp. Med.*, **148**, 223.
247. Heussen, C. and Dowdle, E. B. (1980). **102**, 192.
248. Lomholt, B. (1975). *Anal. Biochem.*, **65**, 569.
249. Christen, P. and Gasser, A. (1980). *Anal. Biochem.*, **109**, 270.
250. Tolley, E. and Craig, I. (1975). *Biochem. Genet.*, **13**, 867.

Reagents for the isotopic labelling of proteins

DAVID RICKWOOD

A wide variety of reagents is commercially available for the isotopic labelling of proteins and these are summarized below. Following the summary, the reagents are listed in alphabetical order, with a description of their properties and selected references to demonstrate their use.

1. Summary

Reacting groups	Labelling reagent
—NH$_2$ Free amino groups (N-terminal or lysine residues)	Acetic anhydride Bolton and Hunter reagent Dansyl chloride Ethyl acetimidate 1-Fluoro-2,4-dinitrobenzene Formaldehyde Isethionyl acetimidate Maleic anhydride Methyl 3,5-diiodohydroxybenzimidate Phenyl isothiocyanate Sodium borohydride/potassium borohydride Succinic anhydride N-Succinimidyl propionate
—SH Thiol groups (cysteine residues)	Acetic anhydride Bromoacetic acid Chloroacetic acid p-Chloromercuribenzenesulphonic acid p-Chloromercuribenzoic acid Dansyl chloride N-Ethylmaleimide Iodoacetamide Iodoacetic acid

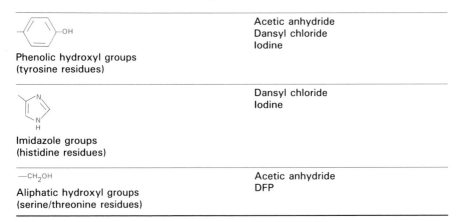

Phenolic hydroxyl groups (tyrosine residues)	Acetic anhydride Dansyl chloride Iodine
Imidazole groups (histidine residues)	Dansyl chloride Iodine
—CH₂OH Aliphatic hydroxyl groups (serine/threonine residues)	Acetic anhydride DFP

2. Reagents

2.1 Acetic anhydride (^{3}H or ^{14}C)

Acetic anhydride is a non-specific acylating agent that may be used to label serine, lysine, threonine, tyrosine or cysteine and N-terminal residues.

References

1. Avivi, P., Simpson, S. A., Tait, J. F., and Whitehead, J. K. (1954). *Proceedings 2nd Radioisotope Conference Oxford,* Vol. 1, p. 313. Butterworths, London.
2. O'Leary, M. H. and Westheimer, F. H. (1968). *Biochemistry,* **7**, 913.
3. Ostrowski, K., Barnard, E. A., Sawicki, W., Chorzelski, T., Langner, A., and Mikulski, A. (1970). *J. Histochem. Cytochem.,* **18**, 490.
4. Heinegård, D. K. and Hascall, V. C. (1979). *J. Biol. Chem.,* **254**, 921.
5. Whitehead, J. K. (1958). *Biochem. J.,* **68**, 662.
6. Barnard, E. A., Wieckowski, J., and Chiu, T. H. (1971). *Nature,* **234**, 207.
7. Brems, D. N. and Rilling, Hans C. (1979). *Biochemistry,* **18**, 860.
8. Gersten, D. M. and Goldstein, L. S. (1979). *Int. J. Appl. Radiat. Isot.,* **30**, 469.

2.2 Bolton and Hunter reagent (^{125}I)

This reagent is specific for free amino groups and has been used extensively for the iodination of proteins under mild conditions.

References

1. Kågedal, B. and Källberg, M. (1977). *Clin. Chem.*, **23**, 1694.
2. Roberts, R., Sobel, B. E., and Parker, C. W. (1978). *Clin. Chim. Acta*, **83**, 141.
3. Culvenor, J. G. and Evans, W. H. (1977). *Biochem. J.*, **168**, 475.
4. Pinder, J. C., Phethean, J., and Gratzer, W. B. (1978). *FEBS Lett.*, **92**, 278.

2.3 Bromoacetic acid (^{14}C)

The haloacetic acids, bromo-, chloro-, and iodoacetic acid, are the most widely used protein alkylating reagents. They react primarily with cysteine residues. The reactivity depends on the halide, and decreases in the order iodoacetic acid > bromoacetic acid > chloroacetic acid.

Bromoacetic acid is a fairly mild alkylating reagent but it is not very selective, reacting with most thiol groups except those in very protected positions.

References

1. Glick, D. M., Goren, H. J., and Barnard, E. A. (1967). *Biochem. J.*, **102**, 7c.
2. Fanger, M. W., Hettinger, T. P., and Harbury, H. A. (1967). *Biochemistry*, **6**, 713.

2.4 Chloroacetic acid (^{14}C)

Chloroacetic acid is less reactive than bromoacetic acid. It is therefore used for very selective alkylations since it will only react with the most reactive thiol groups in the protein.

Reference

1. Gerwin, B. I. (1967). *J. Biol. Chem.*, **242**, 451.

2.5 *p*-Chloromercuribenzenesulphonic acid (^{203}Hg)

Organic mercurials react rapidly and specifically with thiol groups at about pH 5. The mercurials absorb strongly in the ultraviolet and the reaction may be used quantitatively.

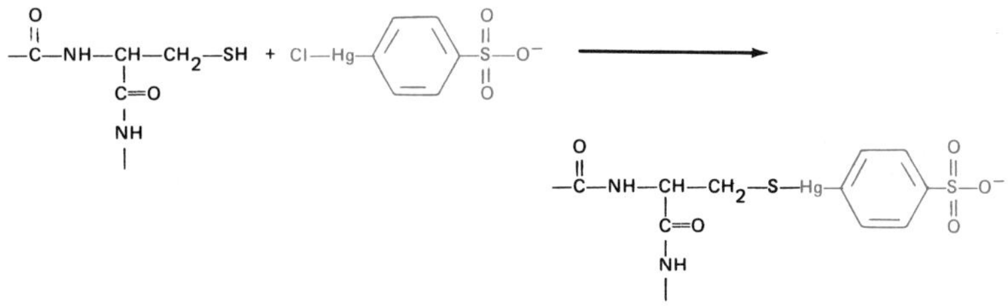

$$-\overset{O}{\underset{\parallel}{C}}-NH-\underset{\underset{NH}{\overset{|}{C=O}}}{\overset{|}{CH}}-CH_2-SH \;+\; Cl-Hg-\!\!\bigcirc\!\!-\overset{O}{\underset{\parallel}{\underset{O}{\overset{\parallel}{S}}}}-O^- \longrightarrow$$

$$-\overset{O}{\underset{\parallel}{C}}-NH-\underset{\underset{NH}{\overset{|}{C=O}}}{\overset{|}{CH}}-CH_2-S-Hg-\!\!\bigcirc\!\!-\overset{O}{\underset{O}{\overset{\parallel}{\underset{\parallel}{S}}}}-O^-$$

Reference

1. Velick, S. F. (1953). *J. Biol. Chem.*, **203**, 563.

2.6 *p*-Chloromercuribenzoic acid (^{203}Hg)

This reagent is used in a similar way to *p*-chloromercuribenzenesulphonic acid for labelling thiol residues with ^{203}Hg.

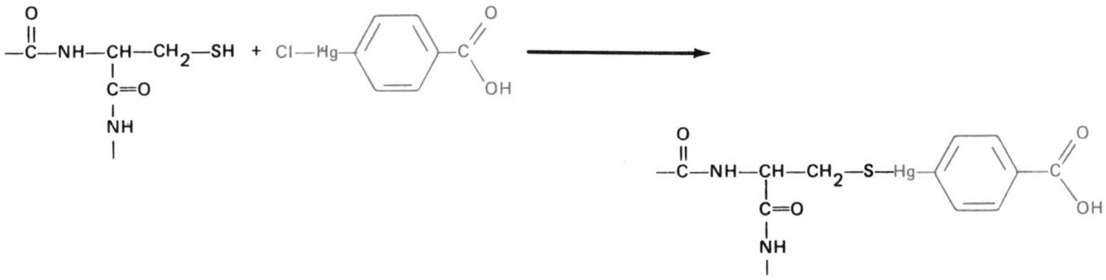

$$-\overset{O}{\underset{\parallel}{C}}-NH-\underset{\underset{NH}{\overset{|}{C=O}}}{\overset{|}{CH}}-CH_2-SH \;+\; Cl-Hg-\!\!\bigcirc\!\!-\overset{O}{\overset{\parallel}{C}}\!\!-OH \longrightarrow$$

$$-\overset{O}{\underset{\parallel}{C}}-NH-\underset{\underset{NH}{\overset{|}{C=O}}}{\overset{|}{CH}}-CH_2-S-Hg-\!\!\bigcirc\!\!-\overset{O}{\overset{\parallel}{C}}\!\!-OH$$

References

1. Waterman, M. R. (1974). *Biochim. Biophys. Acta,* **371**, 159.
2. Boyer, P. D. (1954). *J. Am. Chem. Soc.*, **76**, 4331.
3. Bucci, E. and Fronticelli, C. (1965). *J. Biol. Chem.*, **240**, PC551.
4. Guha, A., England, S., and Listowsky, I. (1968). *J. Biol. Chem.*, **243**, 609.
5. Erwin, V.G. and Pedersen, P.L. (1968). *Anal. Biochem.*, **25**, 477.

2.7 Dansyl chloride (5-dimethylamino-1-naphthalene sulphonyl chloride) (^{3}H or ^{14}C)

Dansyl chloride reacts with amino, thiol, imidazole and phenolic hydroxyl groups. In general, the reaction with aliphatic hydroxyl groups is very slow.

Appendix 3

This reagent is used widely to detect very small amounts of protein by means of the intense fluorescence of the sulphonamide formed when it reacts with the terminal amino group of proteins or peptides. These sulphonamides are stable in hot acid and the assay method using dansyl chloride has nearly 100-fold greater sensitivity than that using 1-fluoro-2,4-dinitrobenzene.

References

1. Chen, R. F. (1968). *Anal. Biochem.*, **25**, 412.
2. Gray, W. R. and Hartley, B. S. (1963). *Biochem. J.*, **89**, 59P.
3. Schultz, R. M. and Wassarman, P. M. (1977). *Anal. Biochem.*, **77**, 25.
4. Venn, R. F., Basford, J. M., and Curtis, C. G. (1978). *Anal. Biochem.*, **87**, 278.
5. Airhart, J., Kelley, J., Brayden, J. E., and Low, R. B. (1979). *Anal. Biochem.*, **96**, 45.

2.8 DFP (Di-isopropyl phosphorofluoridate) (^{3}H)

DFP is a specific reagent for serine residues. It is a pseudosubstrate reacting with active-site serine residues in many proteases and esterases. The electrophilic phosphorus reacts with the hydroxyl group forming a stable modified enzyme.

References

1. Darzynkiewicz, Z. and Barnard, E. A. (1967). *Nature,* **213**, 1198.
2. Budd, G. C., Darzynkiewicz, Z., and Barnard, E. A. (1967). *Nature,* **213**, 1202.
3. Ostrowski, K., Barnard, E. A., Darzynkiewicz, Z., and Rymaszewska, D. (1964). *Exp. Cell Res.,* **36**, 43.
4. Rogers, A. W., Darzynkiewicz, Z., Salpeter, M. M., Ostrowski, K., and Barnard, E. A. (1969). *J. Cell Biol.,* **41**, 665.
5. Fischer, E. P. and Thompson, K. S. (1979). *J. Biol. Chem.,* **254**, 50.

2.9 Ethyl acetimidate (^{14}C)

This reagent reacts specifically with free amino groups under relatively mild conditions. It penetrates cells without impairing membrane function, and labels the protein under physiological conditions.

Ethyl acetimidate is rapidly hydrolysed by water, but the products do not interfere with the main reaction.

References

1. Hunter, M. J. and Ludwig, M. L. (1962). *J. Am. Chem. Soc.*, **84**, 3491.
2. Whiteley, N. M. and Berg, H. C. (1974). *J. Mol. Biol.*, **87**, 541.

2.10 *N*-Ethylmaleimide (^{14}C)

This reagent reacts specifically with more exposed thiol groups and may be used over a wide temperature range at neutral pH. Although *N*-ethylmaleimide may be used quantitatively to determine thiol groups, it has also been widely used to study the effects on enzyme activity of substitution of active-site thiol groups.

References

1. Sekine, T., Barnett, L. M., and Kielley, W. W. (1962). *J. Biol. Chem.*, **237**, 2769.
2. Lai, Tzen-son (1971). *J. Chin. Chem. Soc.*, **18** (3), 145.
3. Barns, R. J. and Keech, D. B. (1968). *Biochim. Biophys. Acta,* **159**, 514.
4. Yamada, S. and Ikemoto, N. (1978). *J. Biol. Chem.*, **253**, 6801.
5. Kielley, W. W. and Barnett, L. M. (1961). *Biochim. Biophys. Acta,* **51**, 589.
6. Riggs, A. (1961). *J. Biol. Chem.*, **236**, 1948.
7. Gadasi, H., Maruta, H., Collins, J. H., and Korn, E. D. (1979). *J. Biol. Chem.*, **254**, 3631.

2.11 1-Fluoro-2,4-dinitrobenzene (^{3}H or ^{14}C)

This reagent is widely used under mildly alkaline conditions to identify terminal amino acids. The N-terminal residue can be separated after hydrolysis of the modified peptide and identified by comparison with standards.

At strongly alkaline pH, 1-fluoro-2,4-dinitrobenzene also reacts with phenolic, thiol and imidazole groups, but these dinitrophenyl groups may be displaced by treatment of the modified protein at pH 8 with 2-mercaptoethanol.

References

1. Whitehead, J. K. (1961). *Biochem. J.*, **80**, 35P.
2. Schultz, R. M., Bleil, J. D., and Wassarman, P. M. (1978). *Anal. Biochem.*, **91**, 354.
3. Travis, J. and McElroy, W. D. (1966). *Biochemistry*, **5**, 2170.
4. Gerber, G. B. and Remy-Defraigne, J. (1965). *Anal. Biochem.*, **11**, 386.

2.12 Formaldehyde (^{14}C)

Formaldehyde is used primarily with a reducing agent such as sodium cyano-borohydride in reductive methylation of free amino groups. However, because of its high reactivity and water solubility it may also be used to bring about crosslinking by reaction with thiol and amino groups.

References

1. Dottavio-Martin, D. and Ravel, J. M. (1978). *Anal. Biochem.*, **87**, 562.
2. Rice, R. H. and Means, G. E. (1971). *J. Biol. Chem.*, **246**, 831.
3. Nelles, L. P. and Bamburg, J. R. (1979). *Anal. Biochem.*, **94**, 150.
4. Peterson, D. T., Merrick, W. C., and Safer, B. (1979). *J. Biol. Chem.*, **254**, 2509.
5. Tolleshaug, H., Berg, T., Fröhlich, W., and Norum, K. R. (1979). *Biochim. Biophys. Acta*, **585**, 71.
6. MacKeen, L. A., DiPeri, C., and Schwartz, I. (1979). *FEBS Lett.*, **101**, 387.

2.13 Iodine (^{125}I)

Usually [^{125}I]iodine is used for labelling proteins because it has a longer half-life than ^{131}I (60 days compared to 8 days for ^{131}I) and is safer to use since the γ-rays are much less penetrating. Iodination is carried out using sodium iodide in an oxidizing environment which favours the formation of the cation I^{+}. The formation of these ions can be catalysed either chemically or enzymatically although the choice of conditions is critical if protein activity is to be preserved (1). Tyrosine is the amino acid most commonly modified though in more alkaline conditions histidine is also iodinated.

An alternative very mild procedure for labelling with ^{125}I uses the Bolton and Hunter reagent which is specific for amino groups (Section 2.2).

References

1. Bolton, A. E. (1977). *Radioiodination Techniques* (Review 18), Radiochemical Centre, Amersham, England.
2. Samols, E. and Williams, H. S. (1961). Nature, **190**, 1211.
3. Greenwood, F. C., Hunter, W. M., and Glover, J. S. (1963). *Biochem. J.*, **89**, 114.
4. Redshaw, M. R. and Lynch, S. S. (1974). *J. Endocrinol.*, **60**, 527.
5. Thorell, J. I. and Johansson, B. G. (1971). *Biochim. Biophys. Acta*, **251**, 299.

2.14 Iodoacetamide (^{14}C)

Iodoacetamide reacts with cysteine residues. This carboxyamidation reaction is often used to convert every cysteine residue within a protein to a derivative that is stable to acid hydrolysis except in the presence of oxygen.

References

1. Truitt, C. D., Hermodson, M. A., and Zalkin, H. (1978). *J. Biol. Chem.*, **253**, 8470.
2. Inagami, T. (1965). *J. Biol. Chem.*, **240**, PC 3453.
3. Inagami, T. and Hatano, H. (1969). *J. Biol. Chem.*, **244**, 1176.
4. Heinrikson, R. L. (1966). *J. Biol. Chem.*, **241**, 1393.
5. Anderson, J. M. (1979). *J. Biol. Chem.*, **254**, 959.
6. Toste, A. P. and Cooke, R. (1969). *Anal. Biochem.*, **95**, 317.
7. Nusgens, B. and Lapiere, Ch. M. (1979). *Anal. Biochem.*, **95**, 406.
8. Kröger, M., Sternbach, H., and Cramer, F. (1979). *Eur. J. Biochem.*, **95**, 341.

2.15 Iodoacetic acid (^{3}H or ^{14}C)

Iodoacetic acid is the most reactive of the haloacetic acids (see bromoacetic acid and chloroacetic acid). It reacts primarily with cysteine residues but is not very selective. With decreasing pH, the reactivity increases and iodoacetic acid will also react with methionine, histidine, lysine, aspartate and glutamate residues.

References

1. Takahashi, K., Stein, W. H., and Moore, S. (1967). *J. Biol. Chem.*, **242**, 4682.
2. Baldwin, G. S., Waley, S. G., and Abraham, E. P. (1979). *Biochem. J.*, **179**, 459.
3. Harris, I., Meriwether, B. P., and Harting Park, J. (1963). *Nature*, **198**, 154.
4. Colman, R. F. (1968). *J. Biol. Chem.*, **243**, 2454.
5. Price, P. A., Moore, S., and Stein, W. H. (1969). *J. Biol. Chem.*, **244**, 924.
6. Li, T. K. and Vallee, B. L. (1965). *Biochemistry*, **4**, 1195.
7. Crestfield, A. M., Stein, W. H., and Moore, S. (1963). *J. Biol. Chem.*, **238**, 2413.
8. Neumann, R. P., Moore, S., and Stein, W. H. (1962). *Biochemistry*, **1**, 68.
9. Weinryb, I. (1968). *Arch. Biochem. Biophys.*, **124**, 285.
10. Baldwin, G. S., Waley, S. G., and Abraham, E. P. (1979). *Biochem. J.*, **179**, 459.
11. Wiman, K., Trägardh, L., Rask, L., and Peterson, P. A. (1979). *Eur. J. Biochem.*, **95**, 265.
12. Holmgren, A. (1979). *J. Biol. Chem.*, **254**, 3664.
13. Anderson, P. J. (1979). *Biochem. J.*, **179**, 425.

2.16 Isethionyl acetimidate (^{14}C)

Unlike ethyl acetimidate (see Section 2.9), isethionyl acetimidate is unable to penetrate intact cells and may therefore be used to label the proteins on the outer surfaces of the membranes. It reacts specifically with free amino groups under relatively mild conditions.

References

For references to the use of this material, see Section 2.9.

2.17 Maleic anhydride (^{14}C)

This reagent may be used for the reversible alkylation of amino groups. Maleyl proteins tend to be water soluble and stable at neutral pH but are rapidly hydrolysed on acidification. This hydrolysis is more rapid than that of the corresponding succinyl derivatives (see Section 2.22).

The much slower maleylation of thiol groups is not reversed by acidification.

Reference

1. Butler, P. J. G., Harris, J. I., Hartley, B. S., and Leberman, R. (1969). *Biochem. J.*, **112**, 679.

2.18 Methyl 3,5-diiodohydroxybenzimidate (^{125}I)

This reagent is used to label lysine residues and terminal amino groups. It is a milder reagent than the Bolton and Hunter reagent, and has the advantage of preserving the charge on the protein.

References

1. Wood, F. T., Wu, M. M., and Gerhart, J. C. (1975). *Anal. Biochem.*, **69**, 339.
2. Ulevitch, R. J. (1978). *Immunochemistry*, **15**, 157.
3. Morgan, J. L., Holladay, C. R., and Spooner, B. S. (1978). *FEBS Lett.*, **93**, 141.
4. Morgan, J. L., Holladay, C. R., and Spooner, B. S. (1978). *Proc. Natl. Acad. Sci. USA*, **75**, 1414.
5. Miller, N. E., Weinstein, D. B., and Steinberg, D. (1978). *J. Lipid Res.*, **19**, 644.
6. Subramani, S., Bothwell, M. A., Gibbons, I., Yang, Y. R., and Schachman, H. K. (1977). *Proc. Natl. Acad. Sci. USA*, **74**, 3777.

2.19 Phenyl isothiocyanate (^{14}C or ^{35}S)

This reagent reacts preferentially with terminal amino groups and is used to sequence peptides by step-wise degradation (Edman degradation). Each phenylthiohydantoin formed can be identified by comparison with standards.

$$-NH-\overset{O}{\overset{||}{C}}-\underset{R^1}{CH}-NH-\overset{O}{\overset{||}{C}}-\underset{R}{CH}-NH_2 \; + \; S=C=N-\bigcirc \xrightarrow{OH^-}$$

$$-NH-\overset{O}{\overset{||}{C}}-\underset{R^1}{CH}-NH-\overset{O}{\overset{||}{C}}-\underset{R}{CH}-NH-\overset{S}{\overset{||}{C}}-NH-\bigcirc$$

$$\downarrow H^+$$

$$-NH-\overset{O}{\overset{||}{C}}-\underset{R^1}{CH}-NH_2 \; + \; \text{(phenylthiohydantoin ring)}$$

References

1. Callewaert, G. L. and Vernon, C. A. (1968). *Biochem. J.*, **107**, 728.
2. Laver, W. G. (1961). *Biochim. Biophys. Acta,* **53**, 469.
3. Laver, W. G. (1961). *Virology,* **14**, 499.
4. Geising, W. and Hornle, S. (1973). *Peptides* (Proceedings of the 11th Peptide Symposium, 1971), published 1973, p. 146 (in German).
5. Levy, N. L. and Dawson, J. R. (1976). *J. Immunol.,* **116**, 1526.

2.20 Potassium borohydride (^{3}H)

This reagent may be used as an alternative to sodium borohydride (see Section 2.21) in reductive methylation of free amino groups. It has the advantage of being water soluble and stable in aqueous solution for short periods, unlike sodium borohydride which is hydrolysed almost instantaneously.

$$-\overset{O}{\overset{||}{C}}-NH-CH(\overset{\underset{C=O}{|}}{\underset{NH}{|}})-(CH_2)_4-NH_2 \; + \; HCHO \longrightarrow -\overset{O}{\overset{||}{C}}-NH-CH(\underset{C=O,NH}{|})-(CH_2)_4-N=CH_2$$

$$\downarrow KBH_4$$

$$-\overset{O}{\overset{||}{C}}-NH-CH(\underset{C=O,NH}{|})-(CH_2)_4-NH-CH_3$$

Reference

1. Kumarasamy, R. and Symons, R. H. (1979). *Anal. Biochem.*, **95**, 359.

2.21 Sodium borohydride (^{3}H)

This reagent is used together with formaldehyde (see Section 2.12) in the reductive methylation of free amino groups. The formaldehyde reacts with the amino group to form a Schiff's base which is then reduced with the borohydride. Using tritiated sodium borohydride the *N*-[^{3}H]methyl derivative is produced.

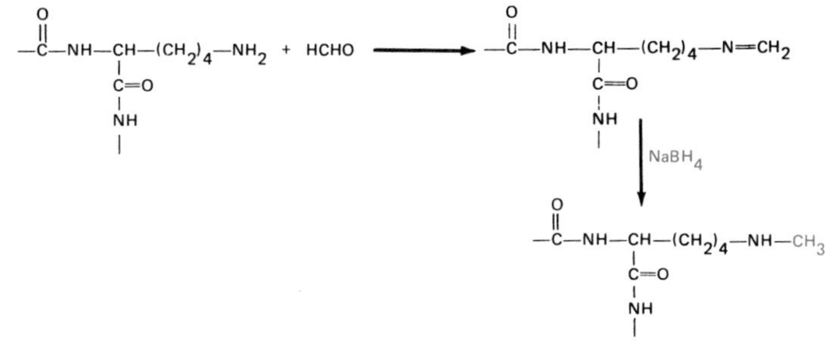

References

1. Biocca, S., Calissano, P., Barra, D., and Fasella, P. M. (1978). *Anal. Biochem.*, **87**, 334.
2. De La Llosa, P., Marche, P., Morgat, J. L., and De La Llossa-Hermier, M. P. (1974). *FEBS Lett.*, **45**, 162.
3. Means, G. E. and Feeney, R. E. (1968). *Biochemistry*, **7**, 2192.
4. Moore, G. and Crichton, R. R. (1973). *FEBS Lett.*, **37**, 74.
5. Chansel, D., Sraer, J., Morgat, J. L., Hesch, R. D., and Ardaillou, R. (1977). *FEBS Lett.*, **78**, 237.
6. Ascoli, M. and Puett, D. (1974). *Biochim. Biophys. Acta*, **371**, 203.
7. Keul, V., Kaeppeli, F., Ghosh, C., Krebs, T., Robinson, J. A., and Retey, J. (1979). *J. Biol. Chem.*, **254**, 843.

2.22 Succinic anhydride (^{14}C)

Succinylation of free amino groups is carried out in mildly alkaline solution using conditions similar to those for acetylations with acetic anhydride. However, whereas acetylation of the cationic group yields an electrically neutral product, succinylation yields an anionic product. For this reason, succinylated proteins may be more soluble than the acetylated ones.

Although succinylation is reversible under acid conditions, the reaction is less facile than hydrolysis of the corresponding maleylated proteins (see Section 2.17).

References

1. Habeeb, A. F. S. A., Cassidy, H. G., and Singer, S. J. (1958). *Biochim. Biophys. Acta,* **29**, 587.
2. Chu, F. S., Crary, E., and Bergdoll, M. S. (1969). *Biochemistry,* **8**, 2890.
3. Frist, R. H., Bendet, I. J., Smith, K. M., and Lauffer, M. A. (1965). *Virology,* **26**, 558.

2.23 *N*-Succinimidyl propionate (^{3}H)

This reagent is specific for free amino groups and sulphydryl groups at neutral pH. It reacts in an analogous manner to the Bolton and Hunter reagent (see Section 2.2). It has the advantage of being a smaller molecule than the Bolton and Hunter reagent and hence causes less alteration to the protein structure.

References

1. Tang, Y. S., Davis, A.-M., and Kitcher, J. P. (1982). *J. Lab. Comps. Radiopharm.,* **XX**, 277.
2. Fink, D. J. and Gainer, H. (1980). *Science,* **208**, 303.
3. Ockleford, C. D. and Clint, J. M. (1980). *Placenta,* **1**, 91.
4. Muller, G. H. (1980). *J. Cell Sci.,* **43**, 319.
5. Dolly, J. O., Nockles, A. V., Lo, M. M. S., and Barnard, E. A. (1981). *Biochem. J.,* **193**, 919.
6. Caras, I. W., Friedlander, E. J., and Bloch, K. (1980). *J. Biol. Chem.,* **255**, 3575.
7. Kummer, U. (1981). *J. Immunol. Meth.,* **42**, 367.

Acknowledgement

I wish to express my thanks to Amersham International for allowing me to base this bibliography on their Technical Bulletin.

A4

Molecular masses and isoelectric points of selected marker proteins

B. D. HAMES

1. Marker proteins for SDS-PAGE in the presence of thiol reagents

Molecular masses of suitable polypeptides for use in SDS-PAGE in the *presence* of thiol reagents are given in *Table 1* below. The polypeptide molecular masses given

---

Table 1. Molecular masses of marker polypeptides in the presence of thiol reagents

Polypeptide	Molecular mass
Myosin (rabbit muscle) heavy chain	212 000
RNA polymerase (*E.coli*) β'-subunit	165 000
β'-subunit	155 000
β-Galactosidase (*E.coli*)	130 000
Phosphorylase a (rabbit muscle)	92 500
Bovine serum albumin	68 000
Catalase (bovine liver)	57 500
Pyruvate kinase (rabbit muscle)	57 200
Glutamate dehydrogenase (bovine liver)	53 000
Fumarase (pig liver)	48 500
Ovalbumin	43 000
Enolase (rabbit muscle)	42 000
Alcohol dehydrogenase (horse liver)	41 000
Aldolase (rabbit muscle)	40 000
RNA polymerase (*E.coli*) α-subunit	39 000
Glyceraldehyde-3-phosphate dehydrogenase (rabbit muscle)	36 000
Lactate dehydrogenase (pig heart)	36 000
Carbonic anhydrase	29 000
Chymotrypsinogen A	25 700
Trypsin inhibitor (soybean)	20 100
Myoglobin (horse heart)	16 950[a]
α-Lactalbumin (bovine milk)	14 400
Lysozyme (egg white)	14 300
Cytochrome c	11 700

[a] Calculated from the sequence data given in ref. 6.

---

here are mainly from refs 1−3 and are the molecular masses in the presence of excess thiol reagent. A more comprehensive list of suitable polypeptides is available from the original sources. The molecular mass range ~ 12 000−68 000 is reasonably well covered but there are few suitable proteins with subunit molecular masses above this range. This can be overcome by using polypeptides which are crosslinked to form an oligomeric series (4, 5). Kits of these are commercially available. Several proteases, for example trypsin, chymotrypsin, and papain, have been used as molecular mass standards by various workers but these may sometimes cause proteolysis of other polypeptide standards and so are omitted here.

2. Molecular masses and isoelectric points of polypeptides in the absence of thiol reagent

The list given in *Table 2* is not intended to be comprehensive but rather a selection of marker proteins which are defined in terms of the number of subunits, subunit molecular mass (in the *absence* of thiol reagent), and isoelectric points, and most of which are commercially available in an essentially pure form. A more comprehensive list is available from the original sources (3, 7). An additional list of protein molecular masses and isoelectric points has been published by Righetti and Caravaggio (8).

Any standard proteins which are found to consist of more than one species should be used with great caution since the protein desired may be only a minor species of the mixture.

Table 2. Molecular masses and isoelectric points of polypeptides in the absence of thiol reagents

Protein	Species	Tissue	Number of subunits	Subunit M_r	Isoelectric point
Adenine phosphoribosyl-transferase	Human	Erythrocyte	3	11 000	4.8
Nerve growth factor	Mouse	Salivary gland	2	13 259	9.3
Ribonuclease	Bovine	Pancreas	1	13 700	7.8
Haemoglobin	Rabbit	Erythrocyte	4	16 000	7.0
Micrococcal nuclease	S. aureus	−	1	16 800	9.6
β-Lactoglobulin	Bovine	Serum	2	17 500	5.2
Ceramide trihexosidase	Human	Plasma	4	22 000	3.0
Adenylate kinase	Rat	Liver (cytosol)	3	23 000	7.5
Trypsinogen	Cow	Pancreas	1	24 500	9.3
Chymotrypsinogen A	Bovine	Pancreas	1	25 700	9.2
Triosephosphate isomerase	Rabbit	Muscle	2	26 500	6.8
Galactokinase	Human	Erythrocyte	2	27 000	5.7
Arginase	Human	Liver	4	30 000	9.2
Deoxyribonuclease I	Cow	Pancreas	1	31 000	4.8
Uricase	Pig	Liver	4	32 000	6.3
Glycerol-3-phosphate dehydrogenase	Rabbit	Kidney	2	34 000	6.4
Malate dehydrogenase	Pig	Heart	2	35 000	5.1

Table 2. (continued)

Protein	Species	Tissue	Number of subunits	Subunit M_r	Isoelectric point
Alcohol dehydrogenase	Yeast	–	4	35 000	5.4
Deoxyribonuclease II	Pig	Spleen	1	38 000	10.2
Aldolase	Yeast	–	2	40 000	5.2
Pepsinogen	Pig	Stomach	1	41 000	3.7
Hexokinase	Yeast	–	2	51 000	5.3
Lipoxidase	Soybean	–	2	54 000	5.7
Catalase	Cow	Liver	4	57 500	5.4
Alkaline phosphatase	Calf	Intestine	2	69 000	4.4
Acetylcholinesterase	Electrophorus	–	4	70 000	4.5
Glyceraldehyde-3-phosphate dehydrogenase	Rabbit	Muscle	2	72 000	8.5
β-Glucuronidase	Rat	Liver	4	75 000	6.0
Lysine decarboxylase	E.coli	–	10	80 000	4.6
Glycogen synthetase	Pig	Kidney	4	92 000	4.8
Phosphoenolpyruvate carboxylase	E.coli	–	4	99 600	5.0
Phosphoenolpyruvate carboxylase	Spinach	Leaf	2	130 000	4.9
Urease	Jack Bean	–	2	240 000	4.9

References

1. Weber, K. and Osborn, M. (1969). *J. Biol. Chem.*, **244**, 4406.
2. Lambin, P. C. (1978). *Anal. Biochem.*, **85**, 114.
3. Fasman, G. D. (ed.) (1976). *Handbook of Biochemistry and Molecular Biology, Proteins.* 3rd edn, Vol. II. CRC Press, Cleveland, Ohio, USA.
4. Payne, J. W. (1973). *Biochem. J.*, **135**, 867.
5. Carpenter, F. H. and Harrington, K. T. (1972). *J. Biol. Chem.*, **247**, 5580.
6. Dautrevaux, M., Boulanger, Y., Han, K., and Biserte, G. (1969). *Eur. J. Biochem.*, **11**, 267.
7. Malamud, D. and Drysdale, J. W. (1978). *Anal. Biochem.*, **86**, 620.
8. Righetti, P. G. and Caravaggio, T. (1976). *J. Chromatogr.*, **127**, 1.

A5

Applications of two-dimensional gel electrophoresis

J. A. ALEC CHAMBERS and DAVID RICKWOOD

1. Applications

1.1. Characterization of cells

System	References
E. coli	1
HeLa cells	2
Brain (rat)	3
Muscle (human)	4
Kidney (human)	5
Liver (rat)	6
Liver and brain (mouse)	7
Plasma proteins (human)	8

1.2 Characterization of organelles

Organelle	System	References
Nucleus	Fibroblast nuclear matrix	9
	Rat non-histone proteins	10
Membranes	*Physarum* plasmodia and amoeba	11
	GTP-binding proteins of rat brain	12
Mitochondria	Rat liver	13
	Pisum sativum (Pea)	14
Chloroplast	*Pisum sativum*	15

1.3 Characterization of proteins and modifications

Modification	System	References
Phosphorylation	Eukaryotic translation initiation factors	16
Hormonally-regulated systems		17

Sulphation	Human hepatoma cell line	18
Acetylation	Flagellar tubulins of *Naegleria gruberi*	19
Glycosylation	Trypanosome variable surface antigens	20
ADP-ribosylation	Nuclear proteins of HeLa cells	21
	Physarum polycephalum nuclear proteins	22
Fatty acid acylation	Membrane proteins	23

1.4. Environmental changes in cell proteins

Stimulus	*System*	*References*
Hormonal or growth factor stimulation	Auxin-stimulated soybean hypocotyls	24
	Pig Sertoli cells	25
	Epidermal growth factor stimulation	26
Light induction	Blue-light induction in *Neurospora crassa*	27
Heat shock	*Drosophila* cells	28
Exposure to toxic, carcinogenic or mutagenic substances	Action of chlorinated hydrocarbon (Aroclor 1254) on rat liver	29
	Effects of ethylnitrosourea	30

1.5 Developmental changes of cell proteins

System	*References*
Lower eukaryotes (*Dictyostelium discoideum* developmental cycle)	31,32
(conidiation in *Neurospora crassa*)	33
Plants (greening of *Pisum sativum*)	34
Invertebrates (sea-urchin embryogenesis)	35
Vertebrates (mouse mammary epithelium)	36
Culture systems (rat muscle cell line L6)	37

1.6 Genetic variation of proteins

Type of study	*System*	*References*
Characterization of mutant proteins	*Aspergillus nidulans* tubulins	38
Effects of mutation on higher-order structures	*Chlamydomonas* flagella	39

1.7 Medical and clinical applications

References

1. Neidhardt, F. C. and Phillips, T. A. (1984). In *Two-Dimensional Gel Electrophoresis of Proteins: Methods and Applications,* (ed. J. E. Celis and R. Bravo), p. 417. Academic Press, New York.
2. Bravo, R. and Celis, J. (1984). In *Two-Dimensional Gel Electrophoresis of Proteins: Methods and Applications,* (ed. J. E. Celis and R. Bravo), p. 445. Academic Press, New York.
3. Heydorn, W. E., Gierschik, P., Creed, C. J., Patel, J., and Jacobowitz, D. M. (1986). *Neurochem. Int.,* **9**, 357.
4. Giometti, C. S., Anderson, N. G., and Anderson, N. L. (1979). *Clin. Chem.,* **25**, 1877.
5. Smith, S. C., Racine, R. R., and Langley, C. H. (1980). *Genetics,* **96**, 967.
6. Iynedjian, P. B., Mobius, G., Seitz, H. J., Wollheim, C. B., and Renold, A. E. (1986). *Proc. Natl. Acad. Sci. USA,* **83**, 1998.
7. Voris, B. P. and Young, D. A. (1980). *Anal. Biochem.,* **104**, 478.
8. Klose, J. and Feller, M. (1981). *Electrophoresis,* **2**, 12.
9. Anderson, N. L. and Anderson, N. G. (1977). *Proc. Natl. Acad. Sci. USA,* **74**, 5421.
10. Fey, E. G. and Penman, S. (1988). *Proc. Natl. Acad. Sci. USA,* **85**, 121.
11. Inoue, A., Higashi, V., Hasuma, T., Horisawa, H., and Yukioka, M. (1983). *Eur. J. Biochem.,* **135**, 61.
12. Pallotta, D., Barden, A., Martel, R., Kirovac-Brunet, J., Bernier, F., Lord, A., and Lemieux, G. (1986). *Can. J. Biochem. Cell. Biol.,* **62**, 831.
13. Heydorn, W. E., Gierschik, P., Creed, G. J., Milligan, G., Spiegel, A., and Jacobowitz, D. M. (1986). *J. Neurosci. Res.,* **16**, 541.
14. Henslee, J. G. and Srere, P. A. (1979). *J. Biol. Chem.,* **254**, 5488.
15. Remy, R., Ambard-Bretteville, F., and Colas des France, C. (1987). *Electrophoresis,* **8**, 528.

16. Dietz, K.-J. and Bogorad, L. (1987). *Plant Physiol.*, **85**, 808.
17. Clemens, M. J., Galpine, A., Austin, S. A., Pannieres, R., Henshaw, E. C., Duncan, R., Hershey, J. W. Z., and Pollard, J. W. (1987). *J. Biol. Chem.*, **262**, 767.
18. Murtaugh, T. J., Wright, L. S., and Siegel, F. L. (1987). *J. Biol. Chem.*, **260**, 15932.
19. Liu, M.-C., Yu, S., Sy, J., Redmann, C. M., and Lipman, F. (1985). *Proc. Natl. Acad. Sci. USA*, **82**, 7160.
20. Shea, D. K. and Walsh, C. J. (1987). *J. Cell Biol.*, **105**, 1303.
21. Pearson, T. W., Kar, S. K., McGuire, T. C., and Lundin, L. B. (1981). *J. Immunol.*, **126**, 823.
22. Adolph, K. W. (1987). *Biochim. Biophys. Acta*, **909**, 222.
23. Poirier, G. G., Cote, S., and Pallotta, D. (1987). *Biochem. Cell. Biol.*, **65**, 81.
24. Wilcox, C. A. and Olson, E. N. (1987). *Biochemistry*, **26**, 1029.
25. Zurfluh, L. L. and Guilfoyle, T. J. (1980). *Proc. Natl. Acad. Sci. USA*, **77**, 357.
26. Perrard-Sapori, M. H., Saez, J. M., and Dazord, A. (1985). *Mol. Cell. Endocrinol.*, **43**, 189.
27. Lamy, F., Roger, P. D., Lecoq, R., and Dumont, J. E. (1986). *Eur. J. Biochem.*, **155**, 265.
28. Chambers, J., Hinkelammert, K., and Russo, V. E. A. (1985). *EMBO J.*, **4**, 3649.
29. Ballinger, D. G. and Pardue, M. L. (1983). *Cell*, **33**, 103.
30. Anderson, N. L., Swanson, M., Gierg, F. A., Tollaksen, S., Gemmell, A., Vance, S., and Anderson, N. G. (1986). *Electrophoresis*, **7**, 44.
31. Marshall, R. R., Raj, A. S., Grant, F. J., and Heddle, J. A. (1983). *Can. J. Genet. Cytol.*, **25**, 457.
32. Sinclair, J. H. and Rickwood, D. (1985). *Biochem. J.*, **229**, 771.
33. Berlin, V. E. and Yanofsky, C. (1985). *Mol. Cell. Biol.*, **5**, 839.
34. Dietz, K.-J. and Bogorad, L. (1987). *Plant Physiol.*, **85**, 816.
35. Bedard, P. A. and Brandorst, H. P. (1983). *Dev. Biol.*, **96**, 74.
36. Asch, H. L. and Asch, B. B. (1985). *Dev. Biol.*, **107**, 470.
37. Garrels, J. I. (1971). *Dev. Biol.*, **73**, 134.
38. Sheir-Ness, G., Lai, M. H., and Morris, N. R. (1978). *Cell*, **15**, 639.
39. Piperno, G., Huang, B., Ramanis, Z., and Luck, D. J. L. (1981). *J. Cell Biol.*, **88**, 73.
40. Bahrman, N., de Vienne, D., Thiellement, H., and Hofmann, J.-P. (1985). *Biochem. Genet.*, **23**, 247.
41. Dunbar, B. D., Bundman, D. S., and Dunbar, B. S. (1985). *Electrophoresis*, **6**, 39.
42. Olsen, I., Rosseland, S. K., Thorsrud, A. K., and Jellum, E. (1987). *Electrophoresis*, **8**, 532.
43. Gahne, B., Juneja, R. K., and Stratil, A. (1987). *Hum. Genet.*, **76**, 111.
44. Kondo, I., Harada, S., Shibasaki, M., Yamakwa, K., Yamamoto, T., and Hamaguchi, H. (1986). *Hum. Genet.*, **72**, 303.
45. Cheung, C., Ingolca, D. E., Bobonis, C., Dunbar, B. S., Riser, M. E., Siciliano, M. J., and Kellums, R. E. (1983). *J. Biol. Chem.*, **258**, 8338.
46. Tracy, R. P. and Young, D. S. (1984). In *Two-Dimensional Gel Electrophoresis of Proteins: Methods and Applications,* (ed. J. E. Celis and R. Bravo), p. 194. Academic Press, New York.
47. Marshall, R. C. and Gillespie, J. M. (1985). *J. Forensic Sci. Soc.*, **22**, 377.
48. Marshall, T. and Williams, K. M. (1987). *Electrophoresis*, **8**, 493.

Production of polyvalent antibodies for immunoelectrophoresis*

R. D. JURD and T. C. BØG-HANSEN

1. Introduction

Monoclonal antibodies are generally not used for immunoelectrophoresis, mainly because immunoelectrophoresis is based upon precipitation of polyvalent antigens or antigens with multiple epitopes. However, this does not exclude the measurement of the affinity constant for a monoclonal antibody or the characterization of a monoclonal antibody or of a mixture of monoclonal antibodies in immuno-electrophoretic systems. Neither do the authors want to exclude the use of a mixture of monoclonal antibodies for immunoelectrophoretic studies. Readers interested in monoclonal antibodies should consult the relevant literature including a companion volume in *The Practical Approach Series* (1).

A number of animal species are suitable for immunization with antigens of interest. The species most often used nowadays are mice, rats, rabbit, sheep, goat, and pigs. Rabbits are commonly used because they are docile animals which can be kept and handled easily. If large amounts of antibody are required, several animals are immunized to minimize animal-to-animal variations and the blood is pooled before antibody preparations. Either the crude antiserum or purified antibodies can be used for immunoelectrophoresis.

Antigens to be used as immunogens should be as pure as possible since it is usually easier to prepare a pure antigen than to remove unwanted cross-reactivity in an antiserum preparation. Immunization schedules vary widely and the ideal schedule for a given antigen will be found by experience. Two schedules in routine use in the authors' laboratories are given in the next section. Well-tested procedures have also been published elsewhere (e.g. ref. 2).

---

*Workers in the United Kingdom are reminded that, under the terms of the 1986 Animals (Scientific Procedures) Act, they will require Home Office permission to immunize and bleed animals. Further advice should be sought from the experimenter's local Home Office Inspector.

2. Immunization procedures

A rabbit should be immunized with about 0.1 mg of protein per immunization. This amount will give a good response in most cases with a multi-component mixture of proteins for production of a multivalent antibody. In the case of a pure protein, much less protein may give a good high-titre antibody. Note that the length of the immunization period is of importance to obtain a good response as well as the source of the antigen (see below). For large animals it is generally recommended to use larger amounts of protein for immunization. However, good results have been obtained with considerably less immunogen.

The antigen is frequently administered to the rabbits as an emulsion with Freund's adjuvant which generally results in a better immune response. Freund's incomplete adjuvant consists solely of mineral oil and emulsifier whilst Freund's complete adjuvant also contains mycobacteria to produce a far stronger immune reaction. However, so strong is the response with Freund's complete adjuvant that it should not be used repeatedly on the same animal in order to limit trauma. In the two general immunization procedures listed below, *Protocol 1* uses only Freund's incomplete adjuvant which (because it can be used for all the antigen injections) allows the preparation of large numbers of samples for immunization to cover the entire period using only one recipe. The second procedure, *Protocol 2*, uses Freund's complete adjuvant for only the initial antigen injection.

Many other adjuvants are available (3), but the authors find that Freund's adjuvant works well and should be tried before anything else is chosen. However, for viral antigens and for peptides as well as with other antigens it could very well pay to use ISCOMS (immune stimulating complexes with Quil A; see refs 4 and 5).

Protocol 1. Immunization using only Freund's incomplete adjuvant

1. It is advisable to prepare the antigen as a ready-to-use vaccine in a number of vials, each containing enough immunogen for one immunization. Therefore prepare the immunogen as follows.

   (a) Using N animals to be immunized M times, pipette $N \times 0.1$ ml of Freund's incomplete adjuvant into M vials.

   (b) Pipette $N \times 0.1$ ml of protein antigen (1mg/ml) dropwise into each vial under agitation on a mixer (Vortex or similar) to make a water-in-oil emulsion.

   (c) Freeze until use.

2. Immunize each animal with 0.2 ml of the emulsion.

Either (a) Immunize each animal every 14 days for about 3 months then bleed and immunize alternately every week.[a]

or (b) Immunize once every three or four weeks then bleed the week after.[a]

Protocol 1. *continued*

In each immunization, inject 0.1 ml of the antigen emulsified in Freund's incomplete adjuvant under the skin as superficially as possible on both shoulders (*Figure 1*). Do not inject the vaccine into the foot pads; there is no proof that this gives a better antibody but it does cause considerable discomfort to the animal. Indeed, it is doubtful whether such a procedure can now be contemplated, at least in the UK under the present law. After repeated immunization, the animal may maintain a good titre of the antibody without further booster injections, so you may omit every other immunization.

[a] When preparing a monospecific antiserum, it may be advisable to take the antibodies very early in the immunization period, before the animal begins to respond to inevitable impurities present in the 'pure' antigen; remember the animal is a very powerful detection system with good biological amplification for impurities in 'pure' substances.

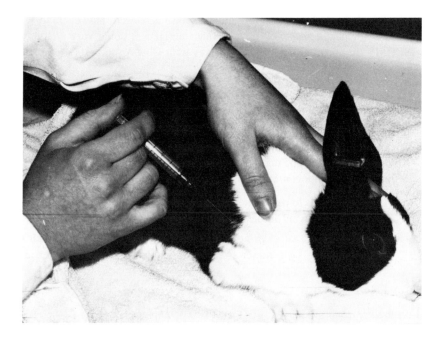

Figure 1. Technique for injecting a rabbit in the upper fore-limb. The rabbit is placed on a towel and is steadied with the left hand.

Protocol 2. Immunization using Freund's complete adjuvant

1. *Day 0.* Inject a 3 kg rabbit with 1 ml of protein antigen solution containing 0.1−5.0 mg of protein in 0.14 M NaCl, buffered to pH 7.3 with 0.02 M sodium phosphate (phosphate-buffered saline) after it has been emulsified with an equal volume of complete Freund's adjuvant (Gibco or Difco Laboratories Ltd), to give 2 ml of emulsion altogether. To emulsify complete Freund's adjuvant with the protein solution, pump the constituents in and out of a 2-ml glass hypodermic syringe fitted with a 19-gauge needle until a thick white emulsion is formed. The stability of the emulsion can be tested by dropping it onto ice-cold water: the second and third drops should float on the surface of the water and not disperse. Inject the emulsion subcutaneously into the upper arms and legs of the rabbit, 0.5 ml of the emulsion at each site (*Figure 1*). It is a safe policy to immunize two rabbits in parallel or, alternatively, in tandem, one rabbit 1 week after the other.

   Smaller amounts of highly purified antigen (down to 50 μg of protein) may be used for the production of monospecific antibody.[a]

2. *Day 28.* Administer a second, booster injection of protein antigen, as on day 0 except use Freund's incomplete adjuvant (Gibco or Difco Laboratories Ltd) instead of complete Freund's adjuvant, and inject subcutaneously between the shoulder blades and above the sacrum (between the hips); inject 1 ml of emulsion at each site.

   Alternatively, 'alum-precipitated' protein antigen may be administered in the booster injection. Mix the protein solution, containing up to 2 mg/ml of protein, with half its volume of 1 M $NaHCO_3$ followed by 1 volume of 0.2 M potassium aluminium sulphate solution. The protein is adsorbed onto the resulting aluminium hydroxide precipitate. Slowly stir the mixture and leave it for 30 min. Then pellet the precipitate by centrifugation for 15 min at 300 *g* and wash it three times with phosphate-buffered saline. Suspend the precipitate in 2 ml of phosphate-buffered saline and inject it intramuscularly into the buttock and shoulder muscles, 0.5 ml at each site.

3. *Day 35.* Collect blood (~25 ml) from the rabbit by nicking a marginal ear vein as described in *Protocol 3*.

4. *Day 42.* Collect a further 25 ml of blood as on day 35.

   If more antiserum is required, the rabbit may be given another injection of antigen, exactly as described for day 28, and bleeding 7 and 14 days later.

---

[a] Even smaller amounts of antigen can be dispersed on the immunosorbent microcrystalline cellulose; 1 μg of protein has been used successfully (6). Proteins separated on polyacrylamide gels may also be used as antigen in which case the polyacrylamide acts as a supplementary adjuvant. The gel containing the protein(s) is homogenized in isotonic saline and then emulsified with complete Freund's adjuvant prior to injection.

3. Collecting blood

Blood can be collected either from a marginal ear vein or by heart puncture. The latter procedure must be left to experienced personnel. Although it yields more blood than ear puncture, the procedure cannot be used if the intention is to keep the animal over a long period of time to supply a steady source of antibody.

To obtain blood from the ear use *Protocol 3*.

Protocol 3. Collection of blood from the ear of a rabbit

1. Place the rabbit on a table and wrap it in a large cloth so that only one ear and the top of the head are exposed. The cloth serves to restrain the animal. Alternatively, use one of the custom-built rabbit restrainers which are commercially available.
2. Rub one ear gently and shave the fur from the posterior margin.
3. Make a longitudinal nick about 0.2 cm long, in the marginal ear vein.
4. Collect the venous blood, dripping from the nick, in a small (100 ml) beaker (*Figure 2*).

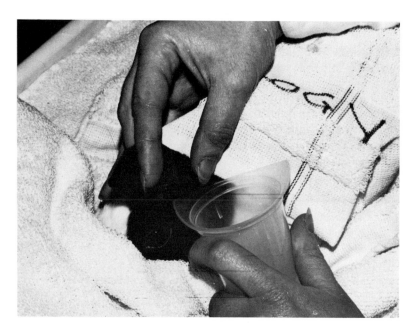

Figure 2. Bleeding a rabbit from the marginal ear vein. The rabbit is wrapped in a cloth or towel. Using the right hand, the ear is held over a small beaker used to collect the blood.

Approximately 25 ml of blood may be collected in this way. The flow of blood may be facilitated by working in a quiet, warm room, by reassuring the rabbit by stroking it, and by warming the ear with a reading lamp placed about 20 cm above it. If the blood flow ceases, it may be restarted by gently wiping the cut with a soft tissue. When finished, bleeding may be arrested by pressing the cut firmly for about 30 sec with a wad of cotton wool.

4. Preparation of antiserum

Preparation of antiserum from the blood involves allowing it to clot and then collecting the serum which remains. A simple procedure is described in *Protocol 4*.

Protocol 4. Preparation of antiserum

1. Let the blood coagulate in the beaker at room temperature. Rim the edges of the clot with a glass pipette to prevent it from adhering to the wall of the beaker. This is important in order to obtain the maximum amount of serum from blood.

2. After about 4 h at room temperature, pipette off the straw-coloured serum. More serum can be obtained by letting the blood clot overnight at 4°C but this usually gives somewhat more haemolysis.

3. Centrifuge the serum (150 g for 15 min) to free it from any residual cells.

4. If only small amounts of antiserum are being prepared, it can now be stored at −20°C in 1 ml aliquots. Alternatively, if several rabbits are being bled, test each antiserum for titre and specificity (see Section 5) and then pool appropriate antiserum preparations before freezing in aliquots. Freezing in aliquots is necessary to avoid multiple freezing and thawing which can lead to reduced antibody activity.

---

5. Assay of antibody activity

Antisera can be titred by simple immunodiffusion as described in *Protocol 5*. However, this method gives very limited data concerning specificity and so other methods should be used to check this (see Chapter 4).

Protocol 5. Determination of antibody titre

1. Prepare a 1% agarose gel (including 0.02% sodium azide) in a Petri dish; the gel should be about 0.2 cm thick.
2. Cut wells in the agar gel in the pattern shown in *Figure 3*. Use a commercial well-cutter (see Chapter 4, *Figure 3*) to do this.
3. Place serially diluted (1:1, 1:2, 1:4, 1:8, etc.) antiserum in a succession of peripheral wells and antigen at a concentration of 1 mg/ml in the central well.
4. Cover the plate and leave it in a humid environment for 48 h. The appearance of a white precipitin line between the antigen and antiserum wells indicates the presence of antibody activity in the serum. The maximum dilution of antiserum giving a visible precipitin line after 48 h gives the 'titre' of the antiserum (7).

6. Purification of IgG antibody

In some cases it is sufficient to use crude antisera for immunoelectrophoresis. However, it is usually advantageous to prepare the IgG fraction of antibodies from rabbit serum since the presence of IgM antibodies in antisera may lead to the formation of a second (IgM−antigen complex) precipitin line in the immunoelectrophoresis gel, rendering analysis of the precipitin patterns more difficult.

Numerous procedures have been described for the purification of antibodies: dialysis (2), salt precipitation and ion-exchange (2), and protein A chromatography (8) to yield the immunoglobulin fraction essentially pure. Immunosorption purifies only the active specifically binding antibody molecules (9).

Two procedures for antibody purification are described here. The first method (*Protocol 6*) is based on chromatography on T-gel (patented by J. Porath, Uppsala, Sweden). This gel matrix is a mercaptoethanol derivative of divinylsulphate-activated agarose. However, it is readily prepared from divinylsulphate-activated agarose as described below. The purification of antibodies using T-gel is rapid (taking less than a single afternoon), easy and inexpensive. Furthermore, the equipment required is generally available in any laboratory; a simple glass column fitted with a UV detector is sufficient. There is no need for either a pump or a fraction collector. Finally, the eluted immunoglobulin fraction is immediately ready for use in an immunoelectrophoresis system.

The second procedure (*Protocol 7*) given is more traditional, relying on ammonium sulphate precipitation and ion-exchange and gel filtration chromatography. Although time-consuming, it does routinely yield pure IgG.

(a)

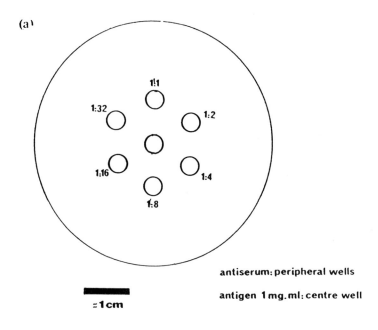

antiserum: peripheral wells

antigen 1 mg.ml: centre well

≏1cm

(b)

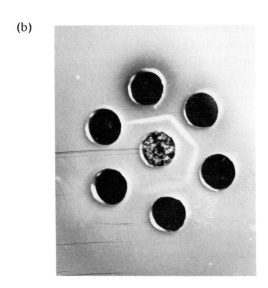

Figure 3. Immunodiffusion plates. (a) Pattern of wells cut in the agarose gel and dilutions of antiserum. (b) A typical immunodiffusion plate. The centre well has antigen (*Limulus* haemocyanin) at a concentration of 1 mg ml^{-1}. Peripheral wells have antiserum to *Limulus* haemocyanin diluted (from top, clockwise) 1:1, 1:4, 1:16, 1:64, 1:256, 1:1024.

Protocol 6. The T-gel method for purification of immunoglobulins

1. To 200 ml of a 1:1 suspension of divinyl sulphone-activated agarose (Minileak, Kem-En-Tec) add 20 ml of 2-mercaptoethanol in a fume cupboard, and titrate the mixture to pH 9 with 2 M NaOH.

2. Let the mixture stand at room temperature overnight. Then wash the gel on a sintered glass filter with 2 litres of water, then 1 litre of 0.75 M $(NH_4)_2SO_4$. The T-gel is now ready to use.

3. Pack a column with 100 ml of T-gel as prepared above.

4. Apply 100 ml of rabbit serum diluted 1:1 with 1.5 M $(NH_4)_2SO_4$ at a flow rate of about 5 ml/min.

5. Wash the column with 300−500 ml of 0.75 M $(NH_4)_2SO_4$ at a flow rate of about 5 ml/min. Follow the elution of protein by UV absorbance at 280 nm.

6. When the UV absorbance reaches the base-line, elute the immunoglobulins with 0.1 M NaCl at a flow rate of 5 ml/min. Collect the peak as monitored by UV absorbance.

7. The T-gel can be reused after washing with 0.75 M ammonium sulphate.

Protocol 7. Preparation of IgG by traditional methodology

1. Mix the serum (100 ml) with 60 ml of saturated ammonium sulphate solution (prepared in 0.02 M Tris−HCl (pH 8.0), 1 mM EDTA) for 30 min at room temperature.

2. Spin down the precipitate at 4000 g for 45 min and redissolve it in 40 ml of the same buffer.

3. Repeat the ammonium sulphate precipitation (using 24 ml of the salt solution) until the precipitate is white.

4. Dissolve the final precipitate in 20 ml of 0.005 M sodium phosphate (pH 8.0), and dialyse the solution against this buffer.

5. The crude IgG solution is further purified by ion-exchange chromatography on DEAE cellulose (DE52) (W & R Balston Ltd). Prepare 3 g for every millilitre of protein solution as follows.

    (a) Add 0.5 M NaH_2PO_4 until the solution reaches pH 8.0. Then dilute the solution with 0.005 M sodium phosphate buffer (pH 8.0) so that there is about 5 ml of buffer for every gram of wet cellulose.

    (b) Stir the slurry up and allow it to settle for about 30 min to give a column 10 cm high.

    (c) Discard the supernatant, add 0.5 volumes of 0.005 M sodium phosphate buffer (pH 8.0) and stir the cellulose into it.

Protocol 7. *continued*

    (d) Pour the slurry into a short, wide column (diameter ∼ 1/5 the height) with the tap open and pack it by pumping 0.005 M sodium phosphate (pH 8.0) through at 60 ml/h/cm^2 internal column cross-section. Typically 5 litres of buffer should be used to wash a 15 cm × 3 cm column.

6. Apply the dialysed protein preparation to the top of the column.

7. Pump the starting buffer (0.005 M sodium phosphate, pH 8.0), contained in the mixing chamber of a gradient mixer, through the column at 60 ml/h and monitor eluant samples for protein content by their UV absorbance at 280 nm. The second chamber contains 0.25 M sodium phosphate (pH 8.0) which is then gradually added to the starting buffer in the mixing chamber of the gradient mixer thus forming a gradient of ionic strength (the shape of which should be convex). The first protein peak eluted contains the IgG.

8. The IgG may be concentrated by ultrafiltration by, for example, using an Amicon apparatus in combination with a PM10 filter.

9. The IgG is further purified by gel chromatography on a column of Sephadex G-200 (Pharmacia Ltd).

    (a) Swell about 20 g (dry weight) of Sephadex G-200 in 0.14 M NaCl, 0.02 M sodium phosphate (pH 7.3) by heating in a boiling water-bath for about 6 h.

    (b) When the swollen gel has cooled, de-gas it for 5 min by stirring it in a Buchner flask attached to a vacuum pump; this removes air bubbles which could distort the elution of the protein bands.

    (c) Pour the gel down a glass rod into a vertical chromatography column (100 cm × 2.6 cm). All the gel slurry must be poured into the column at the same time. Open the outlet at the base of the column and allow the gel to settle over a period of several hours.

    (d) Wash the column by allowing 3 vol. of phosphate-buffered saline to flow through it. The flow rate should be maintained at 15−25 ml/h using a peristaltic pump.

10. Load up to 7.5 ml of the IgG preparation (concentration about 2 mg/ml) onto the column and run it at a flow rate of between 15 and 25 ml/h. Collect 3-ml samples of eluate using a fraction collector. Monitor the samples for protein content; the second peak to be eluted contains the IgG.

---

7. Removal of unwanted antibody activity

Immunodiffusion in agar gels of antisera or IgG preparations using proteins related to the antigen may reveal unwanted cross-reactivity to shared antigenic determinants. Thus, antiserum to protein X may cross-react with proteins Y and Z because all share a given determinant. Absorption may remove this. Using the example above, anti-X can be mixed with proteins Y and Z (the exact amounts being found by experience) and left overnight at 4°C. Next morning the mixture is centrifuged at 5000 *g* for 30 min to remove antibody−antigen complexes. The supernatant can be tested by immunodiffusion to see if any anti-Y or anti-Z activity remains. The minimum amount of Y and Z needed to remove unwanted antibody activity completely can be ascertained by experiment.

8. Commercially available antibody

Commercially prepared antisera of high titre and specificity are available for many common antigens. The catalogues of the following firms are worth inspection, although the list is by no means exhaustive:

Gibco Biocult Ltd
Miles Laboratories Ltd
Nordic Immunological Laboratories Ltd
Wellcome Reagents Ltd

References

1. Catty, D. (ed.) (1988). *Antibodies, Volume 1: A Practical Approach.* IRL Press, Oxford.
2. Harboe, N. M. G. and Ingild, A. (1973). *Scand. J. Immunol.,* Suppl. **2**, 161.
3. Vanselow, B. A. (1987). *Vet. Bull.,* **57**, 881.
4. Lovgren, K., Lindmark, J., Pipkorn, R., and Morein, B. (1987). *J. Immunol. Meth.,* **98**, 137.
5. Dalsgaard, K. (1978). *Acta Vet. Scand.,* Suppl. **69**, 1.
6. Stevenson, G. T. (1974). *Nature,* **247**, 477.
7. Ouchterlony, O. (1949). *Acta Pathol. Microbiol. Scand.,* **26**, 507.
8. Hjelm, H., Hjelm, K., and Sjoquist, J. (1972). *FEBS Lett.,* **28**, 73.
9. Ruoslathi, E. (ed.) (1976). *Immunoadsorbents in Protein Purification.* Universitetsforlaget, Oslo, Norway.

Index

377

Index

Index

Index

MDPF[2-methoxy-2,4-diphenyl-3(2H)-furanone]
   for protein detection 67,326
membrane proteins, analysis by
   'charge-shift' electrophoresis 115−16
   PAGE, using
      chloral hydrate 115
      non-ionic detergents 115−16
      phenol-acetic acid−urea 115
      two-dimensional PAGE 233
mini- and micro-gels
   rod gels 125
   slab gels 125−7
   two-dimensional gels 224−5
Molecular mass determination
   denatured proteins in reducing conditions (by
      SDS−PAGE)
      anomalous behaviour 19−20
      basic method using uniform concentrations
         gels 8,16−21,44−8
      molecular mass marker polypeptides 46−8,
         359−61
      using concentration gradient gels 21,
         117−19
   denatured proteins in non-reducing conditions
      20−1
   native proteins 16−17,119
   oligopeptides 20,103−8

Nigrosin
   as protein stain 282−3
N,N′-bisacrylylcystamine, see BAC
N,N′-methylenebisacrylamide, see bisacrylamide
N,N,N′,N′-tetramethylenediamine, see TEMED
non-histone proteins, analysis of, using
   SDS−PAGE 113−14
   two-dimensional PAGE 256−8

oligopeptides
   molecular mass determination of 20,103−8
Ornstein−Davis discontinuous buffer
   system 10−13,36
OPA (o-phthaldialdehyde)
   for protein detection 67−8,311,326

PAGE, analytical one-dimensional
   analysis of gels after electrophoresis, see
      analysis of polyacrylamide gels; detection
      methods for proteins; rimming; R_f
      determination, etc.
   apparatus
      rod (tube) gels 23−4
      slab gels 24−7
   artifacts and trouble-shooting 136−9
   experimental approach
      choice of buffer system 7−13
      choice of gel concentration 5,14−16

choice of ionic strength 32
choice of pH 13−14
choice of polymerization catalyst 3−5,14
choice of rod or slab gels 6−7
optimal conditions for protein stability and
   resolution 23,27,32,38
pore size, importance of 5
pore size, optimization 14−16
gel preparation
   concentration gradient gels, see gradient
      polyacrylamide gels
   large numbers of gels 127−30
   uniform concentration rod gels 38−41
   uniform concentration slab gels 41−4
molecular mass determination of proteins, see
   molecular mass determination
non-ionic detergents, use of 115−16
sample loading and electrophoresis 49−50
sample preparation 44−9
SDS−PAGE, see SDS−PAGE
urea, use of, see urea
PAGE, preparative
   by gel slice extraction
      localization of protein bands 99−101
      recovery of protein 97−9,101−3
   from electroblots 103
PAGE, two-dimensional
   see two-dimensional PAGE
peptide mapping
   apparatus 303−5
   characterization of proteinases 313−16
   chemical methods for peptide bond
      cleavage 302−3
   detection of peptides after 309−12
   electroblotting after 311−12
   interpretation of data 316−18
   of protein mixtures 312−13
   of radioactive proteins 311
   prelabelling proteins for 309−12
   proteinase specificities for 304
   standard procedure 305−9
   strategies 302−3
   uses of 301−3
   variations to the standard procedure 309−12
Periodic acid Schiff (PAS)
   stain for glycoproteins 81−2,329
phosphoproteins
   detection of 82,330
photography of stained protein bands 55−7
o-phthalaldehyde, see OPA
pI, see isoelectric points of proteins
polyacrylamide gel
   advantages in electrophoretic analysis of
      proteins 1−3
   chemical structure 3
   choice of gel concentration for one-
      dimensional PAGE 14−16
   concentration gradient gels, see gradient
      polyacrylamide gels

Index

Index

multiple probing of blots 97
recovery of proteins from blots 103
specific immunodetection methods
   basic procedure 93−6
   biotinylated antibodies 99,333
   detection after general protein staining
     94,96
   enzyme-conjugated antibodies 95−6,333
   fluorescent antibodies 95−6,333
   immunogold staining 95−6,333
   immunogold silver staining 96
   radiolabelled antibodies 95−6
   using *S.aureus* Protein A 96,333
use of molecular mass standards 92−3

Xylene Cyanine Brilliant G, *see* Coomassie
   Blue G-250

Y_0 15−16

zone electrophoresis
   definition of 1−2
zone-sharpening
   continuous buffer system 48
   discontinuous buffer system 9−13